ELEMENTS OF ENZYMOLOGY

ENCYCLOPAEDIA OF ENZYMOLOGY - I

ELEMENTS OF ENZYMOLOGY

By

Dr. Arvind N. Shukla

School of Studies of Zoology & Biotechnology
Vikram University
Ujjain (India)

DISCOVERY PUBLISHING HOUSE PVT. LTD.
NEW DELHI-110 002

Published by:
Namit Wasan
DISCOVERY PUBLISHING HOUSE PVT. LTD.
4383/4B, Ansari Road, Darya Ganj
New Delhi-110 002 (India)
Phone : +91-11-23279245; 23253475; 43596065
E-mail : discoverybooksindia@gmail.com
discoverypublishinghouse@gmail.com
namitwasan9@gmail.com
web : www.discoverypublishinggroup.com

***First Published:* 2009**
***Reprinted:* 2021**

ISBN: 978-81-8356-357-4 (Set)

ISBN: 978-81-8356-403-8

Elements of Enzymology

© Author

All rights reserved. No part of this publication should be reproduced, stored in a retrieval system, or transmitted in any form or by any means: electronic, mechanical, photocopying, recording or otherwise, without the prior written permission of the author and the publisher.

This book has been published in good faith that the material provided by authors/editors is original. Every effort is made to ensure accuracy of material, but the publisher and printer will not be held responsible for any inadvertent error(s). In case of any dispute, all legal matters are to be settled under Delhi jurisdiction only.

Printed at:
Infinity Imaging Systems
Delhi

Preface

This classic *Encyclopaedia of Enzymology* has been compilated to provide an understanding of enzymology for university degree study and teaching. It comprehensively covers theory and applications of these biological life supporting catalysts in a readable book that will serve as a bridge to more advanced and specialized areas. It not only introduces the fundamentals and biochemistry of enzymes but also provides the molecular and experimental background. This unique piece of work compiled five volumes containing wealth of information for researchers, microbiologists, and biotechnologists. This excellent title is intended mainly for students taking degree course which have a substantial biochemistry component. The chief aim of this particular book is to help student understand the concepts involved in enzymology. An attempt has been made to give a perspective of each topic, and examples are quoted where appropriate.

There can be no claim to originality except in the manner of treatment and much of the information has been obtained from the books and scientific journals available in the different libraries.

The author express his thanks to his friends and colleagues whose continue inspirations have initiated him to bring out this book.

The author is painfully aware of the shortcomings, errors and misprints that have crept in, and shall be grateful to receive suggestion for improvement of the next edition from all the readers.

The author express his gratitude to Mr. Wasan and staff of M/s Discovery Publishing House Pvt. Ltd. for their whole hearted co-operation in the publication of this book.

Author

Preface

This classic *Encyclopaedia of Enzymology* has been compiled to provide an understanding of enzymology for university degree study and teaching. It comprehensively covers theory and applications of these biological life supporting catalysts in a readable book that will serve as a bridge to more advanced and specialized areas. It not only introduces the fundamentals and biochemistry of enzymes but also provides the molecular and experimental background. This unique piece of work comprises five volumes containing wealth of information for researchers, microbiologists, and biotechnologists. This excellent title is intended mainly for students taking degree course which have a substantial biochemistry component. The chief aim of this particular book is to help student understand the concepts involved in enzymology. An attempt has been made to give a perspective of each topic and examples are quoted where appropriate.

There can be no claim to originality except in the manner of treatment and much of the information has been obtained from the books and scientific journals available in the different libraries.

The author expresses his thanks to his friends and colleagues whose continue inspirations have initiated him to bring out this book.

The author is painfully aware of the shortcomings, errors and misprints that may have crept in and shall be grateful to receive suggestions for improvement of this publication from all the readers.

The author expresses his gratitude to Mr. Vikash and staff of M/s Discovery Publishing House Pvt. Ltd. for their whole hearted cooperation in the publication of this book.

Author

Contents

1

INTRODUCTION

It has been our intention that this book should serve a practical purpose, and this aspect has been taken into consideration in the compilation of the preceding chapters. However, details of the individual procedures would have disturbed the uniformity of those chapters, and so they are given separately here.

In selecting the methods to be given it was necessary to take into account the extremely rapid development of protein analysis during the past 20 years. The main trends in development suggested that (i) outdated analytical methods, such as the *chromatography* of proteins and peptides on ion-exchange resin columns, should be omitted and (ii) the methods given should be such that, even when they too become outdated, they will be of help in understanding the future methods.

For all the methods, the material examined is the same: hen egg-white. This permits the *demonstration* of how the various separation methods do not always separate a certain protein sample in the same way and with the same effectiveness; examples are the *chomatography* of egg-white on *CM-Sepharose* and on *DEAE-Sepharose*. This is not to say, of course, that the findings are generally applicable and that separations with similar resolution will be obtained with other protein samples: in the separation of the component proteins of a material examined by the various methods, characteristic effects are displayed by the *physico-chemical* properties of the invidual proteins concerned, such as their molecular weights, charges and steric structures.

The methods described have been carried out by experts who have great experience in laboratory work of this nature.

The whites of 10 eggs not older than one day are combined, thoroughly mixed, and forced through a plastic sieve. The *sieved* material is placed in *Petri dishes* to a depth of about 5 mm, and freeze-dried. The residue after freeze-drying is ground in a *porcelain* mortar.

By subjecting egg-white to chromatography on a *CM-cellulose* column, Rhodes *et al.* isolated the following proteins:

	pI	Amount of protein from N content, mg	Distribution of N in fractions, %
Ovomucoid +flavoprotein	3.9-4.3	304	12.3
Ovomucoid	3.9-4.3	6	0.3
Ovalbumin A_1	4.58	1106	50.8
Ovalbumin A_2	4.65	237	10.8
Ovalbumin A_3	4.75	70	3.2
Globulin	22	1.5	
Conalbumin	6.5-6.8	256	12.4
Conalbumin	6.5-6.8	97	4.7
Globulin	46	2.2	
Globulin	15	0.7	
Globulin	7	0.3	
Avidin	10	6	0.3
Globulin	8	0.4	
Lysozyme	10.7-11.3	11	0.16
		2191	100.0

DETERMINATION OF NITROGEN CONTENT

Through determination of the nitrogen content and knowledge of the nitrogen conversion factor for the protein being examined, reliable data are obtained on the protein content of a sample. If the appropriate nitrogen conversion factor is not known, it may be calculated from the *amino-acid* composition of the *proteins*.

Of the *numerous* methods for determination of the protein contents of samples, the following *procedures* have been selected on the basis that they are reliable and can be carried out in any *laboratory*.

Weigh a sample containing 200-300 mg of protein into a 100-ml Kjeldahl flask, add 10 ml of *concentrated sulphuric acid*, 6 g of *potassium sulphate* and 0.2 g of *copper sulphate* and heat at 320-350

°C for about 90 min. When the solution has cooled, dilute it to about 50 ml, cool it again, transfer it quantitatively to a 100-ml standard flask and make up to volume with distilled water. Pipette 25 ml of this solution into a *Parnas-Wagner* distillation *apparatus*, add 10 ml of 40% sodium hydroxide solution and steam-distil the ammonia, collecting it in either 25 ml of 0.5 *M hydrochloric acid* (accurately measured) or 25 ml of approximately 4% boric acid solution, together with 4 drops of Tashiro indicator (0.2 g of Methyl Red and 0.19 g of *Methylene* Blue dissolved in 100 ml of ethanol). In the Parnas-Wagner method the excess of hydrochloric acid is back-titrated with 0.05 *M sodium hydroxide*, and in the Winkler procedure the ammonium borate formed is titrated with 0.05 *M* hydrochloric acid.

Photometric Determination

This method is suitable for the determination of the nitrogen content of large numbers of samples. There are several advantages to the method: (i) in the case of a *homogeneous* substance with a particle size smaller than 0.2 mm, very small samples are sufficient and a large number of samples may be decomposed simultaneously in a thermal block; *(ii)* the photometry is fast and accurate; (iii) if duplicate samples are decomposed and each resulting solution analysed in duplicate, the mean of the four results gives a satisfactory measure of the nitrogen content of the material.

Reagents

Sulphuric acid for decomposition: 2 g of selenium are heated with 250 ml of sulphuric acid until a. *colourless* solution is obtained; after cooling, this solution is carefully poured into 250 ml of saturated potassium sulphate solution (27.4 of potassium sulphate dissolved in 245 ml of warm distilled water) with cooling.

Nessler reagent

Iodine (22.5 g) is dissolved in a solution of 30 g of potassium iodide in 20 ml of water, and 30 g of *metallic mercury* are added. The vessel is shaken *vigorously* while being cooled under running water, until the solution is practically colourless. A few drops of the solution are then tested with 1 starch solution to check that there is not an *appreciable* amount of unreacted iodine left. The solution is then decanted. If the starch test does not reveal the presence of iodine, a solution of 2.25 g of iodine and 3 of potassium iodide in 2.0 ml of water is added dropwise until a pale yellow colour is restored. The decanted solution is made up to 200 ml with water, and finally 975 ml of 10% *sodium hydroxide* solution are added with stirring.

Standard solutions

Nitrogen concentration 0.1 mg/ml: 47.16 mg of $(NH_4)_2SO_4$ (dried over phosphorus pentoxide) dissolved in 0.2N H_2SO_4 and diluted to looml.

Procedure

The material should have a particle size of at most 0.2 mm. Duplicate samples containing 0.1-0.5 mg of nitrogen are weighed into normal test-tubes and decomposed by heating with 4 drops of H_2O_2 and 0.4 ml of *sulphuric acid* mixture, the test-tubes being inserted in holes drilled in an aluminium block, which is heated to 320 °C. The decomposition requires 1.5-2.5 h (the solutions should become colourless). The operation is best performed in a *fume-cupboard*. The standards (0.1, 0.2, 0.3 ...0.8 ml of 0.1 mg/ml nitrogen standard) are treated similarly.

After decomposition the solution is cooled and made up to 10 ml with distilled water. Then, depending on the expected nitrogen content, two 0.2-4.0 ml portions are taken in graduated 10-m1 test-tubes and made up to 4 ml with distilled water if necessary, then 2.0 ml of Nessler reagent are added with continuous shaking. The absorbances are then measured in 1-cm cells at 440 nm against water. The calibration curve is drawn from the results for the standards, with 4-ml portions of the diluted solution from the decomposition step (the absorbances should be approximately the same as the number of ml of 0.1 mg/ml standard originally taken).

Evaluation of results

The choice of titration or *spectrophotometry* is governed by the purpose of the analysis. The titration procedure must be regarded as the fundamental method, but its use is frequently hindered by sufficient sample not being available; in such a cane, *spectrophotometry* may be of assistance. The larger measurement errors in the spectrophot-ometric procedure necessitate the use of replicate determination. Incorrect results due to faulty sampling or pipetting can be distinguished among the four results obtained. The *spectrophotometric* method is particularly useful for the routine analysis of large numbers of well-homogenized samples.

DETERMINATION OF AMINO-ACID COMPOSITION

Determination of its amino-acid composition is *indispensable* for the general *characterisation* of a protein, elucidation of its primary structure, and nutritional and biological evaluation. The amino-acid composition cannot generally be .found by chroma-tography of a *hydrolysate*, as the individual amino-acids undergo destruction to differing extents during the different hydrolyses. The *hydrolyses* used must therefore permit the exact establishment of the amino-acid composition.

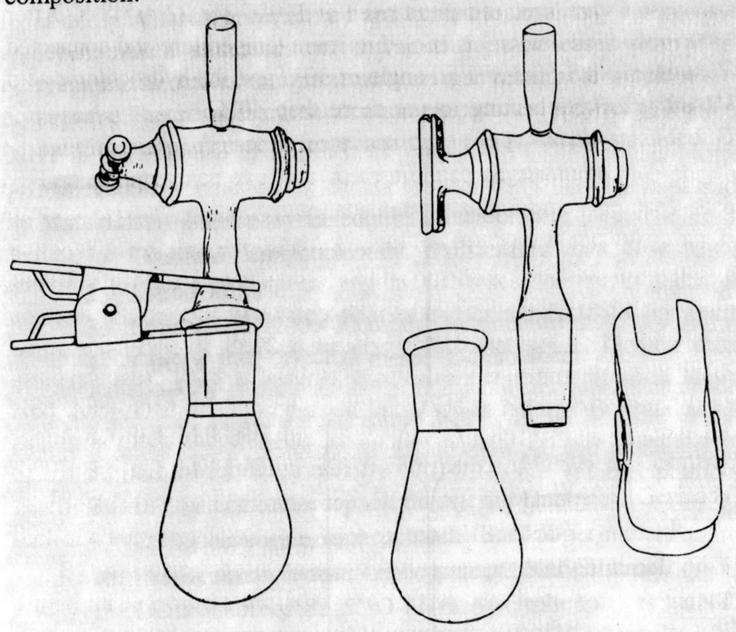

Figure 1.1 : Evacuable vessel for hydrolysis with 6 M HCl.

The following *hydrolyses* are usually performed:

(1) with *6 M* hydro-chloric acid for 22 and 48 h from the results of which the value for zero time can be *extrapolated*

(2) with *5 M* sodium hydroxide, for determination of Trp;

(3) performec acid oxidation, and then hydrolysis with 6 *M hydrochloric acid*, for accurate determination of Cys and Met.

Hydrolysis with 6 *M* HCl

About 13 mg of sample is weighed into the hydrolysis vessel, which is then fitted together and immersed in a "*dry ice*" ethanol cooling mixture; when the solution has frozen, the vessel is evacuated to 7.9 Pa (0.06 mmHg) (measured with a McLeod gauge). To remove traces of air dissolved in the acid, the *hydrolysis* vessel is taken from the cooling mixture and left, after the pump has been switched off, for the frozen solution to melt slowly. As soon as bubbles begin to emerge from the *viscous* solution, the *hydrolysis* vessel is immediately reimmersed in the cooling mixture and evacuation continued until 7.9 Pa (0.06 mmHg) is reached (ca. 10-15 min), but it is not necessary to refreeze the solution. The hydrolysis vessel is then isolated by means of the tap, and disconnected from the vacuum pump. The evacuated vessel is placed in an oven at 110°C for 22 or 48 h. After the hydrolysis, the vessel is cooled to room temperature and connected to a rotary still inclined at an angle of 30°, and the hydrochloric acid is distilled off by heating to not more than 40°C.

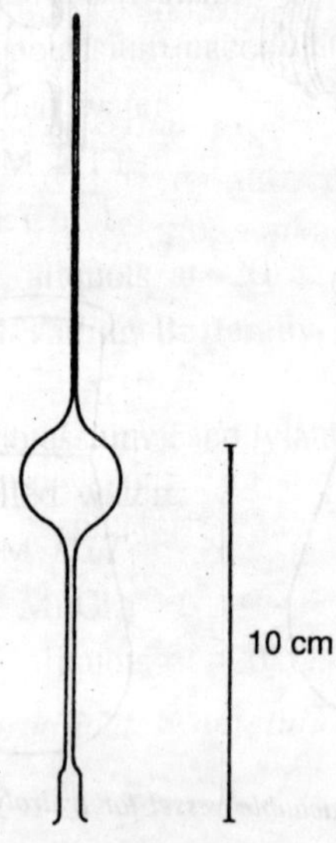

Figure 1.2 : "Ballonki" for filtration of hydrolysates. The bulb is warmed above a low flame, and the open end (packed with scraps of Seitz filter fibre) is immersed to the bottom of the solution to be filtered. As the bulb cools, the solution is filtered upwards; the filter fragments are removed, the end of the capillary is broken off, and the filtered solution can be poured out.

The *hydrolysate* is dissolved in 1 ml of 0.2 *N* sodium citrate buffer (pH 2.2), and any insoluble matter is filtered off with a "*ballonki*".(Solutions not utilized on the same day are stored *frozen* until required.) The optically clear solution (30-μl sample) is chromatographed on a 4 × 23 mm column of Durrum DC 4-A resin with (a) 0.2 *N* sodium citrate buffer (pH 3.25) containing 2.5% ethanol; (b) 0.7 *N* sodium citrate buffer (pH 3.50); (c) 1.6 *N* sodium citrate (pH 3.65) at a flow rate of 20 ml/h. The chromatography takes 75 min and the elution *temperature* is 35-75°C. The eluate is mixed with ninhydrin solution added at a rate of 10 ml/h. The ninhydrin solution consists of 20.0 g of ninhydrin, 750 ml of methylcellosolve, 250 ml of 3.7 *M* sodium citrate (pH 5.5) and 0.4 g of $SnCl_2 \cdot 2\ H_2O$. The absorbance of the ninhydrin/eluate mixture is monitored at 440 and 570 nm.

Hydrolysis with 5 MNaOH for Determination of Trp

A quartz tube containing as starting materials about 13 mg of egg-white, 25 mg of hydrolysed starch and 5 ml of 5 *M* sodium hydroxide is immersed in a "*dry ice*"-ethanol cooling mixture until the materials are frozen, and transferred to a similarly cooled Oehlshlegel hydrolysis vessel. This is then connected to a vacuum pump and evacuated to 23.9 Pa (0.18 mmHg) with cooling, and the glass tap is

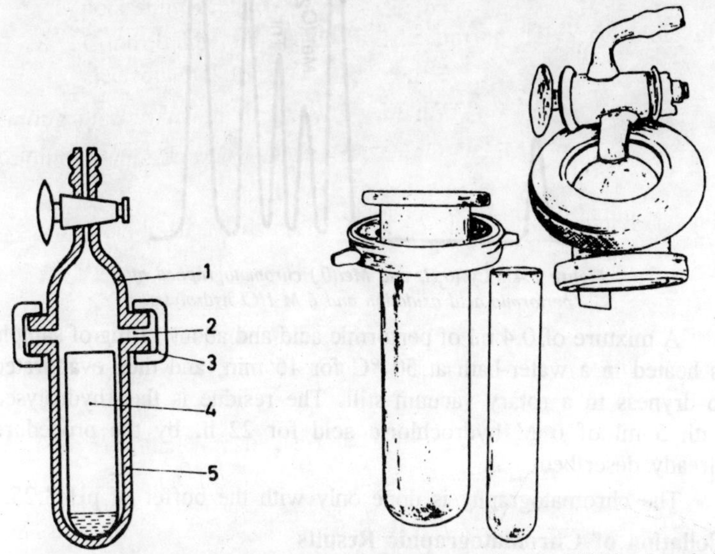

Figure 1.3 : Evacuable vessel for alkaline hydrolysis according to Oehlshlegel et al. 1, upper glass part with tap; 2, position of rubber ring; 3, clamp; 4, quartz tube containing hydrolysate; 5, lower glass vessel.

closed. After being allowed to warm up to room temperature, the *hydrolysis* vessel is placed in an oven at 110 °C for 16 h. After cooling to room temperature, the *quartz tube* is taken out of the *hydrolysis* vessel, 4.6 ml of 6 *M hydrochloric acid* are added to the *hydrolysate*, and the volume is made up to 10 ml with distilled water.

The solution (30 μl) is then *chromatographed* in the same way as the 6 *M* hydrochloric acid *hydrolysates*, and Leu and Trp are determined.

HCL Hydrolysis of Protein Oxidized with Performic Acid

The performic acid is prepared by adding 0.5 ml of 30% H_2O_2 to 4.5 ml of 80% formic acid, and heating in a water-bath at 50 °C for 3 min.

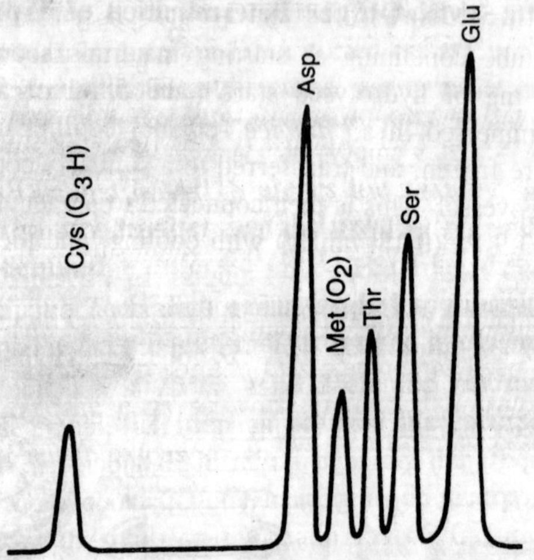

Figure 1.4 : Cys(O_3H) and Met(O_2) chromatographed after performic acid oxidation and 6 M HCl hydrolysis.

A mixture of 0.4 ml of performic acid and about 12 mg of sample is heated in a water-bath at 50 °C for 15 min, and then evaporated to dryness in a rotary vacuum still. The residue is then hydrolysed with 5 ml of 6 M hydrochloric acid for 22 h, by the procedure already described.

The chromatography is done only with the buffer of pH 3.25.

Collation of Chromatographic Results

In the first step, the quantities of amino-acids for.nd by the chromatography of the four hydrolysates are recalculated on the basis of 100 jug of egg-white. Typical quantities found, in ,ug per 100 pg of freeze-dried egg-white, are given below .

The results obtained from the 22-h and 48-h hydrolyses are next extrapolated to zero time to obtain the values before the destruction caused during the acid hydrolysis, except for Cys, Val, Met, lie and Trp. (during the 22-h hydrolysis, the Val and Ile are not liberated quantitatively; the correct values for these are obtained from the 48-h hydrolysis.) These calculations give the following values for the

Table 1.1

Hydrolysis	6MHC1 22 h	6MHC1 48 h	5 M NaOH	6MHCI after performic acid oxidation
Asp	8.92	8.85		
Thr	3.26	3.00		
Ser	5.45	4.96		
Glu	12.34	12.34		
Pro	2.75	2.76		
Gly	2.84	2.82		
Ala	3.06	3.02		
Cys	0.85	0.80		1.67
Val	4.54	4.58		
Met	0.74	0.27		2.03
Ile	3.41	3.56		
Leu	5.85	5.84		
Tyr	2.10	1.95		
Phe	5.07	4.90		
Lys	5.36	5.31		
His	3.36	3.36		
Trp	0	0	1.00	
Arg	3.76	3.75		
NH_2	2.08	2.20		

amino-acid composition (% w/w) of the freeze-dried egg-white:

Asp	9.0%	IIe	3.55%	Val	4.5%	NH_2	1.95%
Thr	3.5%	Leu	5.85%	Cys	1.65%	Arg	3.75%
Ser	5.75%	Tyr	2.2%	Ala	3.1%	Trp	1.0%
Glu	12.35%	Phe	5.15%	Gly	2.85%	His	3.35%
Pro	2.75%	Lys	5.4%	Met	2.05%		

Total amino-acid content: 79.7%

ELECTROPHOETIC METHODS

In contrast to the other methods, the description of the electrophoretic separation methods not only contains data and information on their practical application, but also gives an insight into the decisive extent to which the nature of the support influences the electrophoretic migration of protein molecules with identical charges.

Paper Electrophoresis

Support: Schleicher & Schiill 2043 b, 2 cm wide, 8 cm long (between the fixed points).

Sample applied: 3μl.

Electrode buffer: veronal-sodium acetate, pH 8.6, μ=0.1.

Electrophoresis: 130 V, 4 mA, 160 min.

Staining: 0.4% Ponceau in 7.5% TCA. Destaining: with 5% acetic acid.

Evaluation: 4 distinguishable bands are found on the paper.

Starch Gel Electrophoresis

Support: 35 × 100 × 6 mm 13% starch gel, arranged vertically in triple parallels in a gel holder according to Smithies. Sample applied: 0.05 ml of 1 % freeze-dried egg-white solution. Buffer: 5.4 g of TRIS, 0.4 g of EDTA, 2.75 g of H_3BO_3, 12.5 ml of 1 M NaOH, H_2O to 1000 ml, pH 8.6.

Electrophoresis: 320 V, 18 mA, for 17 h.

Staining: with a 0.2% solution of Amido Black in acetic acid-methanol-water (1 :4: 5) for 30 min.

Destaining: with acetic acid-methanol-water (1 : 4: 5) for 24 h.

Evaluation: Starch gel electrophoresis separation of the proteins of egg-white yields only the main components, in fairly diffuse bands.

Agarose Gel Electrophoresis

Support: 2 ml of 0.8% Koch-Light agarose-containing gel, 26 x 76 × 1 mm. Sample applied: 7μl of 5% egg-white solution. Buffer: 50 mM veronal-veronal-Na, pH 8.6. *Electrophoresis*: 140 V, 5 mA, 20 min.

Fixation: saturated picric acid solution-20% acetic acid (3: 1 v/v). Staining: 0.1% Coomassie BB R-250 in water-ethanol acetic-acid (8: 1 : 1). Destaining: water-ethanol-acetic acid (8: 1 : 1).

Evaluation: in the anodic direction from the position of application of the sample to the gel, strongly diffuse bands can be seen, the number of which is difficult to determine; in the cathodic direction, a scarcely visible band is formed by the *lysozyme* comprising 0.16% of the egg-white.

Polyacrylamide Homogeneous Gel Electrophoresis

Support: 7.5% *(C* = 2.6%) gel, 78 × 78 × 2.7 mm, positioned vertically.

Sample applied: 5 and 10 μl of 2% egg-white solution.

Electrode buffer: TRIS-glycine buffer, pH 8.3 Electrophoresis: 100 V, 80 mA, for 40 min; 200 V, 120 mA, for 180 min. Fixation: in 7% acetic acid for 60 min. Staining: with 0.1% Amido Black solution in 7% acetic acid. Destaining: with 7% acetic acid.

Evaluation: the large number of bands of various widths resulting from the component proteins gives a good reflection of the advantages of PAGE over the previous electrophoretic methods.

Polyacrylamide Linear-gradient Gel Electrophoresis

Support: *Pharmacia Polyacrylamide* Gradient Gel, PAA 4/30.

Apparatus: Pharmacia GE-4 II Gel Electrophoresis Apparatus. Sample applied: 5, 10, 15 and 20 μl of 1 % egg-white solution.

Reference substances (ascending from the bottom on the two edges of the gel): ovalbumin (M.W. 43,000), bovine serum albumin (M.W. 67,000), phos-phorylase b (M.W. 94,000).

Electrode buffer: TRIS-glycine buffer, pH 8.3 Electrophoresis: stable potential 125 V, for 17 h. Fixation: in 7% acetic acid, for 60 min. Staining: with 0.1% Amido Black solution in 7% acetic acid. Destaining: with 7% acetic acid.

Evaluation: the bands obtained in the linear-gradient gel support very well the multiplicity of component proteins in egg-white, as reported by Rhodes *et al.*.

Electrophoresis on SDS-containing, Staged Gradient Polyacrylamide Gel

Support: gels prepared with 40% acrylamide solution (C=5%)

Composition of lower gel (4 cm high):

40% acrylamide solution	10 ml
gel buffer, pH 8.2	20 ml
DMAPN	0.1 ml
10% $(NH_4)_2S_2O_8$ solution	0.1 ml
H_2O	to 40 ml

Composition of middle gel (5 cm high):

40% acrylamide solution	7.5 ml
gel buffer, pH 8.2	20 ml
DMAPN	0.1 ml
10% $(NH_4)_2S_2O_8$ solution	0.1 ml
H_2O	to 40 ml

Composition of upper gel (5 cm high):

40% acrylamide solution	5 ml
gel buffer, pH 8.2	20 ml
DMAPN	0.1 ml
10% $(NH_4)_2S_2O_8$ solution	0.1 ml
H_2O	to 40 ml

Gel buffer: 0.1 M TRIS-acetic acid, pH 8.2, 0.1% SDS.

Electrode buffer: the gel buffer was used.

Samples applied: reference proteins (*bovine serum albumin, serum albumin, egg albumin, catalase, equine myoglobin, cytochrome* C), A: 50μg; B: 100 μg; freeze-dried egg-white: C: 50 μg; D: 100 μg. Electrophoresis: 125 V, 80 mA, for 3.5 h.

Fixation: in methanol-acetic acid-water (5:5: 1), for 60 min.

Staining: 0.25% Coomassie **BB** R-250 solution in methanol-acetic acid-water (5: 5: 1), for 2 h.

Destaining: a mixture of 250 ml of methanol, 375 ml of acetic acid and 4375 ml of water is allowed to flow through overnight.

Evaluation: the separations obtained in cases C and D very clearly illustrate not only the *quantitative* proportions of the egg-white component proteins, but also their distribution according to *molecular* weight.

Rocket Immunoelectrophoresis

Support: 120 × 150 × 1.8 mm gel slab containing 32 ml of 1 % Indubiose A 37 agarose+0.7 ml of antiovalbumin rabbit serum. Buffer: Svendsen buffer (barbiturate-TRIS-Gly), pH 8.6, μ=0.08.

Sites of application of samples: holes 3 mm in diameter, separated by distances of 8 mm.

Volume of sample applied: 5μl. Egg-white solutions were applied in a standard dilution series (the first 5 peaks from the left in Fig. 7.10). with concentrations of 0.0125, 0.025, 0.2, 0.1 and 0.05 μg/ ml; the other peaks originate from 0.90/ NaCl extracts of various pastry products containing egg-white.

Electrophoresis: 4 V/cm standard potential, for 16 h, with the gel-supporting glass plate situated on a cooling plate. Staining: with a 0.1 % solution of *Coomassie* BB R-250. Destaining: according to Laurell. Evaluation: The egg-white quantities in the NaCl solution extracts of the various pastry products can be well evaluated on the basis of the lengths of the *electroendo-osmotic* movements; these are compared with the standard dilution series containing known amounts.

ISOELECTRIC FOCUSING (IEF)

IEF in an ampholyte-containing polyacrylamide gel slab in the pH interval 3-10 provides valuable information for the selection of subsequent separation methods in which the fractionation is based on the charges of the substances to be separated (IEF in a narrower pH interval, various *electrophoresis* techniques, ion-exchange *chromatography*).

If IEF in the pH interval 3-10 is used to establish the limits within which the pl values of the components of the substance under examination are to be found, the pl values of the individual components or of one particularly important protein may be determined accurately by means of a second IEF. For this, an ampholyte with a narrower pH interval is employed. In addition, the focusing may first be performed with the electrodes a large distance apart; the electrodes are then brought nearer to each other, care being taken that the bands of importance always lie between the electrodes.

IEF focusing of egg-white in ampholyte-containing polyacrylamide gel slab in the pH interval 3-10

The purpose of the IEF is to obtain information on the pl values of the egg white proteins, and also to demonstrate that during the IEF the identical proteins migrate from any position to the position with pH corresponding to the pl value.

A Pharmacia FBE 3000 flat bed apparatus and ECPS 2000/300 constant power supply are used.

The acrylamide-BIS solution (T=10%, *C* = 3%) has the composition:

acrylamide	9.7 g
BIS	0.3 g
H_2O	to 100 ml

and when mixed is shaken with 1.5 g of Elgalite or Amberlite MB 1, and stored with this at 4°C.

The TCA-sulphosalicylic acid solution is made from:

trichloroacetic acid	20 g
sulphosalicylic acid	10g
H_2O	to 200 ml

A 0.2% Coomassie BB R-250 solution in *methanol-water-acetic acid* (9: 9 : 2) is used for staining. The destaining solution is *methanol-water-acetic acid* (9:9:2).

The sample is 10 mg of freeze-dried egg-white in 1 ml of water.

The gel slab is prepared with:

acrylamide solution	18 ml
glycerol	4 ml
Pharmalyte, pH 3-10	1.8 ml
H_2O	6 ml

This solution is placed in a 100-ml suction flask, which is then closed with a rubber stopper and connected to a water-pump, and the solution is degassed. *Ammonium persulphate* solution (2.28%, 0.2 ml) is added to the *degassed* solution, and after mixing the solution is poured into the 230 × 115 × 1 mm glass plate chamber of the FBE 3000. *Gelation* occurs about 30 min later. The plastic plate and the glass plate (previously treated with a 0.4% solution of Silane A-174, Union Carbide Co.) are carefully removed from the gel layer adhering to the glass plate, and the 1 mm thick gel slab adhering to the glass plate is placed in the FBE 3000 apparatus. The sample applicator is placed on the gel in three positions, with three excisions at each position, into which 5, 10 and 15 μl samples are added side by side.

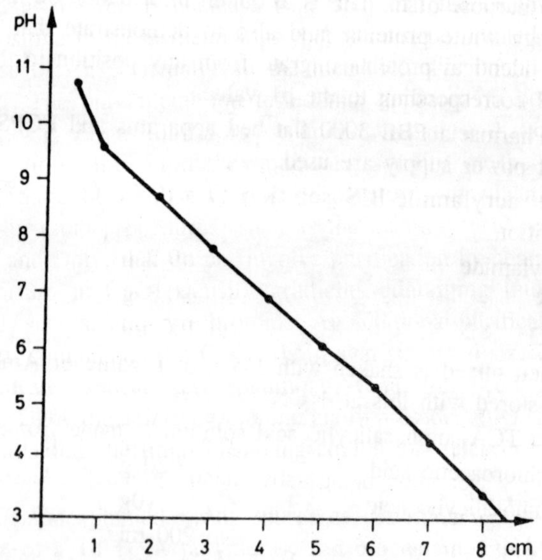

Figure 1.5 : pH gradient produced on gel slab during isoelectric focusing of egg-white.

During IEF the current strength is maintained at almost the same level: initially, a potential of 900 V is necessary for 33 mA, and at the end of the IEF 2000 V is necessary for 35 mA. The IEF lasts for 90 min; the pH distributions developing in the gel slab during this time are measured with a surface *electrode*.

The gel slab adhering to the glass is fixed in *TCA-sulphosalicylic acid* solution for 1 h, then stained for 1 h, and destained overnight. Evaluation: depending on their pI values, the proteins of the sample solutions applied in the various positions migrate to the gel band of corresponding pH. The egg-white proteins are predominantly situated in the pH interval 4-5. A characteristic deviation from this is displayed by the migration of lysozyme; after IEF, this is found on the cathodic edge of the gel. The experimental results depicted in show that the pH distribution of the gel was uniform.

IEF of Egg-white in a PAG Slab Containing an Ampholyte of pH 4-6.5

The purpose of the IEF is to determine the pI values of the proteins occurring in larger amounts in egg-white.

The apparatus and reagents are the same as for IEF in the pH interval 3-10, plus *Pharmalyte* (pH 4-6.5), 0.03 *M glutamic acid* to wet the anode strip, and 0.1 *M histidine* to wet the *cathode strip*.

The IEF technique is the same as for IEF at pH 3-10. A standard current strength of 36 mA is applied for 2 h.

Evaluation: the separations achieved with this IEF clearly demonstrate the importance of the quantity of sample applied, insofar as the bands of the ovalbumins (present in predominant amounts) can be well observed in the cases of the 5 and 10 μl samples. The many types of egg-white proteins are very well separated on the basis of their pI values in the pH interval 4-6.5.

IEF of Egg-white in PAG Containing an Ampholyte of pH 4-6.5, in a Narrowed Electric Field

The aim is to attain very high resolution in the separation of the proteins with *isoelectric* points lying in the pH interval 4.5-4.8. Accordingly, the IEF is begun with the *electrodes* 220 mm apart, the distance of separation being subsequently gradually decreased.

The apparatus is the same as for IEF at pH 3-10, and the chemicals those for IEF ar pH 4-6.5.

The gel slab is 230 × 115 × 1 mm, 5 and 10 μl samples of 10-mg/ml freeze-dried egg-white solution are applied.

Before application of the samples, with the electrodes 220 mm apart, the gel slab is subjected to prefocusing in the longitudinal direction for 1 h at 2000 V and 30 W. Next, the samples are applied *midway* between the *electrodes* (still 220 mm apart), and the IEF is begun at 2000 V. After 100 min, the cathode is brought 80 mm closer to the anode, and the IEF continued for a further 120 min. Following this stage, both the anode and the cathode are moved 36 mm towards each other, so that the distance of separation is 50 mm. The IEF is continued under these conditions at 2000 V for a further 45 min. The field of the IEF is then 400 V/cm, and pH interval between the *electrodes* is 1.

The *fixation*, *staining* and *destaining* are done as already described above. Evaluation: IEF performed with electrodes separated by a decreasing distance very, clearly *demonstrates* that the bands of the proteins with pl values in the interval pH 4.5-5.5 are separated with very high resolution. This method appeared to be more sensitive than all the others attempted.

Preparative IEF of Egg-white in a Sephadex G-200 Flat Bed Containing an Ampholyte of pH 4-6.5

The aim was to obtain preparative quantities of the fractionated proteins present in greatest amount in egg-white and focused in the pH interval 4-6.5, for possible subsequent analyses.

The apparatus was the same as before, used with Whatman 1 MM filter paper, 217 × 217 mm.

Preparation of gel bed and application of sample

Sephadex IEF (8 g) is *swollen* with 225 ml of water, and then is degassed in a suction flask. The supernatant liquid is poured off, 20 ml of *Pharmalyte* (pH 4-6.5) are mixed with the gel, and the slurry is *poured* into the FBE 3000 preparative tray. Next, to bind the excess of liquid in the gel, sufficient *Sephadex* IEF must be sieved uniformly onto the surface of the gel bed so that this no longer moves when the bed is inclined at an angle of 45°. The tray containing the gel is placed on the cooling plate of the FBE 3000 apparatus, the preparative sample applicator is positioned on roughly the middle of the gel slab, and the gel enclosed by the two discs of the applicator is lifted out with a spatula. The freeze-dried egg-white solution (13 mg/ml) and 9 ml of aqueous 5% *Pharmalyte* solution (pH 4-6.5) are mixed into the gel removed. This sample-containing gel is next replaced in the empty site of the preparative sample *applicator*, and the two discs of the *applicator* are removed. The *electrodes* are positioned on the two edges of the gel bed, parallel with the line of the sample (anode strip wetted with 0.1 M *phosphoric acid*, *cathode strip* with 0.1 M sodium hydroxide), and the IEF is carried out at 650 V and 100 mA for 3 h, and then at 1150-1180 V and 50 mA for 75 min. After completion of the IEF, the gel bed is covered with rolled-up Whatman 1 MM filter paper, without air bubbles. Two min later, the paper (containing absorbed protein) is placed in 10% TCA + 5 % sulphosalicylic acid solution for fixation, the fixation solution is washed out of the paper (with the following destaining solution), and staining is performed with 0.2% *Coomassie* BB R-250 solution in methanol-acetic acid-water (9:2:9), and destaining with *methanol-acetic* acid-water (9:2:9).

Evaluation: on the basis of the paper *fingerprint* by appropriate positioning of the discs used with the preparative sample *applicator*, the gel band containing the desired fraction may be lifted out with a spatula, the gel packed into a column, and the protein fraction eluted.

COLUMN-CHROMATOGRAPHIC METHODS

Although of great value in certain special cases, adsorption and partition *column-chromatography* and the *column-chromatographic* methods performed with ion-exchange resins have been virtually completely replaced by the use of *cellulose-*, *dextran-* and *agarose-*based ion-exchange gels, which have a higher capacity and give a better resolution. Accordingly, we have not carried out separations by the three above-mentioned *chromatographic* procedures.

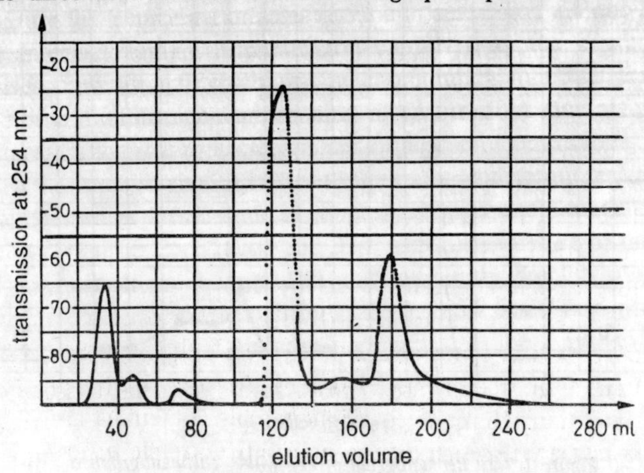

Figure 1.6 : Chromatogram of egg-white, chromatographed with linear gradient elution on CM-cellulose

The various types of ion-exchange *chromatography* actually performed do not always give uniformly good separations, but for just this reason they are important for illustrative purposes.

Ion-exchange Chromatography on a CM-cellulose Column

The CM-cellulose (0.7 meq/g) was prepared for *chromatography* as described in Chapter and packed in a column 16 mm in diameter and 120 mm long. The sample applied was 100 mg of freeze-dried egg-white, dissolved in 4 ml of the starting buffer (0.04 M sodium acetate, pH 4.4). The limiting buffer was 0.5 M sodium acetate, pH 9.5, and the flow rate was 40 ml/h. The eluate was monitored at 254 nm.

The elution started from the acidic region; the sodium ion concentration and pH were increased with a linear gradient.

Evaluation: On the *chromatogram* there are six well separated peaks.

Ion-exchange Chromatography on a CM-Sepharose CL-6B Column

The CM-Sepharose CL-6B column was 16 mm in diameter and 100 mm long and the sample was 100 mg of freeze-dried egg-white, dissolved in 4 ml of starting buffer. The linear *gradient* elution started with 0.04 M sodium acetate buffer, pH 4.4, and ended with 0.5 M sodium acetate buffer, pH 9.5, the flow rate being 40 ml/h, and the

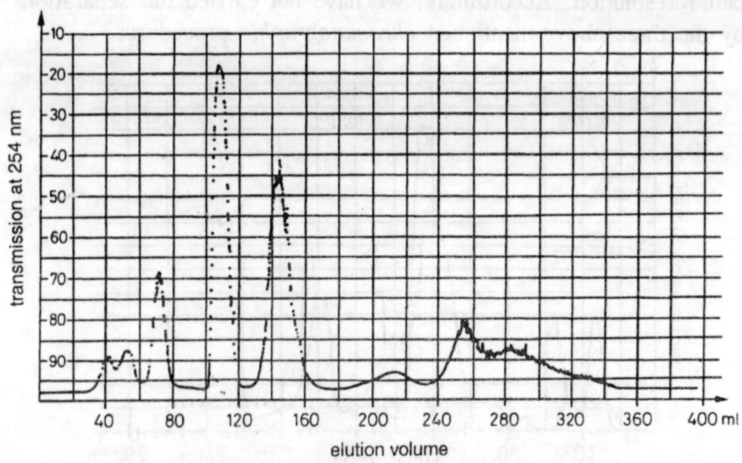

Figure 1.7 : Chromatogram of egg-white, chromatographed with linear gradient elution on CM-Sepharose CL-6B.

eluate was monitored at 254 nm. Evaluation: Chromatography led to nine fractions here, in contrast to six fractions obtained with the CM-cellulose column, due to the fact that CM-Sepharose CL-6B is a better sorbent.

Ion-exchange Chromatography on a DEAE-Sephacel Column

Elution with a buffer of constant TRIS concentration and decreasing pH

The DEAE-Sephacel column was 16 mn in diameter, and 100 mm long and the sample was 100 mg of freeze-dried egg-white, dissolved in 4 ml of starting buffer. The linear *gradient elution* started with 0.04 *M* TRIS buffer, pH 8.6, and ended with 0.04 *M* TRIS buffer, pH 3.4, the flow rate being 40 ml/h, and the eluate was monitored at 254 nm.

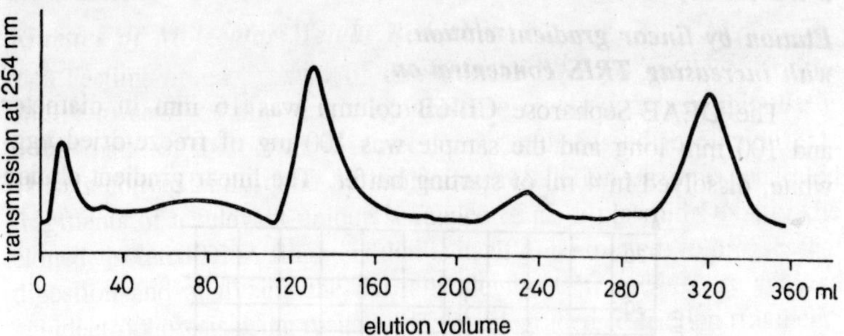

Figure 1.8 : Chromatogram of egg-white, chromatographed with a buffer of constant TRIS concentration and decreasing pH on DEAE-Sephacel.

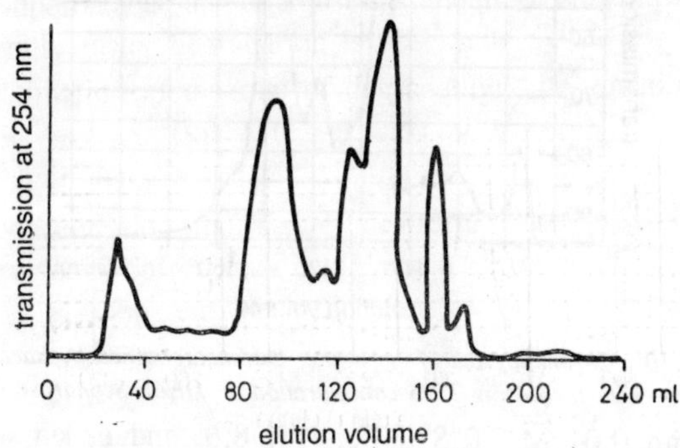

Figure 1.9 : Chromatogram of egg-white, chromatographed with a buffer of increasing TRIS concentration and decreasing pH on DEAE-Sephacel.

Elution with a buffer of increasing TRIS concentration and decreasing pH

The DEAE-Sephacel column was 16 mm in diameter and 100 mm long and the sample was 100 mg of freeze-dried egg-white, dissolved in 4 ml of starting buffer. The linear gradient elution started with 0.04 M TRIS buffer, pH 8.6, and ended with 0.5 M TRIS buffer, pH 3.4, the flow rate being 40 ml/h, and the eluate was monitored at 254 nm.

Evaluation: The two chromatographic separations demonstrate clearly the importance of the ion-concentration of the buffer in ion-exchange chromatography: the elution with a buffer of increasing ion-concentration required a smaller volume of buffer, but yielded more protein fractions.

Ion-exchange Chromatography on a DEAE-Sepharose CL-6B Column

Elution by linear gradient elution, with increasing TRIS concentration

The DEAE-Sepharose CL-6B column was 16 mm in diameter and 100 mm long and the sample was 100 mg of freeze-dried egg-white, dissolved in 4 ml of starting buffer. The linear gradient elution

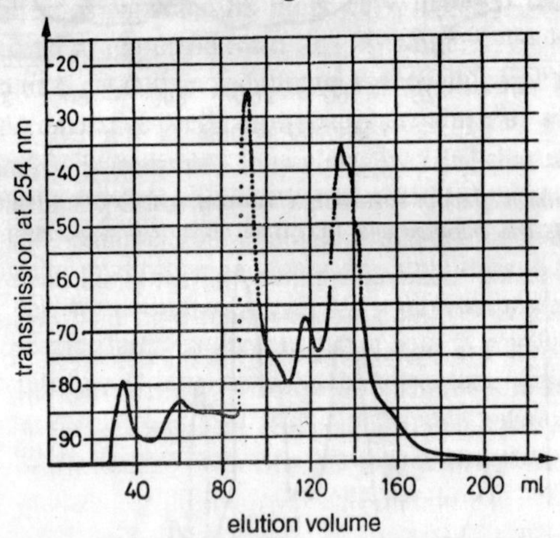

Figure 1.10 : Chromatogram of egg-white, chromatographed by linear gradient elution, with increasing TRIS concentration on DEAE-Sepharose CL-6B.

started with 0.04 M TRIS buffer, pH 8.6, and ended with 0.5 M TRIS buffer, pH 3.4, the flow rate being 40 ml/h, and the eluate was *monitored* at 254 nm.

Elution by stepwise gradient elution with increasing sodium ion concentration

The intention here was to compare the effects of linear *gradient* and *stepwise* gradient elution.

The DEAE-Sepharose CL-6B column was 25 mm in diameter and 100 mm long and the sample was 100 mg of freeze-dried egg-white, dissolved in 4 ml of *starting buffer*. The stepwise gradient elution started with 0.01 *M* TRIS-HCI, pH 8.4, the flow rate being 150 ml/h and the eluate was monitored at 254 nm. The stepwise concentrations of the second gradient component (NaCI) were:

1. 0
2. 0.09 *M*
3. 0.16 *M*
4. 0.185 *M*
5. 0.21 *M*
6. 0.265 *M*
7. 0.35 *M*
8. 0.5 *M*

Evaluation: Stepwise increase of the *ion-concentration* of the buffer gave a better separation then elution with a linear gradient. The differences between the chromatograms given by the two elution methods, especially at the beginning of elution, are caused by the different concentrations of the starting buffers.

(Work by L. Szabo, Department of Biochemistry, Jozsef Attila University, Szeged.)

Conclusion on the Results of the Chromatographic Methods Above

The significance of the quality of the ion-exchanger is shown well in the separations on CM-derivatives, where better separations are achieved with better ion-exchanger (CM-Sepharose CL-6B). As regards the three basic factors in ion-exchange chromatography (temperature of the column, ion-concentration and pH of the eluent), the importance of the ion-concentration is demonstrated well by the chromatography on the DEAE-Sephacel column .

With respect to the chromatography carried out by linear and stepwise gradient elution with increasing ion-concentration, the stepwise procedure gave more fractions. This is caused mainly by the increase in the ion-concentration, and to a smaller degree by the different *chromatographic dimensions*.

From the negatively charged egg-white proteins, more fractions may be obtained on a *DEAE-Sephacel* column, eluted with a buffer with ion-concentration increased by linear gradient, than under the same chromato-graphic conditions on DEAE-Sepharose CL-6B.

In order to establish definitely whether the sequences of the protein fractions eluted from *CM-Sepharose* CL-6B and *DEAE-Sephacel* columns are mirror images of each other, it is necessary to make control examination of the separated fractions by IEF.

GEL FILTRATION BY A COLUMN-CHROMATOGRAPHIC TECHNIQUE

The column form of gel filtration provides a possibility for separation of various proteins into fractions of proteins with the same molecular weights, by elution of a solution containing them. Since the basis of the fractionation is separation according to molecular weight, if the solution applied to the gel contains different proteins of nearly the same molecular weight, these will be contained in one *eluted fraction*. In such cases, therefore, *gel filtration* yields group separation. The composition or homogeneity of each fraction can only be assessed after further treatment of the fractions resulting from preparative-scale gel filtration .

The gel filtration of the freeze-dried egg-white was facilitated by the fact that the *molecular weights* of the component proteins were known from the literature data. On this basis, Sephadex G-75 and G-100, Sepharose 6-B and Sephacryl S-200 were used for the gel filtration.

Gel Filtration on Sephadex G-75 and G-1.00 Columns

Gel preparation: 12 g of Sephadex G-75 or G-100 Fine were hydrated with 200 ml of boiling 0.9% sodium chloride solution.

Gel column dimensions: 25 mm diameter, 400 mm long. Sample applied: 200 mg of freeze-dried egg-white, dissolved in 5 ml of 0.9% sodium chloride solution. (The membranes remaining undissolved after 6 h, comprising 4-6% of the total proteins, were filtered off with a densely-woven nylon sieve.)

Eluent: 0.9% sodium chloride solution.

Flow rate: 1 ml/min.

Detection: continuously at 254 nm.

Gel Filtration on Sepharose 6B and Sephacryl S-200 Columns

Gel column: 25 mm diameter, 350 mm long, Sepharose 6B or Sephacryl S-200 column.

Evaluation of the Gel Filtrations

The *gel filtrations* performed gave very good illustrations of the *molecular-sieve* properties of the various gel beds.

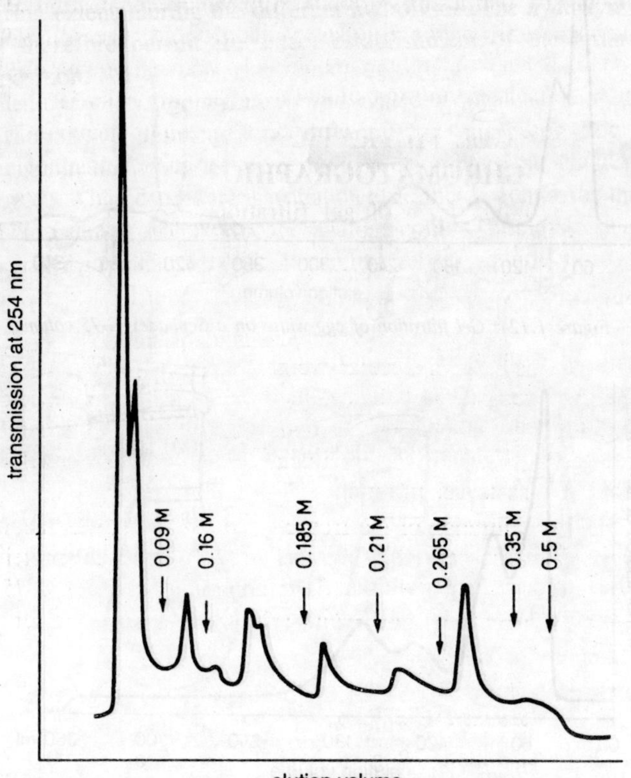

Figure 1.11 : Chromatogram of egg-white, chromatographed by stepwise gradient elution, with increasing sodium ion concentration on DEAE-Sepharose CL-6B.

With the Sephadex G-75 gel column, which is recommended for the separation of protein molecules in the M.W. range 3000-80,000, the ovalbumin and the proteins with M.W. above 45,000 (the first two peaks) were well separated. In the chromatogram obtained with Sephadex G-100, this separation was even more marked, in contrast to the proteins with lower M.W. (peaks 3 and 4). Proteins with M.W. higher than those of the *ovalbumins* were separated very well with the Sephacryl S-200 gel column. Understandably, much poorer separations were achieved with the Sepharose 6B column, which is recommended for the gel filtration of proteins with M.W. in the range 10^4-4×10^6.

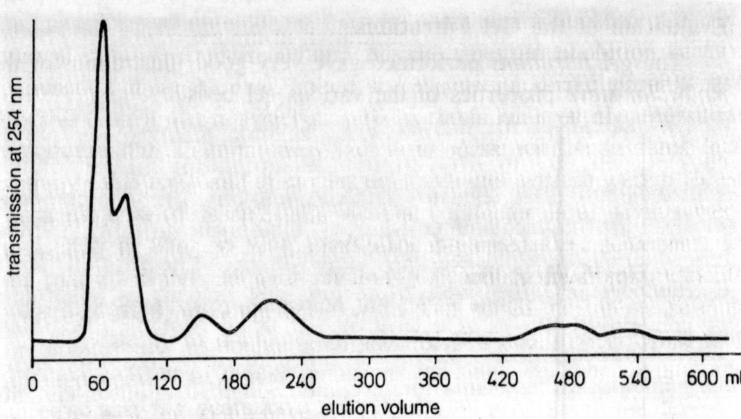

Figure 1.12 : Gel filtration of egg-white on a Sephadex G-75 column.

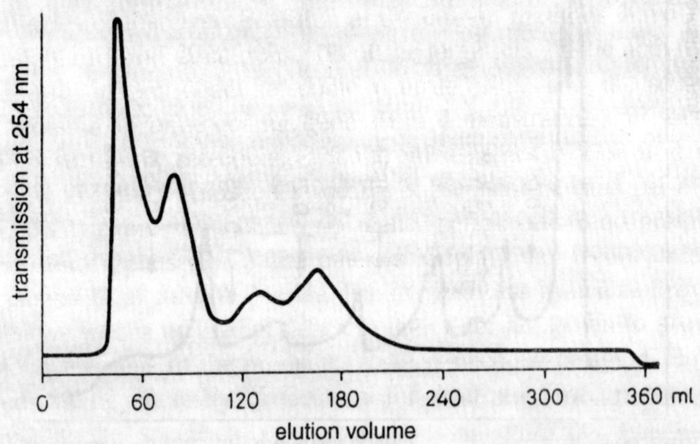

Figure 1.13 : Gel filtration of egg-white on a Sephadex G-100 column.

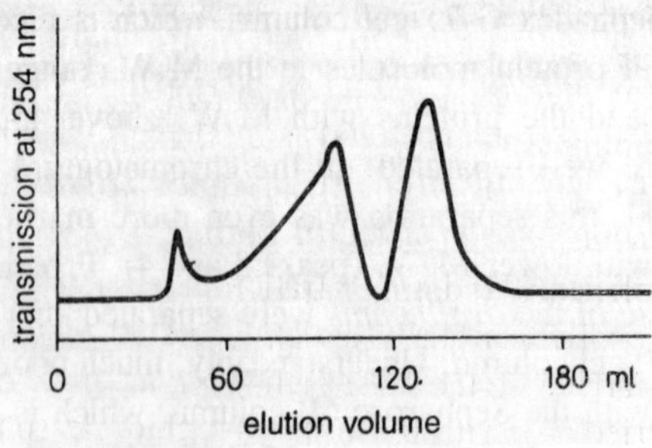

Figure 1.14 : Gel filtration of egg-white on a Sepharose 6B column.

THIN-LAYER GEL FILTRATION

In thin-layer gel filtration, separation in the gel layer also occurs on the basis of the sizes of the sample substances; accordingly, this technique may be employed for the determination of the approximate molecular weights of the *dissolved proteins*. *Pharmacia* TLG apparatus is used with Whatman 3 MM filter paper, 200 × 200 mm.

Reference solutions: cytochrome-C solution, 20 mg/ml, M.W. 12400; *soya-bean trypsin* inhibitor solution, 20 mg/ml, M.W. 21,500; ovalbumin solution, 20 mg/ml, M.W. 45,000. Sample solution: freeze-dried egg-white, 20 mg/ml. (The reference substances and the sample are dissolved in 0.9% sodium chloride solution.)

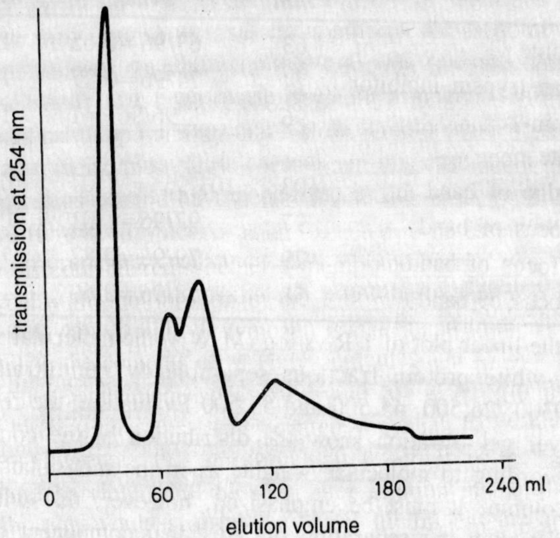

Figure 1.15 : Gel filtration of egg-white on a Sephacryl S-200 column.

Preparation of gel: with *gentle stirring*, 150 ml of boiling 0.9% *sodium chloride* solution are poured onto 8 g of Sephadex G-100 superfine in a 300 ml beaker. *Swelling* of the *Sephadex* is complete within 5-6 h. Preparation of gel layer: with the TLG spreader, a 0.6 mm thick layer is spread on a 200 × 200 mm glass plate. The gel layer is washed overnight with 0.9% *sodium chloride* solution, with the apparatus inclined at an angle of 15°.

Sample application and gel filtration: the sample and reference substances are applied in quantities of 5 and 10 μl to the gel layer, about 15 mm apart. With the plate inclined at an angle of 30°, the gel filtration is carried out with 0.9% sodium chloride solution for 4 h.

The gel layer is lifted out of the apparatus and positioned horizontally, and a Whatman 3 MM *filter paper* of the same size is rolled onto it. After 1 min, the moist filter paper is removed and immersed for 1 min in 0.1% *Bromophenol* Blue solution (in 9:1 v/v methanol-acetic acid mixture) and the excess of dye is then dissolved out by treating the filter paper with 5% acetic acid solution as in the procedure of Radola.

Results and calculations: the reference basis for the calculations was taken as the distance covered by the ovalbumin (96 mm, R= 1).

	Distance covered, mm	*R ovalb.*	*1/R ovalb.*
1. cytochrome C	63	63/96=0.65	1.53
2. soya-bean trypsin inhibitor	74	74/96=0.77	1.3
3. ovalbumin	96	96/96 =1	1
egg-white			
4. lower edge of band	103	103/96 = 1.07	0.93
5. lower focus of band	97	97/96=1.01	0.99
6. middle focus of band	79	79/96=0.82	1.21
7. upper focus of band	58	58/96=0.60	1.67

From the linear plot of 1/R vs log (M.W.), the molecular weights of the egg-white protein fractions separable by gel filtration are obtained: 9300, 26,500, 45,500 and 53,500.Evaluation: the results of the thin-layer gel filtration show the distribution of the component proteins according to molecular weight, as in the molecular sieving on a gel column. It must be emphasized, however, that with thin-layer gel filtration the separation of individual component proteins has much lower resolution than that in molecular weight determinations by means of electrophoretic procedures.

CONCLUSIONS

The procedures described are not all the best within the groups of methods in question, but even so, in our view they permit various *comparisons*.

A guide towards the selection of the most appropriate procedure is given by *analytical isoelectric focusing*. in particular to the choice of pH for the *eluents* and the choice of the sorbents in ion-exchange chromatography.

When our model protein preparation, freeze-dried egg-white, was separated by means of the different analytical methods, very good agreement was found with regard to the distribution of the component proteins. Although the resolutions proved different, the characteristic fractions were obtained by the *electrophoretic* methods, the liquid-*chromatographic* procedures, and with column and thin-layer gel filtration alike. Depending on the methods used, these fractions showed up as a large number of bands, or sometimes only as *tailing* (e.g. in thin-layer *gel filtration*).

2

BIOMECHANICAL MOVEMENT

A *vector* is a quantity that possesses the attributes of both magnitude and direction. In the transport of substances across planar surfaces two vector quantities exist, one describing movement in each direction. When we talk of cellular transport, we generally refer to the vector for the process (or processes) that carries material into a cell as the influx (J_i) and the one that carries material out as the efflux (J_e). The net vector quantity, or net flux (J_{net}), is the algebraic sum of the two vectors whose orientations differ by 180°: $J_{net} = J_i - J_e$. Thus, we assign two vectors to the transport of substances across surfaces; no more than two because only two are possible, that is, they describe the system's potentialities fully-no less than two, because transport processes are generally reversible and require both vectors.

Saying that a transport process is reversible does not mean, however, that things move equally in quantity or facility in both directions. Indeed this often is not the case. Moreover, biological transport processes tend to be at least operationally irreversible in that time passes and entropy grows. The point is merely that such processes are generally *capable* of reversible or equilibrium behaviour, even if in the normal course of events a particular substance is moved preferentially and irreversibly in one direction or another. This concept of transport as a reversible process that involves two *vector quantities* is different from that proposed for *exocytosis*. Although exocytosis is often called a "*vectorial*" transport process, it is actually a proposal for a single vector process for the movement of substances across surfaces. That is, a single vector, efflux, accounts fully for the movement of the substance, and therefore the net vector quantity or net flux is equal to, indeed identical with, efflux: Jnet = J_e because $J_i = 0$.

In concrete terms the exocytosis model proposes that concentrated material in the secretion granule enters the dilute extracellular environment through a hole, produced as a result of membrane fusion, either by diffusion or as a result of fluid flow. This is the efflux vector and the only vector, and fusion and hole formation are its underlying mechanisms. Therefore, in the exocytosis model the efflux (and net flux) of material, that is, the rate of its secretion, is determined by the frequency with which granule and cell membranes fuse and holes form, other things being equal. The greater the frequency of fusion events, the greater the rate at which the substance is secreted. Thus, in this model J_e or J_{net} is strictly proportional to the rate of fusion.

Can we devise tests of this "*univector*" hypothesis that can judge its goodness of fit to a particular secretion process? In a series of experiments in pancreas, Isenman, Ho, applied one simple test. If it is true that we are dealing with a transport process that has only an efflux vector, then increasing the concentration of the secreted substance or substances in the extracellular medium would not alter the rate of secretion or the net flux of material out of the cell, because it would not affect J_e, the rate of which is set by intracellular events (fusion), nor could it enter the cell because an influx process does not exist ($J_i = 0$). As a result, J_{net} or the rate of secretion would be unchanged and independent of the concentration of material outside of the cell. This, of course, is different from the behaviour that we would expect if both *influx* and *efflux vectors*, and as a result the potential for equilibrium behaviour, did exist; in this case, increasing the concentration of a secretion product in the extracellular medium would increase its movement into the cell. If there had been a net flux of material out of the cell, namely, secretion, before we increased the extracellular concentration, then it would be reduced because $J_{net} = J_e - J_i$ and J_i would have been increased.

Indeed, in a reversible system if we increased the concentration of the transported substrate in the extracellular medium sufficiently, then we should be able to reduce the net flux out of the cell to zero (by increasing J_i to equal J_e so that Let = 0). In a system in which there is only a single vector, this could not happen, and, as I have said, secretion should continue at essentially the same unchanging rate whatever the external concentration of the transported substrate. Therefore, we have a relatively clear predictive difference between the *exocytosis model* and an equilibrium model for secretion, at least for exocytosis as it was originally proposed and has generally been viewed: a transport process for which only an efflux vector exists. In any event, if the secretion process is equilibrium dependent or at least equilibrium capable, then we should be able to reduce the rate of secretion merely by letting secreted material accumulate in the external medium; in this way *influx* or *backflux* would increase over time. If only an *efflux vector* existed, then such an accumulation of product should have no discernible effect on the rate of secretion, which in this case should continue unabated.

BLOCKAGE OF DUCTS

In an *exocrine gland* such as the pancreas, the simplest way to test this idea is merely to block the secretory duct and to prevent the outflow of material contained in the duct system. In this way, the removal of secreted product from the immediate environment of the secretory cell can be prevented. If exocytosis is the operating mechanism, then under these circumstances we would expect to see secretion continue at a constant rate into the duct system, and as a result we would expect the concentration of product in the duct lumen to increase in a roughly linear fashion during the period of blockage as secreted protein continues to accumulate in the blocked duct.

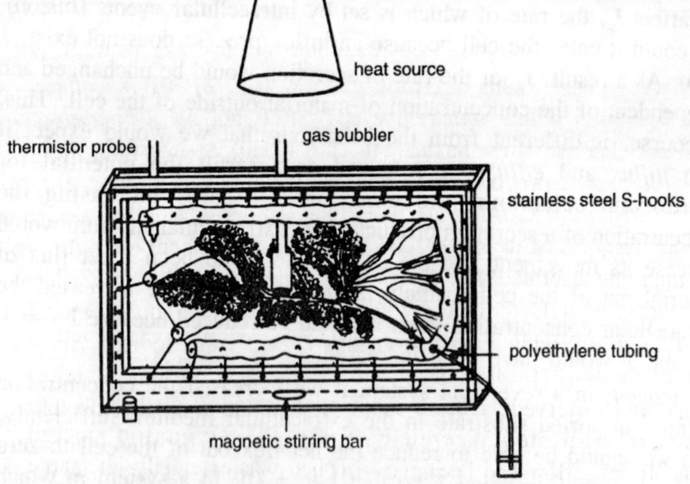

Figure 2.1 : Schematic diagram of rabbit pancreas and duodenum suspended in chamber.

If, on the other hand, an equilibrium-based process is involved, then preventing the removal of material from the site of its secretion and letting it accumulate in the immediate vicinity of the secretory cell should reduce the rate of secretion (as J_i increases) and eventually prevent it altogether (when $J_i = J_e$).

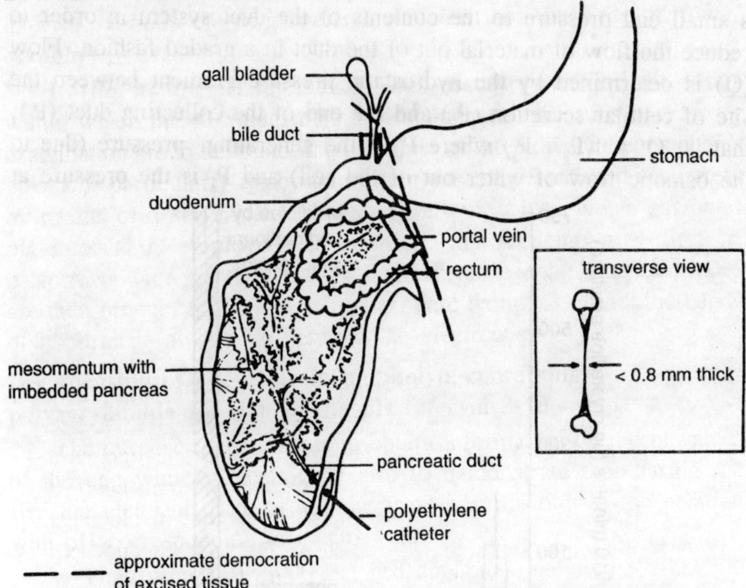

Figure 2.2 : Schematic diagram of rabbit pancreas in situ interposed between ascending and descending limbs of the mesomental loop of duodenum.

This experiment was performed using the rabbit pancreas in organ culture, an experimental system in which the whole pancreas is placed in a chamber, bathed by a physiological salt solution, and secretion collected directly from the cannulated pancreatic duct. Blockage is achieved merely by closing off the indwelling catheter. When the catheter is opened, the accumulated material can then leave the confines of the duct (be excreted and collected). A period of blockage was chosen that would lead to a roughly magnitudinal increase in the concentration of protein in the duct system if secretion continued at a constant rate during blockage.

What was observed? After a substantial period of duct blockage, the amount of protein secreted during the collection period immediately after flow had been reestablished was essentially the same as seen prior to duct blockage; that is, no substantial accumulation of material had occurred during a period in which a tenfold increase in intraductal concentration should have developed if a univector event accounted for secretion. Blockage had led to an essentially complete cessation of secretion.

Thus, the results fit an equilibrium and not a "*univector*" exocytosis view. We explored these relationships further by applying a small end pressure to the contents of the duct system in order to reduce the flow of material out of the duct in a graded fashion. Flow (Q) is determined by the hydrostatic pressure gradient between the site of cellular secretion (P_g) and the end of the collecting duct (P_e); that is, $Q = f(P_g - P_e)$ where P_g is the generating pressure (due to the osmotic flow of water out of the cell) and P_e is the pressure at

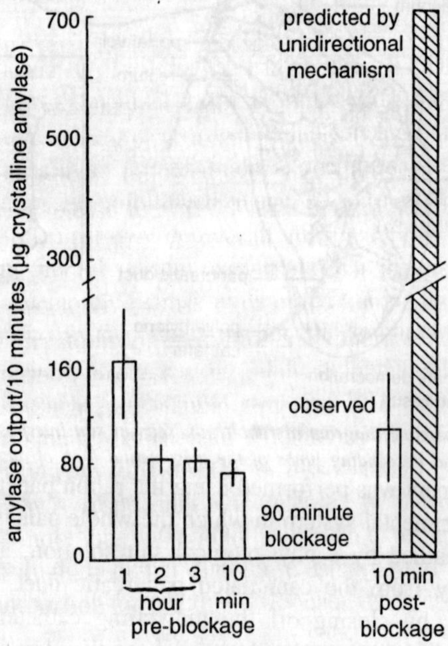

Figure 2.3 : The effect of ductal blockage on am,ylase secretion (open bars). Data are presented as mean amylase output per 10 min + SE (n = 4).

the end where the material flows out of the duct system. Therefore, if we elevate P_e, then we decrease the gradient and decrease Q as well. As we decrease flow, we also reduce the rate at which secreted material leaves the immediate extracellular environment of the secretory cell. The greater the reduction in flow, the less readily material enters the large conduits of the duct'system. When such studies were performed, we found that as flow was reduced, so was protein' secretion and in a directly proportional fashion. Specifically, the concentration of protein in the duct remained essentially unchanged, and protein "*output*," or the amount secreted, decreased in proportion to the reduction in flow. When the process was reversed and flow was gradually increased, protein secretion increased in proportion as well.

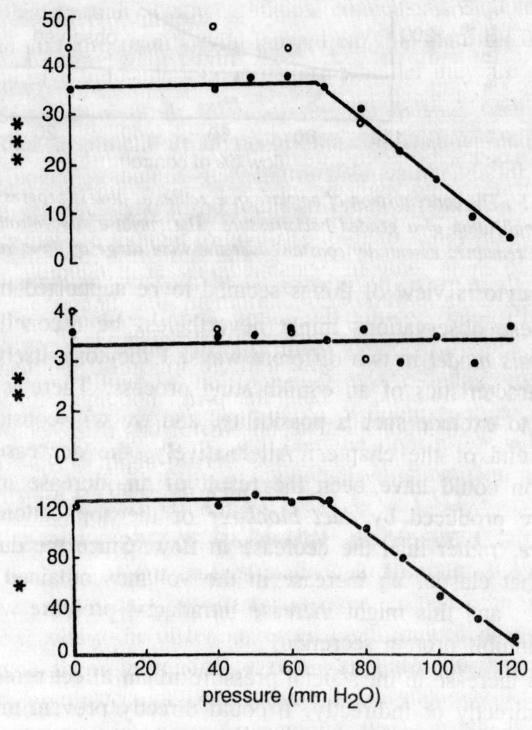

Figure 2.4 : Fluid secretion (flow) amylase concentration, and amylase output as a function of backpressure applied to the column of fluid int eh pancratic duct for a single experiment.

These observations could not be ascribed to changes in the rate of transit of already secreted material through the duct system, and a true reduction in the rate of release of protein from the cell appeared to be involved. The fact that inhibition was completely reversible indicated that there was no permanent damage to the duct system produced by blockage, and, as I have said, the equilibrium and not

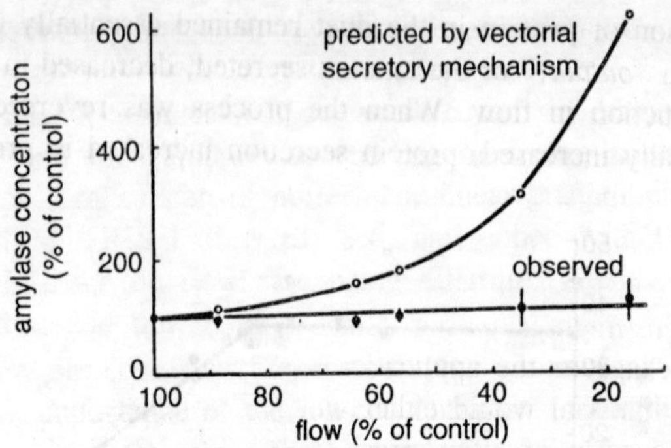

Figure 2.5 : The concentration of amylase insecretion as flow is progressively decreased by the application of a graded backpressure. The amylase concentration observed in secretion remained essentially constant over the wide range of flows studied.

the exocytosis view of things seemed to be supported by the data.

These observations might nevertheless be reconciled with the *exocytosis model* in two different ways. Exocytosis itself might have the characteristics of an equilibrating process. There is no a priori reason to exclude such a possibility, and we will consider this idea at the end of the chapter. Alternatively, the decrease in protein secretion could have been the result of an increase in intraductal pressure produced by *duct blockage* or the application of a *back-pressure*, rather than the decrease in flow. Since the duct system is somewhat elastic, an increase in the volume contained within it is possible, and this might increase intraductal pressure which in turn might inhibit protein secretion.

An increase in intraductal pressure might affect protein secretion either directly or indirectly. It could directly prevent the exocytosis of granules. In order to explain the observations in this way:

(1) exocytosis would have to be very sensitive to low pressures (the end-pressure required to abolish protein secretion totally is only about 10 mmHg);

(2) increases in intraductal pressure would have to produce proportional reductions in the rate of exocytosis (because protein secretion decreases in a linear fashion with increasing back pressure); and most significantly,

(3) two seemingly independent processes, exocytosis and fluid secretion, would have to be altered to the same degree, over the same time course, by identical pressure increments, by chance (because protein secretion and fluid flow changed to

the same proportional extent at the same applied pressure over the same period of time).

An increase in *intraductal pressure* might also indirectly reduce protein secretion, or at least appear to reduce it. If duct blockage and back-pressure led to the development of a sufficient transmural pressure gradient to open paracellular shunts through which the bulk movement of material out of the duct system could occur, then secretion by the acinar cells might actually continue at the same "*univector*" rate despite duct blockage or the application of an end-pressure, but the bulk of secreted material would either wholly (blockage) or in proportion to the applied pressure (backpressure) be diverted from the confines of the duct into interstitial spaces. Such a shunt would have to be responsive to low pressures, open gradually with increasing pressure, and close reversibly as pressure was reduced in order to explain the observations. There is no indication that such a system exists.

FLOW REDUCTIONS

Whatever one may think of these pressure-dependent possibilities, the issue is best resolved by reducing flow in ways that avoid the potential for increasing *intraductal pressure*. This can be done by reducing flow metabolically, that is, by reducing P_g, rather than increasing P_e. In this case, decreases in flow are produced by decreases in intraductal pressure. If the same effect were observed under such circumstances, then flow, and not pressure, could be assumed to be the variable directly responsible for the reduction in protein secretion. We used two different techniques to perform this test. In the first, the concentration of sodium in the medium bathing the gland was lowered. Water secretion by the pancreas is a sodium-dependent process, and a sufficient reduction in the external sodium concentration reduces the transepithelial osmotic force (by reducing the transport of sodium and associated anions) and hence the transepithelial movement of water. The second method, the addition of the Na+, K+-ATPase inhibitor ouabain, inhibits fluid flow for the same osmotic reason, although in a somewhat different manner.

Would a reduction in fluid flow produced in this way also reduce protein secretion? The answer was not only affirmative, but both low-Na^+ media and ouabain treatment produced essentially the same effect as each other and that seen with *mechano-physical methods*, namely, reductions in protein output proportional to reductions in fluid flow, thereby leaving the concentration of protein in the duct essentially unchanged.

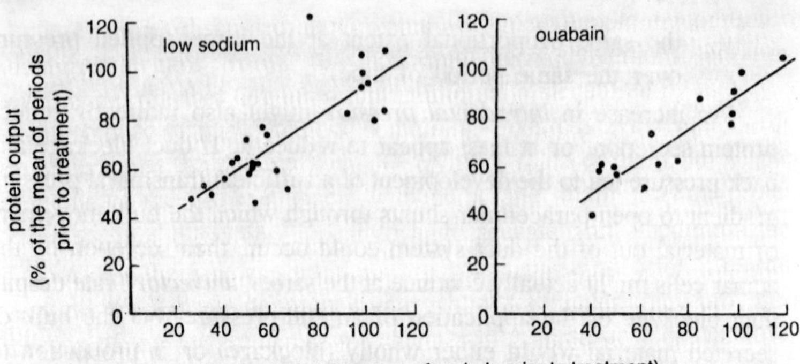

Figure 2.6 : Regression of protein output against flow in unstimulate state Both protein output and flow are expressed as percent of mean of three 15-min collection periods before either replacement of medium by one containing 50 mM sodium (n – 5) or addition of 10^{-5} M ouabain (n – 3).

The same result was obtained when the concentration of protein in the duct system was elevated initially by stimulating protein secretion with the gastrointestinal hormone, *cholecystokinin-pancreozymin*. That is, reductions in flow were accompanied by roughly proportional reductions in protein secretion, even though in this case, the concentration of protein in the duct system was about an order of magnitude higher than in unstimulated tissues. Thus, even when exocytosis is presumably most active, during augmented secretion, one can reduce protein secretion in a proportionate manner merely by reducing fluid flow.

Even so, these observations, at least if considered alone, might still be reconciled with the "*univector*" exocytosis model. For example, the reduction in protein secretion may not have been due to a true reduction in cellular secretion, but may merely have reflected a transient decrease in the *excretion* of already secreted material from the duct system. That is, the effect could merely be due to a reduction in the rate of passage of secreted protein through a ductal dead space, as noted above. If this were the case, then once the dead space of the conduit had been cleared, there would be a compensatory increase in protein output, followed by recovery to normal rates over time, reflecting the continued secretion of protein at the cellular level.

This was not observed. In the low-Na+ situation, the initial reduction in protein secretion was maintained for many hours during which time the functional, as well as the anatomical dead space, of the duct system had been cleared many times over.

Another possibility is that altering ion gradients might affect the

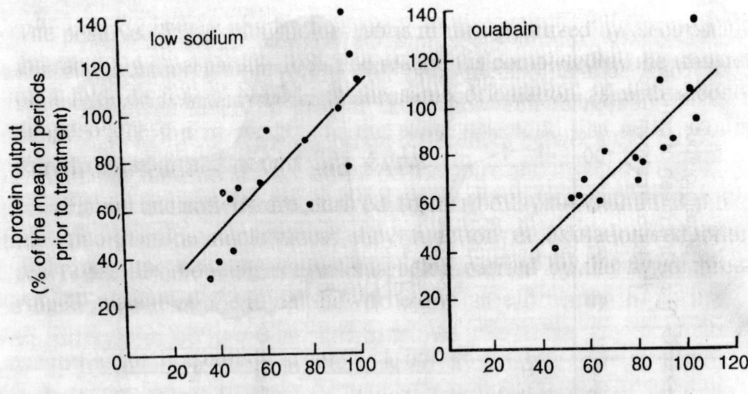

Figur 2.7 : Regression of protein output against flow after stimulation with cholecystokinin-pancreozymin. Both protein output and flow are expressed as percent of mean of four 5-min collections before bath replacement or addition of ouabain (low sodium, n - 3; ouabain, n = 3).

activation of a putative exocytosis mechanism. A low-Na^+ medium of course reduces the sodium gradient into the cell, and this might alter the frequency with which exocytosis occurs in some fashion or another. Alternatively, ouabain, even in the short run, might increase intracellular sodium locally which might also in some way reduce the rate of exocytosis. We chose these two particular metabolic interventions in part to help deal with these possibilities. Ouabain leaves the Na^+ gradient into the cell intact (as opposed to a low-Na^+ medium), and conversely, a low-Na^+ medium is likely to reduce, not increase, intracellular Na+ (unlike ouabain). Nevertheless, both treatments produced almost identical effects, reinforcing the idea that flow was indeed the relevant independent variable.

Whatever questions one may be able to raise, it must be borne in mind that not only did two different metabolic methods produce the same effect, but they were essentially identical to that seen when flow was reduced by mechano-physical methods. Moreover, the same effect was seen in both the unstimulated state and during highly augmented secretion produced by a secretagogue. In all cases, reductions in flow were accompanied by proportional reductions in protein secretion. Although different methods reduced flow in different ways for different reasons, they all affected protein secretion in a remarkably similar fashion. Seemingly, the sole common link between the various treatments was the reduction in flow itself.

We tried a final means of reducing flow metabolically. In this

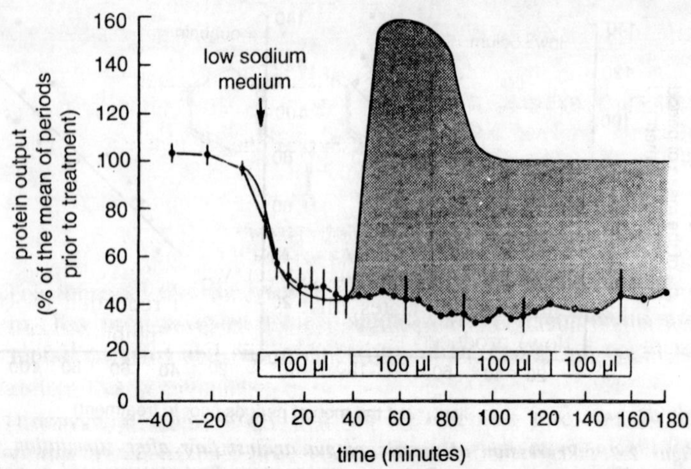

Figure 2.8 : Effects of low-Na⁺ medium on protein output in unstimulated state.

case we merely reduced the temperature of the medium bathing the gland. Fluid flow is dependent, as I have said, on ion transport, and ion transport is a temperature-sensitive process. Although it would certainly not be surprising to find that a reduction in temperature reduced both flow and protein secretion even if they were totally independent processes, one would not expect to find that their

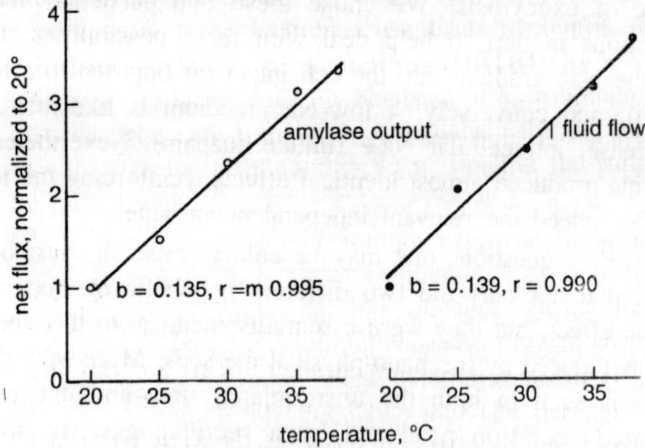

Figure 2.9 : Temperature dependence of fluid and enzyme secretion by whole rabbit pancreas in vitro.

temperature sensitivity was almost identical. This is what was observed, that is, over a temperature range of 20-37°C both processes exhibited the same temperature dependence, once again suggesting a connection between the two.

EQUILIBRIUM BEHAVIOUR

With whatever attendant uncertainty, it seemed reasonable to draw the conclusion that flow was the independent variable that altered protein secretion in these studies. If so, its effect could have been produced in two ways. We have discussed the first, namely, altering the rate of removal of product from the *site of secretion*. It is also possible that the reduction in ion transport itself might have reduced protein secretion if the two processes were linked in some fashion. The fact that the same proportional effect was observed in both stimulated and unstimulated states, that is, a given reduction in ion transport (and flow) produced very different absolute inhibitory effects on stimulated as opposed to unstimulated secretion, argues against this possibility. However, this question, as well as some of the others that I raised above, could best be dealt with if we were able to alter protein secretion in a volume-dependent fashion that did not require chemically or physically altering the secretion of fluid and electrolyte (and that might also bypass some of the problems related to the passage of secreted material through a ductal conduit). That is, one might still argue that although these results are certainly consistent with an equilibrium explanation for the process and do not appear consistent with the exocytosis model, each test involved an artificial circumstance (end-pressure, duct blockage, the addition of a chemical, removal of sodium, or altering temperature), and it is possible that each in some other unknown fashion could have reduced the rate of protein secretion in a proportional manner to the reduction that they produced in fluid flow.

Partly to deal with this possibility, Ho and I used a system in which one can test the equilibrium behaviour of secretion without introducing such "*artificial*" variables. It has been known for some time that if one places pieces of pancreatic tissue in a bathing medium and follows the release of an enzyme (usually amylase) into that medium, that the rate of secretion falls off exponentially with time and that after a few hours approaches zero. This decrease has usually been attributed to metabolic fatigue, indicative of the time-dependent failure of the tissue in vitro. We observed the same phenomenon with pieces of rabbit pancreas. Over a period of about 3 hr, the rate of secretion fell off by some 80% relative to that seen initially [from 51.3 ± 7.5 (SE) jig-min-1-g-1 tissue wet weight for the first 15 min to 8.7 ± 5.1 for the last (n = 10). However, when the same tissue was studied in the whole organ preparation discussed above and

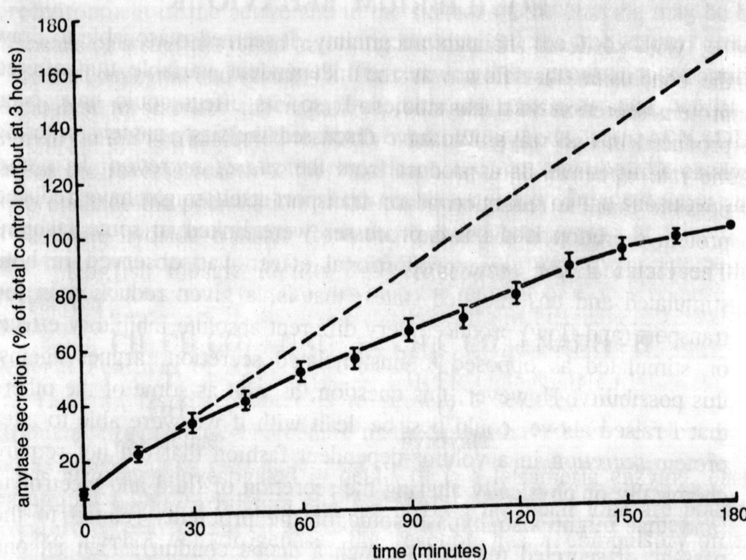

Figure 2.10 : Secretion of amylase by strips of rabbit pancreas. Amylase secretion is normalized relative to total amylase output into bath at end of 3 hr. for each individual experiment.

secretion collected directly from the duct system, such a falloff was not observed, and indeed the rate of amylase secretion was maintained relatively constant over the same 3-hr period. Why should the difference

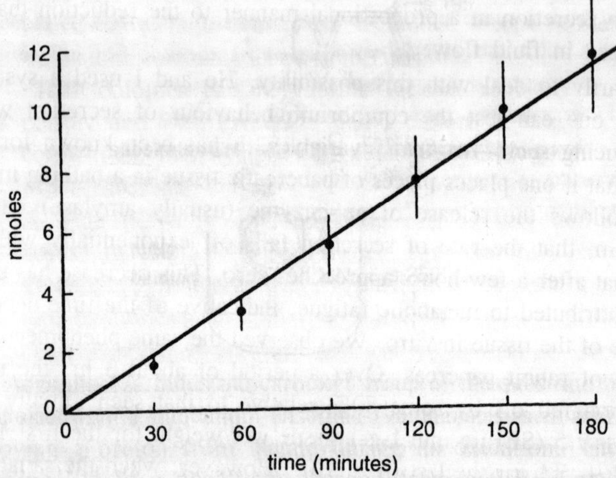

Figure 2.11 : Cumulative amylase output into ductal secretion. Secretion was collected by means of a catheter in the pancreatic duct of the organ in vitro.

between these two very similar situations exist? It occurred to us that the reason for the falloff might be related to the equilibrium nature of secretion and not tissue fatigue. In the organ preparation, secreted material was removed (by fluid flow) from the site of its secretion into a collecting vial, whereas when pieces of tissue were incubated in media, the secretory product accumulated in the bathing medium. Therefore, it was possible that in the second case, the accumulation of product led to the development of a backflux or influx (J) of already secreted material into the tissue, in this way reducing the net efflux (Jnet) over time.

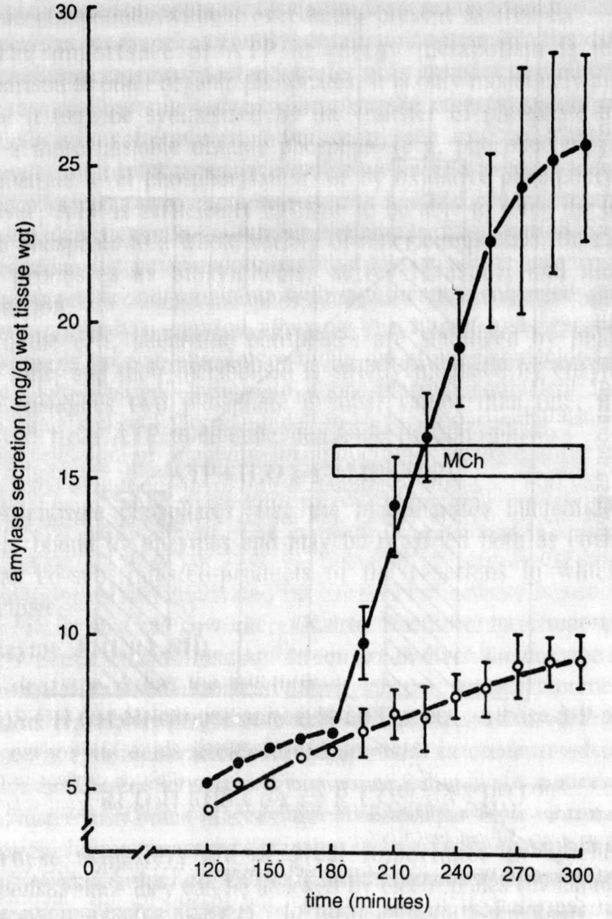

Figure 2.12 : Effect of addition of colinergic drug acetyl-β-methylcholine chloride (MCh) to tissue strips after 3-hr incubation in vitro.

We first determined whether or not the pieces of tissue were actually viable after a 3-hr incubation period, by adding a stimulant of protein secretion at this time (when the rate of amylase release was negligible). The addition of a *cholinergic* agonist led to a prompt and full response, and protein secretion was increased to about the same extent as when the stimulant was added immediately after the tissue was incubated in vitro. Thus, the tissue was viable and able to respond to an appropriate stimulus.

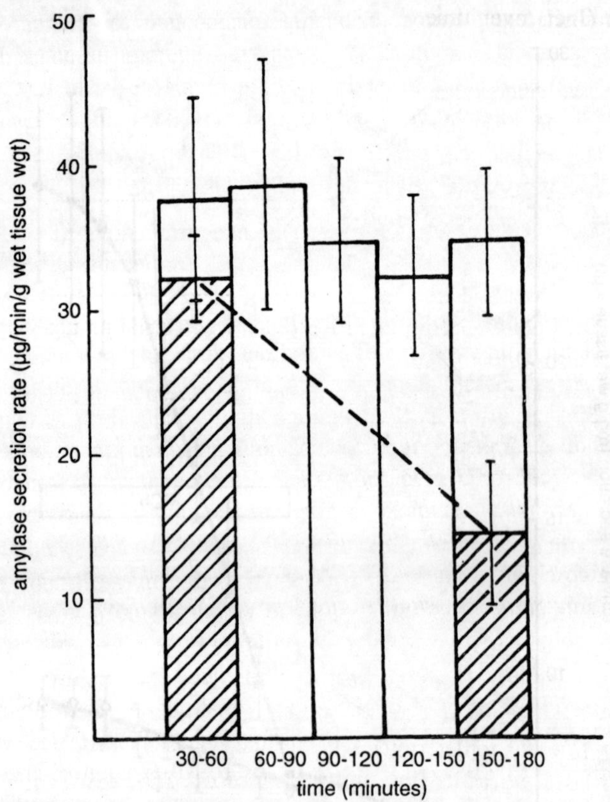

Figure 2.13 : Effect of periodic replacement of bathing medium on amylase secretion by pancreas strips. Medum bathing tissue strips was replaced by an equal volume of fresh medium every 30 min (open bars).

Next we attempted to determine whether we could prevent the falloff in secretion. If it was true that the reason for the decrease in *amylase secretion* over time was its accumulation in the bathing medium, then we should be able to prevent the falloff merely by

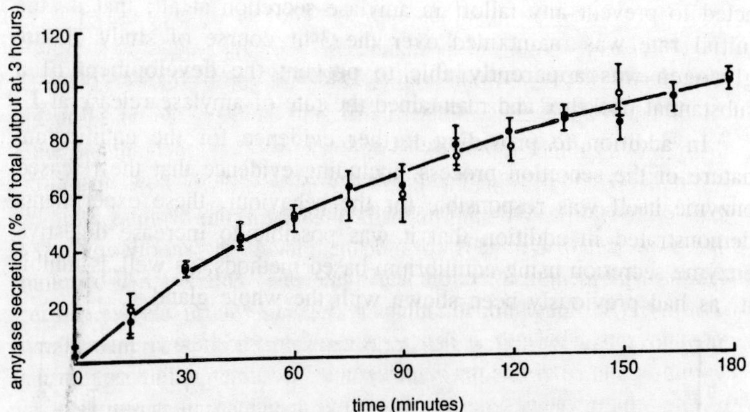

Figure 2.14 : Effect of periodic replacement of an external bathing medium amylase secretion by tissue in a dialysis bag.

replacing the medium periodically, thereby preventing the development of a substantial J_i and maintaining secretion at or close to the initial rate of efflux J_e in the absence of J_i). When this was done, secretion was indeed maintained at a high rate over the 3-hr period of study, and the falloff was no longer observed.

In order to determine whether the cause of the falloff was the release of a large molecule from the tissue, rather than the result of the release of a small, dialyzable substance or the rapid utilization of some component of the incubation medium, all of which were dialyzable, we placed the tissue pieces and media in a dialysis sac, which was then bathed by a large external medium. We changed the external medium periodically, but left the medium within the sac unchanged over the 3-hr course of study. When this was done, the falloff in amylase secretion was still seen, as if the medium had not been changed. Thus, it seemed clear that the cause of the falloff was the release of a nondialyzable compound from the tissue.

In order to determine whether the released compound was indeed amylase, we added the amylase substrate glycogen to the medium. We reasoned that if amylase accumulation in the medium had been the cause of the falloff (due to the development of a back-flux), then perhaps an amylase substrate would bind amylase and compete with the influx process, and we might be able to competitively inhibit the development of a back-flux. In this case, we should be able to maintain the rate of amylase release at a higher rate than seen in the absence of the substrate. Not only was this observed, but glycogen acted to prevent any falloff in amylase secretion at all; that is, the initial rate was maintained over the 3-hr course of study. Thus, glycogen was apparently able to prevent the development of a substantial backflux and maintained the rate of amylase release at J_e.

In addition to providing further evidence for the equilibrium nature of the secretion process, including evidence that the released enzyme itself was responsible for the behaviour, these experiments demonstrated in addition that it was possible to increase digestive enzyme secretion using equilibrium-based methods, as well as inhibit it, as had previously been shown with the whole gland.

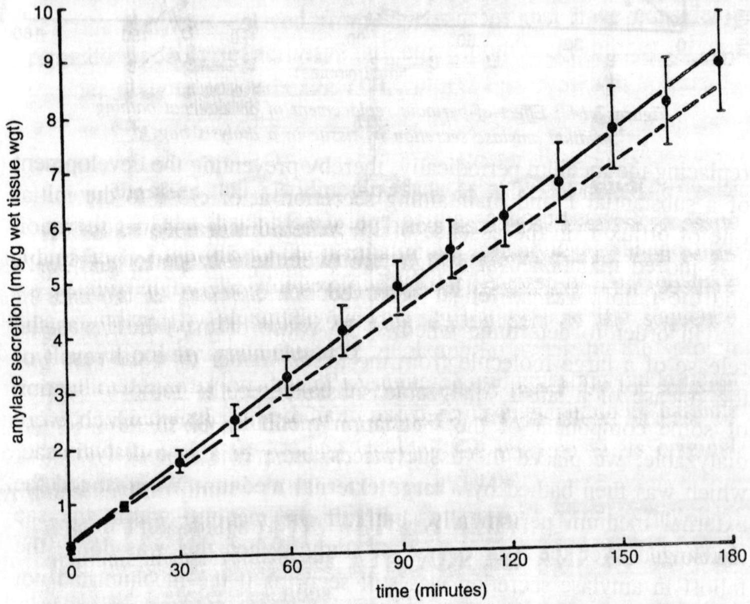

Figure 2.15 : Effect of addition of 0.5% glycogen to medium on amylase secretion.

We followed this line of investigation one step further. We attempted to inhibit the secretory response of strips of pancreatic tissue to a cholinergic agonist by limiting the volume of medium in which secretion was allowed to accumulate. To do this, it was necessary to reduce the volume of the medium to approximate the volume of the tissue itself, so that the sizes of the compartments were roughly comparable. Towards this end, we placed pieces of tissue and a small amount of medium in dialysis sacs. These sacs were then incubated in a much larger external bathing medium. In this way, the tissue had access to the large volume of external medium and all of its dialyzable contents, but *nondialyzable substances* were of course retained within the sac. Using such an experimental system, we were able to inhibit the response to a cholinergic stimulus by about 90%; that is, merely by reducing the nondialyzable volume of the medium, we were able to abolish almost completely the response to a powerful secretagogue. Moreover, we were able to establish a relationship between the volume of the internal incubation medium and the magnitude of the response, indicating its volume dependence; the larger the volume, the greater the response.

The fact that inhibition was indeed due to the volume restriction for nondialyzable compounds was confirmed in the following manner. Tissue was incubated for several hours in a small sac volume that inhibited the response to a cholinergic drug by about 90%. The dialysis sac was then perforated with numerous holes through which nondialyzable material could pass readily. When this was done, amylase secretion escaped from the inhibition and rapidly increased to highly augmented rates.

PERMEABILITY OF THE DIGESTIVE ENZYMES

Another means of examining the equilibrium nature of *digestive enzyme* transport that can avoid some of the problems of dealing with a ducted gland is to look for movement across other membranes that might be permeable to these proteins. For a number of reasons it seemed likely that the basolateral membrane of the acinar cell might be, and Isenman attempted to demonstrate such a permeability. If it existed, studying transport across this surface might be particularly useful, because there were no preconceived notions about such a process, nor was there evidence of vesicular transport across this membrane.

Exocytotic figures had not been reported, nor was there any indication that granules were normally oriented in an endocrine direction in the cell, although they were at times found in the basolateral region. We were able to demonstrate amylase release across the basolateral surface into interstitial space, and our studies suggested, as later work confirmed, that there is a natural endocrine secretion of at least certain digestive enzymes that is responsive to stimulants and is distinct from ductal secretion.

Amylase transport had the characteristics of an equilibrating process. Isenman followed the appearance of amylase in the medium bathing the rabbit pancreas in culture. Using this preparation, ductal

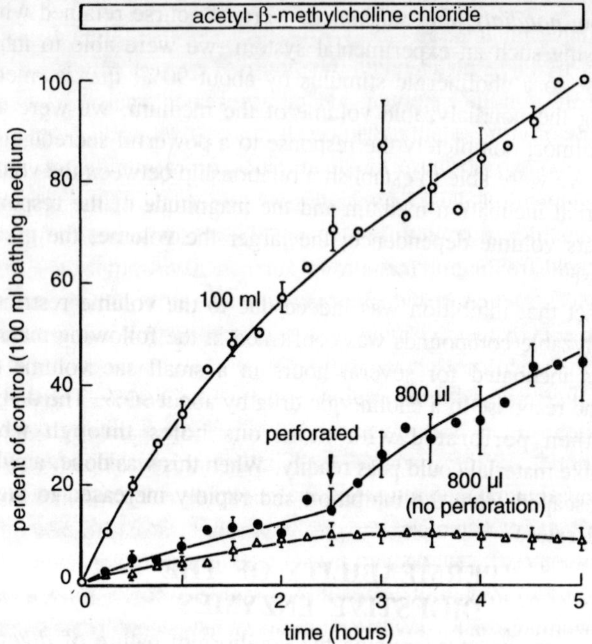

Figure 2.16 : The effect of bathing medium volume on the rate of amylase secretion by strips of rabbit pancreas stimulated by a cholinergic agonist (acetyl-β-methylcholine, 1.0 mg/100 ml medium).

and basolateral secretions can be followed and collected separately. She found much the same kinetic pattern of release as was seen several years later in the tissue strip experiments that I have just discussed. Release was greatest initially and the rate decreased over time in a roughly exponential fashion, establishing a steady-state or stationary concentration in the bathing medium in several hours. As with tissue strips, the addition of a cholinergic agonist after equilibration, elicited a prompt and substantial secretory response. Release was also found to be temperature sensitive, with a "Q_{10}" for transport of 2.3, suggesting a reaction-dependent process. Furthermore, the steady-state concentration of amylase in the medium was very similar to normal levels found in rabbit plasma. Isenman was able to demonstrate the equilibrating nature of the process by periodically changing the medium bathing the tissue and reestablishing the initial rate of release. That is, the rate of release was not a function of some event endogenous to or within the tissue that occurred over time but rather was determined by the concentration of amylase in the bathing medium. If the medium

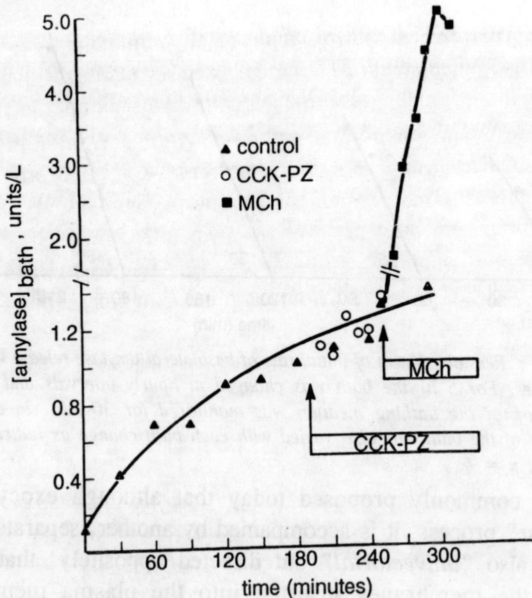

Figure 2.17 : Effect of pancreozymin (cholecystokinin) (CCK-PZ) and acetyl-fi-methylcholine chloride (MCh) on the concentration of amylase in the bathing medium.

concentration was reduced by replacing it, then the initial rate of release was reestablished. Thus, digestive enzyme appears to be transported across the basolateral surface of the acinar cell in an equilibrium-dependent fashion in much the same way as across its apical membrane, across the membrane of the zymogen granule.

NATURE OF THE EQUILIBRIUM

The experiments provide compelling evidence for the equilibrium nature of the secretory process although its details, such as the number and size of the involved pools, the maintenance of steady-state concentration gradients, the effect of different stimulants on the distribution, and so forth, remain to be worked out.

This having been said, however, the existence of equilibrium-based transport does not necessarily mean that individual protein molecules pass through the relevant membranes of the acinar cell accounting for this behaviour, although the observations are certainly consistent with this point of view. It is possible to imagine that vesicular processes are not truly "(uni)*vectorial*," as has been thought, but are themselves capable of equilibrium behaviour. This can be considered in two ways.

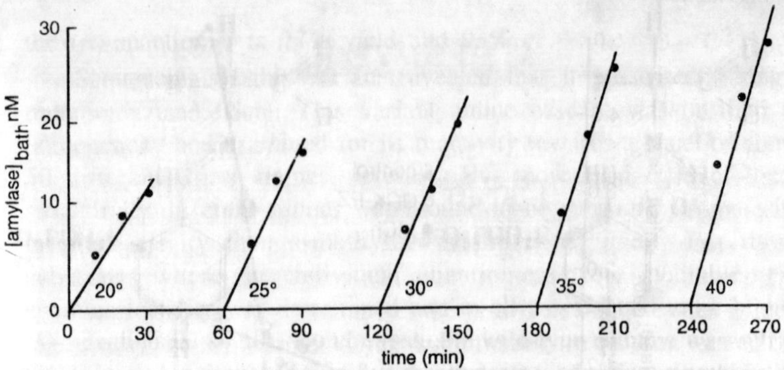

Figure 2.18 : Reestablishment of initial rate of basolateral amylase release with repetitive bath changes. For 5 hr the bath was changed at hourly intervals and the amylase concentration of the bathing medium was monitored for 30 min. In addition, the temperature of the bath was also varied with each bath change as indicated. For all data points, n = 6.

It is commonly proposed today that although exocytosis is a "*univector*" process, it is accompanied by another, separate vesicular process, also "*univectorial*" but directed oppositely, that exists to recover the membranes inserted into the plasma membrane by *exocytosis*. This is the idea of endocytosis in which invaginations in the cell membrane form intracellular vesicles. Such vesicles could carry secreted material back into the cell, ergo, the appearance of equilibrium behaviour.

We could also propose that exocytosis itself is reversible and that the amount of material released by a given fusion event is not an all-or-none phenomenon but can be graded, being responsive to the concentration of secreted material in the extracellular environment. This must not be confused with varying the rate at which material leaves the vesicle environment after fusion, which is not a steady-state effect and would not alter the steady-state rate of secretion, but only the rate at which a new steady-state is achieved. To alter the steady-state rate of secretion, partially emptied granules would have to reseal after having released only a portion of their contents.

The evidence from the experiments poses three problems for either of these hypotheses (*endocytosis or resealing*) insofar as their being able to account for the decreased rate of protein secretion seen during flow reductions, in which circumstance exocytosis itself, determined by intracellular events, presumably would continue at a constant rate. First, increases in reuptake, by either means, require a driving force. The only obvious possibility is the released material itself. For example, if the concentration of enzyme were increased in the immediate vicinity of the cell, then this might act to increase the rate of endocytosis or resealing. But in the studies discussed above, the concentration of protein in the secreted fluid was not significantly increased as output was reduced. Indeed, it was possible to reduce secretion by some 90% without any observable increase in concentration. Under such cir-cumstances we would have to propose a 10-fold increase in endocytosis or resealing, in the apparent absence of a force to produce such a change.

The second problem is also a curious one. If either mechanism accounted for our observations, then how would we be able to vary the rate of protein secretion under normal circumstances? As more material is secreted, an equivalent increase in reuptake would be expected. Also, at least in the unstimulated state, there would be a considerable dilution of the secreted material released by exocytosis into the watery environment of the duct lumen (or bathing medium for tissue strips) from concentrated stores in the secretion granule. In such a circumstance, reuptake would seemingly have to be a considerably more active process, perhaps magnitudinally so in terms of membrane utilization, than the original *exocytosis* event.

Moreover, the reduction in enzyme secretion by tissue strips over time occurs in a constant volume of medium; that is, in this case the rate of protein secretion is time dependent, not volume dependent, and likewise so would a putative reuptake process have to be viewed as time dependent, and not volume dependent. That is, the process would have to be volume dependent in one situation and time dependent in another, a seemingly unlikely possibility although one that is explained simply in terms of a diffusive equilibrium.

3

Enantioselectivity of Enzymes

Academic and industrial interest in the catalytic asymmetric preparation of enantiomerically pure or enriched organic compounds continues to grow rapidly, biocatalytic routes drawing increasing attention. Indeed, already now a significant number of industrial enantioselective processes are based on the use of enzymes. In addition to progress in *biotechnological engineering* (e.g. reactor design), two important developments in the 1990s resulted in even greater interest in the use of enzymes in *asymmetric* catalysis, namely directed evolution of *enantioselective* enzymes and metagenome DNA panning. The evolutionary approach is based on the proper combination of *molecular biological* methods for random gene mutagenesis and gene expression coupled with appropriate high-throughput screening systems to assess the enantiopurity of the chiral products. Typically, thousands of samples arising from the catalytic action of the evolved enzyme variants on a given substrate of interest need to be assayed within a day or two. A similar analytical problem arises in the case of *metagenome* DNA panning in which large numbers of genes are collected in the environment followed by expression of the encoded potentially *enantioselective* enzymes in recombinant *microorganisms*.

The *enantioselectivity* of an enzyme catalyzing a given transformation is traditionally determined by the *enantiomeric* excess of the product (*% ee*) or, in the case of kinetic resolution of a racemate, by the selectivity factor *E*, which reflects the relative reaction rates of the two *enantiomers*. Normally gas *chromatography* (GC) or high-performance liquid *chromatography* (HPLC) based on chiral columns is employed, but the conventional forms of these analytical tools can handle only a few dozen samples per day. For this reason medium- or preferably high-throughput *ee-assays* had to be developed, as in the new field of combinatorial asymmetric transition metal catalysis.

This chapter focuses on the most efficient and practical high-throughput *ee*-screening systems developed specifically to evaluate *enantioselective* enzymes.

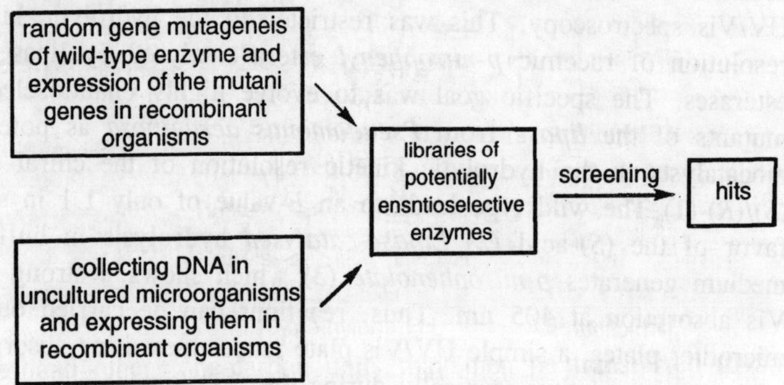

Figure 3.1 : Two sources of large libraries of potentially enantioselective enzymes.

A few other rapid *ee-assays* originally developed for combinatorial *asymmetric* transition metal catalysis, but adaptable to *biocatalysis*, are included as well.

Due to space limitation only a few of the currently available *ee-assays* are illustrated by detailed *protocols*. Many of the *ee-screening* systems are complementary, and no single assay is truly universal. For general information concerning high-throughput *ee-screening* systems, the reader is referred to recent reviews. Selection, as opposed to screening, has not been developed to date in the directed evolution of *enantioselective* enzymes. However, a screening system based on differential cell growth has been described, which is a rare case of a colony-based *ee-assay*. All of the assays described in this chapter are carried out in the spatially addressable wells of *microtiter plates* following bacterial colony picking. If possible, it is advisable to apply an efficient prescreen for activity, which eliminates inactive mutants and thus makes it easier to perform the *ee-assay*.

UV/VIS SPECTROSCOPY-BASED ASSAYS

UV/Vis-based assays have a number of advantages, including the possibility of visual *prescreening* for activity on *microtiter plates*.

More importantly, if a reliable UV/Vis signal arises as a consequence of an *enzymatic* reaction, commercially available UV/Vis plate readers can be used in *screening* thousands of mutant *enzymes catalyzing* reactions in the wells of *microtiter plates*.

Assay for Screening Lipases or Esterases in the Kinetic Resolution of Chiral *p*-Nitrophenyl Esters

The first, though somewhat crude, high-throughput *ee-assay* used in the directed evolution of enantioselective enzymes was based on UV/Vis spectroscopy. This was restricted to the hydrolytic kinetic resolution of racemic *p-nitrophenyl* esters catalysed by lipases or esterases. The specific goal was to evolve highly enantioselective mutants of the *lipase* from *Pseudomonas aeruginosa* as potential biocatalysts in the hydrolytic kinetic resolution of the chiral ester (*S*)/(*R*)-(**1**) The wild type leads to an *E*-value of only 1.1 in slight favor of the (*S*)-acid (**2**) *Lipase-catalysed* hydrolysis in buffered medium generates *p-nitrophenolate* (**3**) which shows a strong UV/Vis absorption at 405 nm. Thus, reactions can be carried out on microtiter plates, a simple UV/Vis plate reader measuring absorption as a function of time (typically during the first 8 min).

However, since the racemate delivers only information concerning the overall rate, the (*S*)- and (*R*)-substrates **1** were prepared and studied separately pairwise on 96-well microtiter plates. If the slopes of the absorption/time curves differ considerably, a hit is indicated (i.e. an enantioselective lipase variant has been identified which is then studied in detail in a lab-scale reaction using traditional chiral GC).

H_2O, lipase-variants

rac-1 ($R = n\text{-}C_8H_{17}$)

(*S*)-2 + (*R*)-2 + 3

Figure elsewhere in this chapter shows two experimental plots, illustrating the presence of a nonselective lipase and a hit. About 500–800 plots of this kind are possible per day. Using error-prone *polymerase* chain reaction (epPCR), saturation *mutagenesis*, and DNA

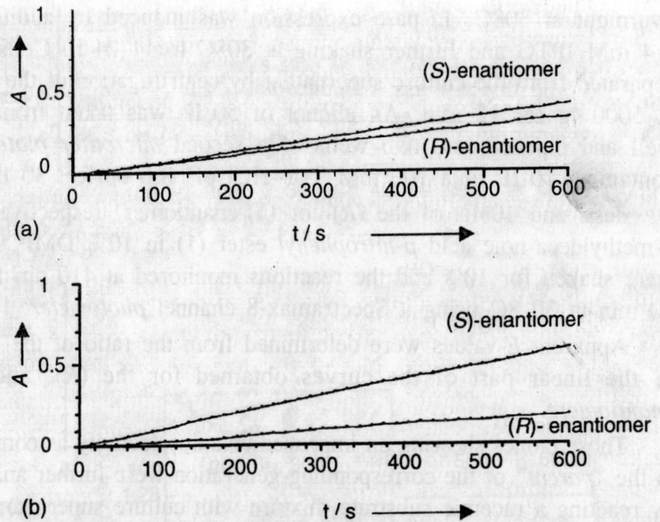

Figure 3.2 : Time course of the lipase-catalysed hydrolysis of the (R)- and (S)-ester 1 measured with a UV/Vis plate reader [5 a, 5 h]. (a) Wild-type lipase from P. aeruginosa, (b) improved mutant in the first generation.

shuffling, a total of 40 000 lipase-variants were generated and screened in the model reaction. Several *enantioselective* lipases variants were obtained, the best one showing an *E*-value of > 51. Moreover, reversal of *enantioselectivity* was achieved (*E*= 30 in favor of (*R*)-2), a process which again required only about 40 000 mutants.

The disadvantage of this assay has to do with the fact that a built-in chromophore is required (*p-nitrophenol*), yet *p-nitrophenyl* esters will never be used in real (industrial) applications. If directed evolution is being applied, the final out-come may be influenced by the fact that the *p-nitrophenyl* ester and not the usual methyl or ethyl ester was chosen. Moreover, since the (*S*)- and (*R*)-substrates are tested separately pairwise, the enzyme does not compete for the two substrates, introducing some uncertainty. Nevertheless, the assay is useful in proof-of-principle studies.

UV/Vis-based Screening of Lipases or Esterases in Kinetic Resolutions

Each colony of *P. aeruginosa* PABST7.1/pUCPL6A was picked with a colony picker and resuspended in a well of a 96-deep-well microtiter plate filled with 1 mL of 2´ LB (20 g L^{-1} trypton, 10 g L^{-1} yeast extract, 10 g L^{-1} NaCl) containing 200 lg mL^{-1} *carbenicillin* and 50 lg mL^{-1} tetracycline. Cells were grown by shaking the plates overnight at 308C. Li-pase expression was induced by addition of 0.4 mM IPTG and further shaking at 308C for 4-24 h. Cells were separated from the culture supernatant by centrifugation of the plates at 5000×*g* for 15 min. An aliquot of 50 lL was taken from each well and pipetted into two wells of a second *microtiter plate* each containing 10 lL of a 100 mM Tris–HCl pH 8.0 buffer, 30 lL A.

dest. and 10 lL of the (*R*)- or (*S*)-enantiomer, respectively, of 2-methyldeca-noic acid *p-nitrophenyl* ester (**1**) in 10% DMF. Plates were shaken for 10 s and the reactions monitored at 410 nm for 6–10 min at 30 8C using a Spectramax-8 channel *photometer*.

Apparent *E*-values were determined from the ratio of the slopes of the linear part of the curves obtained for the (*R*)- and (*S*)-*enantiomers*.

Those clones showing an improved enantioselectivity as compared to the "*parent*" of the corresponding generation were further analysed by reacting a racemic substrate mixture with culture supernatant and subsequent analysis of the reaction products by conventional chiral GC. The genes encoding *lipases* with improved enantioselectivity were sequenced. For this, the *Bam*HI/*Apa*I-fragments from pUCPL6A were cloned into pBluescriptII KS giving pMut5 which also served as the template for the next round of epPCR.

Enzyme-coupled UV/Vis-based Assay for Lipases and Esterases

In those cases in which the product of an enzymatic reaction under study can be transformed by another enzyme into a secondary product which gives rise to a *spectroscopic* signal, an enzyme-coupled assay is possible. This was first demonstrated using *fluorescence* as the spectroscopic detection method. In a UV/V is approach using this idea the hydrolase-catalysed kinetic resolution of chiral acetates was studied using a high-throughput *ee-assay* based on an enzyme-coupled test, the presence of a fluorogenic moiety or a UV/V is chromophore not being necessary. The assay is based on the idea that the acetic acid formed in the hydrolysis of a chiral acetate can be transformed *stoichiometrically* into NADH via a series of coupled enzymatic reactions using commercially available enzyme kits. The NADH is then easily detected *UV-spectroscopically* in the wells of microtiter plates using a standard plate reader. It has been estimated that about 13 000 samples can be evaluated per day, although in practise this has not been demonstrated. Using various commercially available *hydrolases* the *kinetic* resolution of (*S*,*R*)-1-methoxy-2-

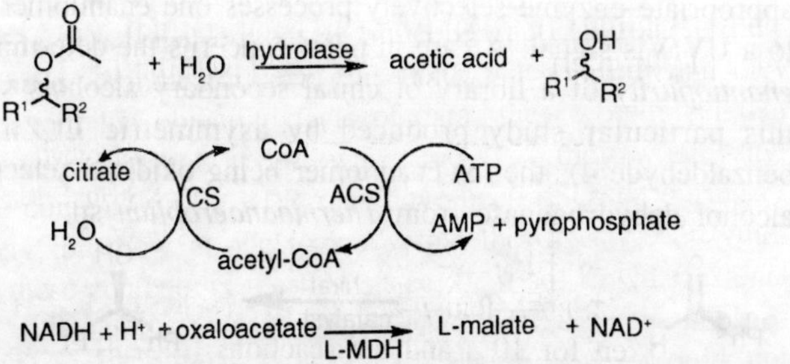

O
O ⟋ + H_2O —hydrolase→ acetic acid + OH
R^1 R^2 R^1 R^2

citrate CoA ATP
CS ACS
H_2O acetyl-CoA AMP + pyrophosphate

NADH + H^+ + oxaloacetate —L-MDH→ L-malate + NAD^+

Figure 3.3 : Enzyme-coupled assay in which the hydrolase-catalysed reaction releases acetic acid.

propylacetate was studied. The agreement between the apparent selectivity factor E_{app} and the actual value *E*true determined by GC turned out to be excellent at low enantioselectivity (*E* = 1.4–13), but less so at higher *enantioselectivity* (20% variation at *E* = 80). This may cause problems when attempting to enhance enantioselectivity beyond *E* = 50 in directed evolution studies.

Enzyme-coupled UV/Vis-based Assay for Lipases and Esterases

Spectrophotometric determination of NADH concentration is performed at 340 nm in the milliliter scale on an Ultrospec 3000 Photometer and on the microliter scale using a fluori- meter (e.g. FLUOstar, BMG LabTechnologies, Offenburg, Germany). In order to determine the amount of acetic acid formed, a solution of *Pseudomonas fluorescens* esterase (PFE) (20 lL, 2 mg mL^{-1}, unless stated otherwise) or the lipase B from *Candida antarctica* (CAL-B) (2 mg mL^{-1}) is added to a mixture of the test-kit components (150 lL). The reactions are started by adding a solution of a chiral acetate (20 lL) in sodium phosphate buffer (10 mM, pH 7.3). Mixtures of the test-kit with buffer or cell lysates of non-induced *E. coli* harboring the gene encoding recombinant PFE serve as controls. In a similar manner, reaction rates are determined by using optically pure (*R*)- and (*S*)-acetates separately. For reactions with crude cell extract, PFE is produced in microtiter plates similar to the published protocol for shake-flask cultivation. However, the cultivation volume is 200 lL per well and cells are disrupted by two freeze–thaw cycles. Finally, cell debris is removed by *centrifugation* and the *supernatants* are used for the assay.

Enzymatic Method for Determining Enantiomeric Excess (EMDee)

A different UV/Vis-based approach is based on the idea that an appropriate enzyme selectively processes one enantiomer, giving rise to a UV/Vis signal. A case in point concerns the determination of the *enantiopurity* of a library of chiral secondary alcohols such as **6** (in this particular study produced by asymmetric Et_2Zn-addition to benzaldehyde **4**), the (*S*)-enantiomer being oxidised selectively by the alcohol dehydrogenase from *Thermoanaerobium* sp..

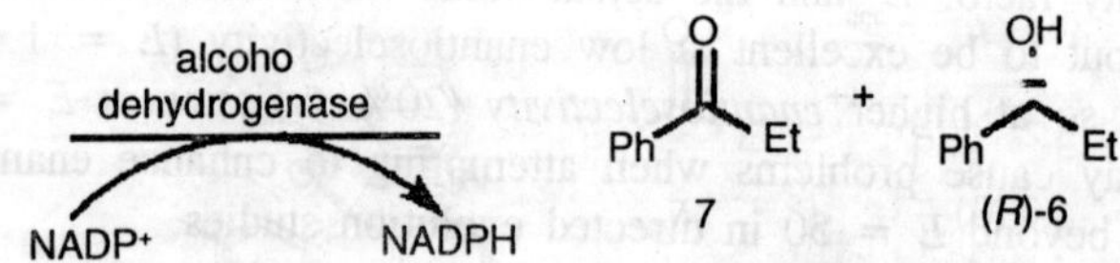

Figure 3.4 : Enzymatic method for determining enantiomeric excess (EMDee) in the case of 1-phenylpropanol produced by asymmetric addition of diethylzinc to benzaldehyde.

The rate of this process can be monitored by a UV/Vis plate reader due to the formation of NADPH (absorption at 340 nm), which relates to the quantity of the (*S*)-enantiomer present in the mixture. Although the screen was not specifically developed to evaluate chiral alcohols produced by an *enzymatic* process (e.g. reduction of ketones), this should be possible, provided an appropriate extraction process can be developed.

The success of the assay depends upon the finding that the rate of oxidation constitutes a direct measure of the *ee*-value. High throughput was demonstrated by analyzing 100 samples in a 384-well format by using a UV plate reader. Each sample contained 1 lmol of 1-phenylpropanol in a volume of 100 lL. The accuracy of the *ee*-value amounts to ± 10%, as checked by independent GC determinations. About 100 samples can be processed in 30 min, which works out at 4800 *ee* determinations per day. Of course, for each new alcohol to be assayed, the alcohol dehydrogenase needs to be selective. If this is not the case, a different (selective) alcohol dehydrogenase has to be found in order for the assay to function properly. Moreover, an accuracy of ± 10% may not be sufficient in the late stages of an evolutionary project when attempting to increase the *ee*-value beyond 90%. Of course, this problem applies to any high-throughput *ee-assay* having such accuracy.

UV/Vis-based Enzyme Immunoassay as a Means to Measure Enantiomeric Excess

UV/Vis-based high-throughput *screening* of *enantioselective* catalysts is possible by enzyme *immunoassays*, a technology that is routinely applied in biology and medicine. As in the case of some of the other screening systems, this new assay was not developed specifically for *enzyme-catalysed* processes. In fact, it was illustrated by analyzing (*R*)/(*S*)-mixtures of *mandelic acid* generated by enantioselective *Ru-catalysed hydrogenation* of benzoyl formic acid. By employing an antibody that binds both enantiomers, it is possible to measure the concentration of the reaction product, thereby allowing the yield to be calculated. The use of an (*S*)-specific antibody then makes the determination of the *ee* possible.

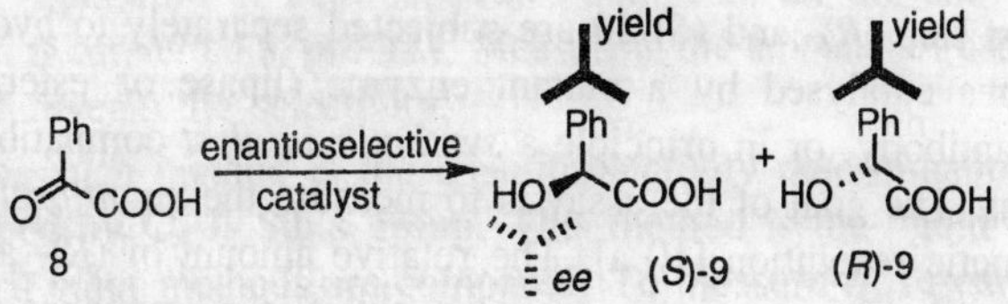

Figure 3.5 : High-throughput screening of enantioselective catalysts by competitive enzyme immunoassays. The solid antibody recognizes both enantiomers, and the hatched antibody is (S)-specific, enabling th determination of yield and ee.

The success of this assay depends upon the availability of specific antibodies, and indeed these can be raised to almost any compound of interest. Moreover, simple automated equipment comprising a plate washer and plate absorbance reader is all that is necessary. About 1000 *ee* determinations are possible per day, precision amounting to ± 9%.

Other UV/Vis-based *ee*-Assays

Several other high-throughput *ee*-assays based on the use of UV/ Vis spectroscopy have been described. Most prominent are colourimetric tests used in the evaluation of enantioselective lipases and esterases. Since the *hydrolysis* of an ester leads to a change in pH, quantification is possible by the use of an appropriate pH indicator.

ASSAYS USING FLUORESCENCE

Assays based on fluorescence, either in the measurement of enzyme activity or in the quantification of *enantioselectivity*, all have a very useful property, namely a high degree of sensitivity. This allows the use of very dilute substrate concentrations and extremely small amounts of catalysts. Basically, there are two different approaches. One involves the use of a substrate of interest to which a fluorescent-active (or potentially active) moiety is covalently attached. The second approach makes use of a fluorescence-based sensor which gives rise to a signal as a consequence of the enzyme-catalysed reaction of a substrate of interest.

Umbelliferone-based Systems

A *fluorescence*-based *ee*-assay using umbelliferone as the built-in fluoro-phore has been developed which works for several different types of enzymatic reactions. In an initial study the system was used to monitor the hydrolytic kinetic resolution of chiral acetates, for example. It is based on a sequence of two coupled enzymatic steps that converts a pair of enantiomeric alcohols formed by the *asymmetric hydrolysis* under study (e.g. (*R*)-and (*S*)-**11**) into a fluorescent product. In the first step (*R*)- and (*S*)-**10** are subjected separately to hydrolysis in reactions catalysed by a mutant enzyme (lipase or esterase), a catalytic antibody, or in principle a *synthetic catalyst* compatible with the system. The goal of the assay is to measure the *enantioselectivity* of this kinetic resolution [10 a]. The relative amount of (*R*)- and (*S*)-**11** produced after a given reaction time is a measure of the *enantioselectivity* and can be ascertained rapidly, but not directly. Two subsequent chemical transformations are necessary. First the

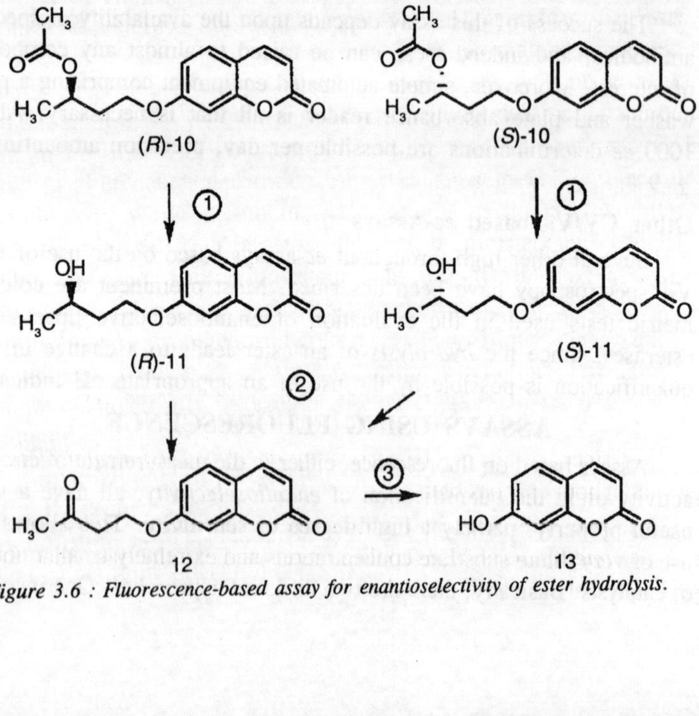

CH_3 O O H_3C O O O (*R*)-10 ① OH H_3C O O O (*R*)-11 ② O H_3C O O O 12 ③ HO O O 13

CH_3 O O H_3C O O O (*S*)-10 ① OH H_3C O O O (*S*)-11

Figure 3.6 : Fluorescence-based assay for enantioselectivity of ester hydrolysis.

enantiomeric alcohols (*R*)- and (*S*)-**11** are *oxidised* separately by horse liver alcohol *dehydrogenase* (HLDH) to the ketone **12**, from which the *fluorescent* final product umbelliferone is released in each case by the catalytic action of bovine serum albumin (BSA). Thus, by measuring the *fluorescence* of **13** for the (*R*)- and the (*S*)-substrate separately, the relative amounts of (*R*)- and (*S*)-**11** can be determined.

Thirty different esterases and lipases were tested. The rate of release of **13** in the wells of standard microtiter plates was monitored by *fluorescence*. Control experiments ensured that the apparent rate of umbelliferone release is directly proportional to the rate of acetate hydrolysis. The predicted and observed *E* and *ee*-values (as checked by standard HPLC assay on a chiral phase) were found to lie within ± 20%. Only in one case was a larger discrepancy observed, a result that was believed to be caused by the occurrence of an unusually low K_m for one of the *enantiomers*.

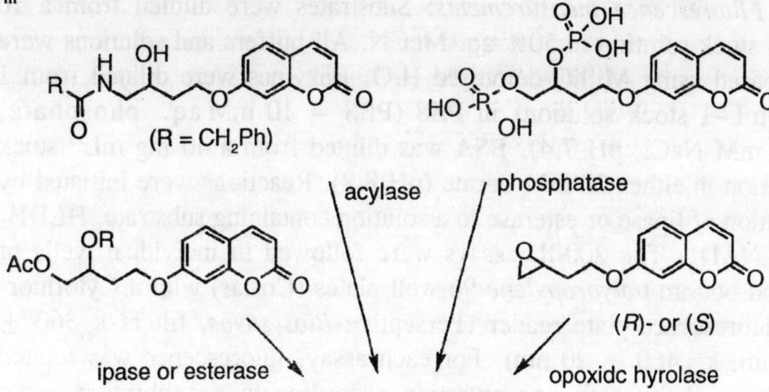

Figure 3.7 : Periodate-coupled fluorogenic assay for hydrolases based on the generation of umbelliferone.

Thus, since the test can be carried out on 96-well *microtiter plates*, high throughput is possible.

Of course, the inherent disadvantage noted earlier for some of the *colourimetric* tests also applies here, namely the fact that the optimisation of a potential catalyst is focused on a specific substrate

10 modified by incorporation of a probe, in this case the *fluorogenic moiety* **13**. However, one can expect the test to be useful in directed *evolution projects* in which proof-of-principle is the goal.

Moreover, this kind of approach can be used in very practical applications, namely as a pretest for the activity of enzymes.

The basic idea has been extended beyond the *ee*-assay for lipases and *esterases* to include such enzymes as *acylases*, *epoxide hydrolases*, *alcohol dehydrogenases*, *aldolases* and *phosphatases*. In many cases the assay makes use of $NaIO_4$-mediated oxidation of a diol-moiety followed by base-catalysed *b-elimination* to generate the fluorescent umbelliferone. Again, the most likely area of application is prescreening, although adaption to include high-throughput *ee*-determination may be possible in some cases.

Umbelliferone-based Assay

Fluorescence measurements: Substrates were diluted from a 10 mM stock solution in 50% aq. MeCN. All buffers and solutions were prepared using MilliQ-deionised H_2O. Enzymes were diluted from 1 mg mL–1 stock solutions in PBS (PBS = 10 mM aq. phosphate, 160 mM NaCl, pH 7.4). BSA was diluted from a 40 mg mL^{-1} stock solution in either 20 mM borate (pH 8.8). Reactions were initiated by addition of lipase or esterase to a solution containing substrate, HLDH, and NAD^+. The 200-lL assays were followed in individual wells of round-bottom *polypropylene* 96-well plates (Costar) with a Cytofluor-II fluorescence-plate reader (Perseptive *Biosystems*, filters k_{ex}360 ± 20 nm, k_{em} 460 ± 20 nm). For each assay, fluorescence was related to the *umbelliferone* concentration according to a calibration curve with pure in the same buffers containing BSA. The reference maximum rate with alcohol substrates was taken from the linear portion of the curves (data points 0–5000 s). For assays with acetates, the linear portions of the curves (data points 15 000–30 000 s) were used to derive the reaction rates.

Preparative assays: Solid enzyme (1 mg) weighed into a 1.5-mL Eppendorf tube was dissolved in 20 mM aq. borate (pH 8.8; 200 lL). Then, a solution of racemic acetate (50 lL; 50 mg mL–1 in MeCN) was added and the resulting suspension agitated on a plate stirrer at room temperature. To follow the reaction, 2 lL aliquots were taken and diluted with hexane/*i*-PrOH 1 : 1. Then, this solution (20 lL) was injected on a Chiralpak AS column (Daicel, 0.45 × 22 cm, flow 1.0 mL min–1, hexane/*i*-PrOH 1 : 2): t_R10.9 ((*R*)-**10**), 23.8 ((*S*)-**10**), 12.1 ((*R*)-**11**), 13.1 ((*S*)-**11**) min. The integral of the peaks recorded by UV at 325 nm was used to calculate the conversion and *ee* of released alcohol.

Fluorescence-based Assay Using DNA Microarrays

A very different fluorescence-based *ee*-assay makes use of DNA microarrays.

This type of technology had previously been employed to determine relative gene expression levels on a genome-wide basis as measured by the ratio of fluorescent reporters. In the case of the *ee*-assay, chiral amino acids were used as model compounds. Mixtures of a racemic amino acid are first subjected to acylation at the amino function with formation of *N*-Boc-protected derivatives. The samples are then covalently attached to amine-functionalised glass slides in a spatially arrayed manner. In a second step the uncoupled surface amino functions are acylated exhaustively. The third step involves complete de-protection to afford the free amino function of the amino acid. Finally, in a fourth step two *pseudo*-enantiomeric fluorescent probes are attached to the free *amino groups* on the surface of the array. An appreciable degree of kinetic resolution in the process of amide coupling is a requirement for the success of the *ee-assay*. In the present case the *ee-values* are accessible by measuring the ratio of the relevant fluorescent intensities. About 8000 *ee-determinations* are possible per day, precision amounting to ± 10% of the actual value. Although it was not explicitly demonstrated that this *ee-assay* can be used to evaluate enzymes (e.g. proteases), this should in fact be possible. So far this approach has not been extended to other types of substrates.

Other Fluorescence-based *ee*-Assays

Several other *ee-assays* relying on use of fluorescence have been described, although their general utility in the high-throughput evaluation of *enantioselective* enzymes remains to be demonstrated in most cases. A high-through-put *ee-assay* based on capillary *electrophoresis* (CE), fluorescence being the detector system is described in this chapter.

ASSAYS BASED ON MASS SPECTROMETRY (MS)

Since *enantiomers* have identical mass spectra, the relative amounts of the (*R*)- and (*S*)-forms present in a given sample and therefore the *ee-value* cannot be measured by conventional MS techniques. However, special MS-based approaches have been described which do in fact allow *ee-determination*. In one method two conditions have to be met: (1) A mass-tagged chiral *derivatisation* agent is applied to the mixture, and (2) in the process of derivatisation a significant degree of kinetic resolution needs to occur (*Horeau's principle*). The relative amounts of mass-tagged *diastereomers* can then be measured by MS simply by integrating the appropriate peaks, the uncertainty in the *ee*-value amounting to $\pm$ 10%. High-throughput application (e.g. in *enzyme catalysis*) has not been demonstrated, but this should be possible. Isotope labeling forms the basis of another MS-based *ee*-*assay* which has been used several times in the directed evolution of *enantioselective enzymes*.

MS-based Assay Using Isotope Labeling

In this MS-based approach derivatisation reactions are not necessary, which makes the overall process easy. It makes use of isotope-labeled *pseudo*-enantiomers or *pseudo-meso*-compounds. This practical method is restricted to studies involving kinetic resolution of racemates and desymmetrisation of prochiral compounds bearing reactive *enantiotopic* groups.

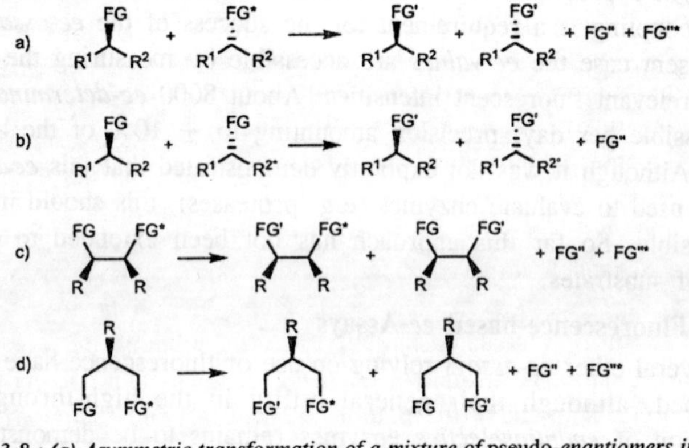

Figure 3.8 : (a) Asymmetric transformation of a mixture of pseudo-*enantiomers involving cleavage of the functional groups (FG) and labeled FG*. (b) Asymmetric transformation of a mixture of* pseudo-*enantiomers involving either cleavage or bond formation at the FGs; isotopic labeling at* R^2*is indicated by the asterisk. (c) Asymmetric transformation of a* pseudo-meso *substrate involving cleavage of the FGs and labeled FG*. (d) Asymmetric transformation of a* pseudo- *prochiral substrate involving cleavage of the FGs and labeled FG*.*

The products of these transformations are *pseudo*-enantiomers differing in absolute configuration and in mass, integration of the MS peaks and data processing affording the *ee*- or *E*-values.

Any type of ionisation can be employed, but *electrospray* ionisation

(ESI) is used most commonly. In some cases an internal standard is advisable if the determination of %-conversion is necessary.

The uncertainty in the *ee*-value amounts less than ± 5%. In the original version about 1000 *ee*-values could be measured per day, but this has been increased to about 10 000 samples per day as a result of a second-generation system based on an eight-channel multiplexed sprayer system. The method is illustrated here using lipase variants from *Bacillus subtilis,* produced by the methods of directed evolution and employed as catalysts in the desymmetrisation of *meso*-1,4-diacetoxy-cyclopentene. The goal is to obtain *enantioselective* variants of this lipase which are expressed in *E. coli*.

This particular substrate and the MS-based *ee*-assay have also been used in another study concerning the assembly of designed *oligonucleotides* (ADO) as a new recombinant method in directed evolution. Accordingly, the D_3-labeled *pseudo-meso*-compound **14** is used as the substrate, the two products of the asymmetric transformation being non-labeled and the D_3-labeled *pseudo*-enantiomers **15** and D_3-**15** which are easily distinguished by ESI/MS because they differ by three mass units.

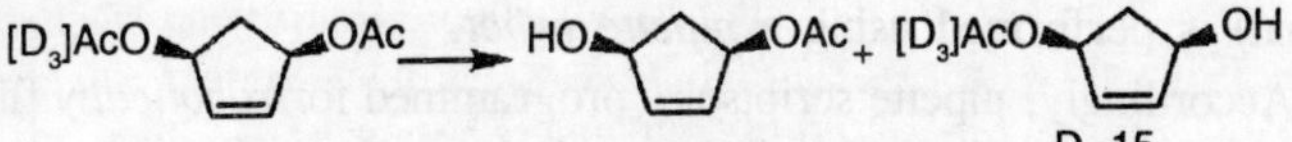

[D$_3$]AcO–(cyclopentene)–OAc ⟶ HO–(cyclopentene)–OAc + [D$_3$]AcO–(cyclopentene)–OH

D_3-14 15 D_3-15

The MS assay has also been applied successfully in the directed evolution of *enantioselective* epoxide hydrolases acting as catalysts in the *kinetic* resolution of chiral epoxides. Moreover, Diversa has applied the MS-based method in the desymmetrisation of a prochiral dinitrile (1,3-dicyano-2-hydroxypropane) catalysed by mutant nitrilases. In this industrial application one of the nitrile moieties was labeled with^{15}N as in^{15}N-**16**, which means that the two *pseudo*-enantio-meric products (*S*)-^{15}N-**17** and (*R*)-**17** differ by one mass unit. This is sufficient for the MS system to distinguish between the two products.

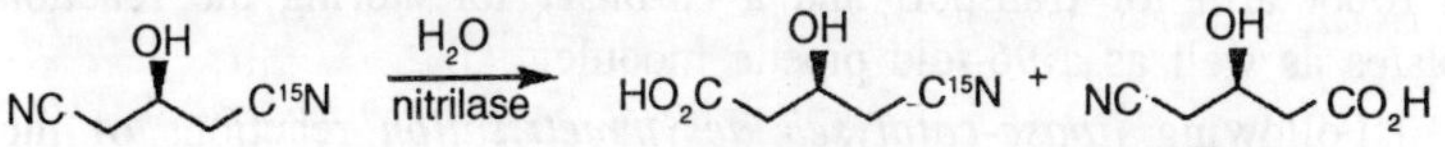

NC–CH$_2$–CH(OH)–CH$_2$–C^{15}N $\xrightarrow[\text{nitrilase}]{H_2O}$ HO$_2$C–CH$_2$–CH(OH)–CH$_2$–C^{15}N + NC–CH$_2$–CH(OH)–CH$_2$–CO$_2$H

^{15}N-16 (*S*)-^{15}N-17 (m/z = 130) (*R*)-17 (m/z = 129)

As a note of caution, in the case of kinetic resolution the MS measurements must be performed in the appropriate time window (near 50% conversion). If this is difficult to achieve due to possibly different amounts or activities of the mutants from well to well, the system needs to be adapted in terms of time resolution. This means that samples for MS evaluation need to be taken as a function of time. Finally, it is useful to point out the possibility of multi-substrate *ee-screening* using the MS-based assay, which allows for enzyme *fingerprinting* with respect to the *enantioselectivity* of several substrates simultaneously.

MS-based Assay Using Isotope Labeling

In this protocol the desymmetrisation of *meso*-1,4-diacetoxy-cyclopentene using labeled D_3-**14** is described. Bacterial colonies on agar plates are collected and placed individually in the deep wells of *microtiter plates* (96-format) containing LB medium with the aid of an appropriate robot (e.g., Colony Picker Q-Pix commercially available from Genetix). Up to 10 000 colonies can be handled per day. The LB medium is added using an eight-channel dispenser (e.g. Dispenser Multidrop DW from Labsystems).

The preparation of the reaction solutions and the process of carrying out the lipase-catalysed reactions in the wells of 96-format microtiter plates are also fully automated. The preparation of the reaction solutions in the deep wells (2.2 mL) of microtiter plates (96-format) is performed using a *pipette robot*.

Accordingly, pipette scripts are programmed for *robotically* filling the wells with *buffer* and *substrate* solutions.

This will vary, depending upon the particular *substrate* and *enzyme*. In the present case, which is typical for lipase-catalysed desymmetrisation, the reaction solutions are composed of 125 lL phosphate buffer (10 mM; pH 7.5), 50 lL supernatant and 25 lL substrate D_3-**14** solution (0.1 M in *dimethyl-sulfoxide*). The reactions are allowed to take place in the wells of 96-format microtiter plates at room temperature in an incubator while shaking for 24 h. In order to activate all of the modules of the robot, the software Facts is used. The pipette robot consists of a workstation with spaces for 12 microtiter plates, a robot arm for transport and a *carousel* for storing the reaction plates as well as a 96-fold pipette module.

Following *lipase-catalysed desymmetrisation* reactions of the substrate (e.g. D_3-**14**) in the wells of 96-format microtiter plates, an extraction step is necessary prior to ESI/MS analysis. In the present case ethyl acetate was used in the extraction step (200 lL reaction solution and 200 lL solvent).

The samples need to be diluted prior to ESI/MS measurements using standard solution (methanol/10 mM NaOAc, 4 : 1 plus undeuterated *meso*-1,4-diacetoxy-cyclopentene as an internal standard necessary for the determination of conversion). This process is controlled by the software Facts. Four modules are controlled simultaneously: The robot arm (RoMa), the carousel for storing the microtiter plates, the 96-pipette system (TeMo) as well as the eightfold pipette head (Gemini). Iteration occurs within 12 min.

The control of the HPLC pump as well as the autosampler and the MS is ensured by the Software Masslynx 3.5 (Micromass). Following optimisation of the measurement conditions, a list of process measurements is set up sample list) and the desired HPLC and MS steps are called upon. Following a measurement the ESI source is automatically brought to room temperature (shut down). Using a 96-microtiter plate 576 samples can be processed per measurement. The *chromatograms* are integrated by the software package Quanlynx and Openlynx (Micromass) and transformed into an Excel (Microsoft, Unterschleissheim, Germany) table. The use of a macro allows the calculation of the absolute intensities and therefore the *ee* as well as the conversion. The *E*-values in the case of kinetic resolution are automatically calculated by the formula derived by Sih. Data processing is possible by the use of the Openlynx Browser (Micromass). The overall process occurs continuously and makes possible the analysis of up to 10 000 samples per day, provided the eight-channel multiplexed sprayer system (e.g. from Micromass) is used. It is also possible to use 384-format microtiter plates.

ASSAYS BASED ON NUCLEAR MAGNETIC RESONANCE SPECTROSCOPY

For decades nuclear magnetic resonance (NMR) measurements were considered to be slow processes, but recent advances in the design of flow-through NMR cells have allowed the method to be applied in *combinatorial chemistry*.

These technological improvements were then applied to the development of two different NMR-based high-throughput *ee*-assays for evaluating the products of enzyme- or transition metal-catalysed reactions. In one version classical derivatisation using a chiral reagent or NMR shift agent is parallelised and automated using flow-through cells, about 1400 *ee-measurements* being possible per day with a precision of ± 5%. In the second embodiment, illustrated here in detail, a principle related to that of the MS system described in Section 2.4 is applied. Chiral or *meso*-substrates are labeled so as to produce *pseudo*-enantiomers or *pseudo-meso*-compounds which are then used in the actual screen. Application is thus restricted to kinetic resolution of racemates and *desymmetrisation* of prochiral compounds bearing reactive *enantiotopic* groups as illustrated previously for the MS system.

A particularly practical form of this assay utilizes ^{1}H NMR *spectroscopy*, ^{13}C-labeling being used to distinguish between the (*R*)- and (*S*)-forms of a chiral compound under study. Essentially any carbon atom in the compound of interest can be labeled, but methyl groups in which the ^{1}H signals are not split by ^{1}H,^{1}H coupling are preferred because the relevant peaks to be integrated are the singlet arising from the CH_3-group of one enantiomer and the doublet of the $^{13}CH_3$-group of the other. A typical example which illustrates the method concerns the lipase- or esterase-catalysed hydrolytic kinetic resolution of *rac*-1-phenylethyl acetate, derived from *rac*-1-phenyl ethanol (**19**). However, the acetate of any chiral alcohol or the acetamide of any chiral amine can be used. Labeling can be carried out in any position of a compound, as in (*S*)-^{13}C- **18**. The synthesis is straightforward, since it simply involves acylation of the (*S*)-alcohol using commercially available ^{13}C-labeled acetyl chloride. Then a 1 : 1 mixture of labeled and nonlabeled compounds (*S*)-^{13}C-**18** and (*R*)-**18** is prepared which simulates a racemate. It is used in the actual *catalytic hydrolytic kinetic*

(*S*)-^{13}C-18 + (*R*)-18 ⟶

(*S*)-19 + (*R*)-19 + $^{13}CH_3COOH$ + CH_3COOH
^{13}C-20 20

resolution, which affords a mixture of true enantiomers (*S*)-**19** and (*R*)-**19** as well as labeled and non-labeled acetic acid ^{13}C-**20** and **20**, respectively, together with unreacted starting esters. At 50% conversion (or at any other point of the reaction) the ratio of (*S*)-^{13}C-**18** to (*R*)-**18** correlates with the *enantiomeric* purity of the nonreacted ester, while the ratio of ^{13}C-**20** to **20** reveals the relative amounts of (*S*)-**19** and (*R*)-**19**, respectively.

An excerpt of the ^{1}H NMR spectrum of a "*racemic*" mixture of

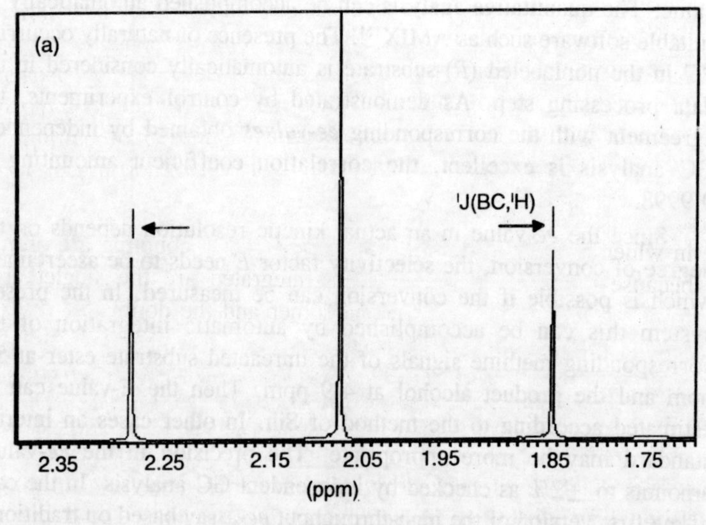

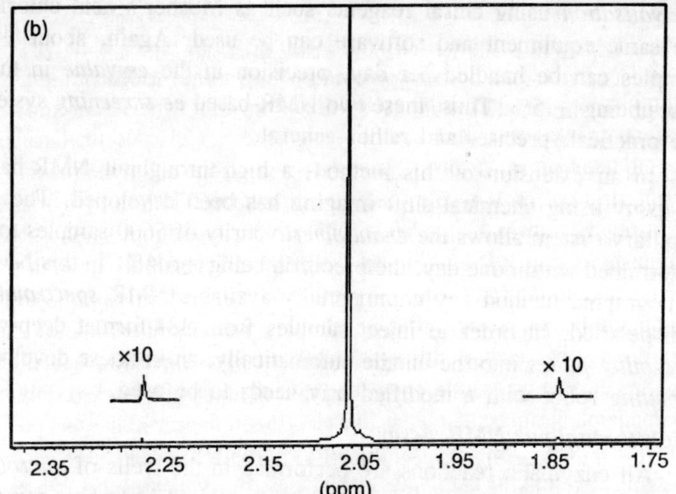

Figure 3.9 : Expanded region of the ^1H NMR spectra of (a) racemic mixture of (S)-^{13}C-18/(R)-18 and (b) (R)-18 alone.

(*S*)-^{13}C-**18** and (*R*)-**18** featuring the expected doublet of the ^{13}C-labeled methyl group and the singlet of the nonlabeled methyl group. The singlet of the non-labeled methyl group of (*R*)-**18**, including the ^{13}C-satellites due to the presence of natural ^{13}C in the sample.

The exact ratio of the two *pseudo*-enantiomers is accessible by simple integration of the respective peaks, which provides the *ee*-

value. The quantitative analysis can be accomplished automatically by suitable software such as AMIX™. The presence of naturally occurring ^{13}C in the nonlabeled (*R*)-substrate is automatically considered in the data processing step. As demonstrated by control experiments, the agreement with the corresponding *ee-values* obtained by independent GC analysis is excellent, the correlation coefficient amounting to 0.9998.

Since the *ee*-value in an actual kinetic resolution depends on the degree of conversion, the selectivity factor *E* needs to be ascertained, which is possible if the conversion can be measured. In the present system this can be accomplished by automatic integration of the corresponding methine signals of the unreacted substrate ester at 5.9 ppm and the product alcohol at 4.9 ppm. Then the *E*-value can be estimated according to the method of Sih. In other cases an internal standard may be more appropriate. The precision in the *ee*-values amounts to ±2% as checked by independent GC analysis. In the case of the first version of the high-throughput *ee*-assay based on traditional *derivatisation* using chiral reagents such as Mosher's acid chloride, the same equipment and software can be used. Again, about 1400 samples can be handled per day, precision in the *ee-value* in these cases being ± 5%. Thus, these two NMR-based *ee-screening* systems are practical, precise, and rather general.

In an extension of this method, a high-throughput NMR-based *ee-assay* using chemical shift imaging has been developed. The 19-capillary system allows the *enantiomeric* purity of 5600 samples to be determined within one day, the precision being ±6.4%. In this NMR-tomographic method any commercially available NMR *spectrometer* can be used. In order to inject samples from 384-format deep-well *microtiter plates* into the bundle automatically, an in-house developed *pipetting* robot with a modified tray needs to be used.

High-throughput NMR Assay

All enzymatic reactions are performed in the wells of *microtiter plates* (96-format) in water (as in *lipase-catalysed hydrolytic* reaction of (*S*)-^{13}C-**18**/(*R*)-**18**), which is followed by a standard automatic extraction step. Depending upon the particular substrate to be assayed and upon the type of solvent used, it may be necessary to remove the solvent. However, this is often not necessary. In the case of enzymatic reactions in organic medium, solvent extraction is not required. For NMR analysis such solvents as $CDCl_3$, D_6-DMSO or D_2O are used. A minimum of about 6 lmol of substrate/product per mL of solvent is needed. Although the flow-through cell system does not need too much solvent (about 1 L in 24 h), the solvents can be mixed with the undeuterated form in 1 : 9 ratios in order to reduce costs.

Several flow-through NMR cells are commercially available, for example, the BEST™ NMR system (Bruker Biospin GmbH) as described here or the VAST™ NMR system (Varian). In addition to the flow-through cell and the NMR spectrometer (300 MHz), the system requires an autosampler, for example, a Gilson 215 autosampler.

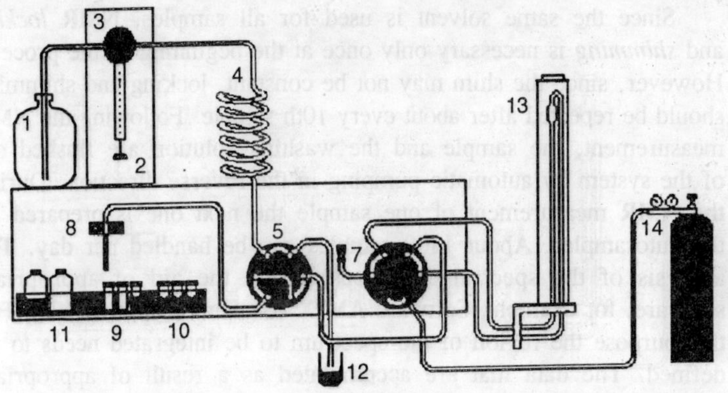

Figure 3.10 : Schematic representation of the BEST system (Bruker Biospin GmbH). (1) Bottle with transporting liquid; (2) Dilutor 402 single syringe (5 mL) with 1100 lL tube; (3) Dilutor 402 three-way valve; (4) Sample loop (250–500 lL); (5) Six-way valve (standard version) loading sample; (6) Six-way valve (standard version)injecting sample; (7) Injection port; (8) XYZ needle; (9) Rack for sample vials; (10) Rack for recovering vials; (11) Rack for washing fluids and waste bottle (three glass bottles); (12) External waste bottle; (13) Flow probe with inner lock container; (14) Inert gas pressure bottle for drying process.

In the present case the hydrolytic kinetic resolution of the acetate of racemic 1-phenyl-ethanol using a 1 : 1 mixture of the *pseudo*-enantiomers (*S*)-^{13}C-**18** and (*R*)-**18** in water is carried out in the wells of microtiter plates (e.g. 96-format), followed by extraction (pipette robot) using 300 lL of $CDCl_3$. For storage the resulting organic layers are placed in the wells of microtiter plates (e.g. 96-format). The samples are then transferred to the *autosampler* of the BEST NMR system and analysed using the high-speed mode as described below. The samples are taken by the movable needle and transferred into the first valve system. At the same time washing solution ($CDCl_3$) is introduced by the Dilutor 402 into the six-port selection valve. Injection occurs via the injection port, whereby the washing solution is first fed into the second six-port selection valve followed by the sample to be measured. The washing solution is then transferred rapidly via tubings into the flow-through cell by a *hydraulic impulse*. Immediately thereafter the sample follows which is separated from the washing solution by a small air gap. The washing solution is pumped through the flow-through cell, and once the sample has entered it, pumping is stopped and the NMR measurement is automatically initiated (maximum of four scans). During this time the washing solution is stored behind the cell.

Since the same solvent is used for all samples, NMR *locking* and *shimming* is necessary only once at the beginning of the process. However, since the shim may not be constant, locking and shimming should be repeated after about every 10th sample. Following the NMR measurement, the sample and the washing solution are flushed out of the system by automatic pumping in the reverse direction. During the NMR measurement of one sample the next one is prepared by the autosampler. About 1400 samples can be handled per day. The analysis of the spectral data occurs with the aid of appropriate software, for example, Software AMIX (Bruker Biospin GmbH). For this purpose the region of the spectrum to be integrated needs to be defined. The data that are accumulated as a result of appropriate NMR peak integration are transferred onto Excel spreadsheets. The *ee*- or *E*-values are readily tabulated with the help of a macro.

FOURIER TRANSFORM INFRARED SPECTROSCOPY

The basic idea of utilizing isotopic labeling for distinguishing *pseudo*-enantiomers in the kinetic resolution of chiral compounds and in the desymmetrisation of prochiral substrates bearing reactive enantiotopic groups can also be applied using Fourier transform infrared spectroscopy (FTIR) as the detection system in some cases. Since FTIR spectroscopy is a cheap analytical technique available in almost all laboratories, the *ee*-assay has great potential. Moreover, it is of particular interest in the analysis of enzymatic reactions because the *ee*- or *E*-values can be measured directly in culture supernatants without time-consuming workup procedures. Of course, it is restricted to substrates of the type which, however, must contain infrared (IR)-active functional groups. If all *prerequisites* are fulfilled, up to 10 000 *ee-values* can be measured per day with an accuracy of ± 7%, making this a particularly practical and economical method for evaluating large libraries of potentially enantioselective enzymes.

When applying this *ee-assay*, the "*best*" position at which isotopes are introduced needs to be determined first. For the purpose of illustration the lipase- or esterase-catalysed kinetic resolution of esters is considered here, although the method is not restricted to this type of reaction. ^{13}C-labeling of carbonyl groups is ideal for several reasons: Carbonyl groups provide intensive vibrational bands in an IR spectrum, allowing for easy and precise determination of the concentration of the compounds by applying Lambert-Beer's law. In the spectral region between 1600 and 1800 cm^{-1}, which is typical for carbonyl stretching vibrations, almost no absorptions of other functional groups appear, eliminating interferences with other vibrational bands. The ^{13}C-labeled compounds can be easily prepared because reactive reagents with ^{13}C-labeled carbonyl groups like 1- ^{13}C-acetyl chloride are commercially available. Finally, the absorption maxima of the carbonyl stretching vibration is shifted by 40–50 cm–1 to lower wave numbers by introducing a ^{13}C-label, which prevents the overlap of the two carbonyl bands.

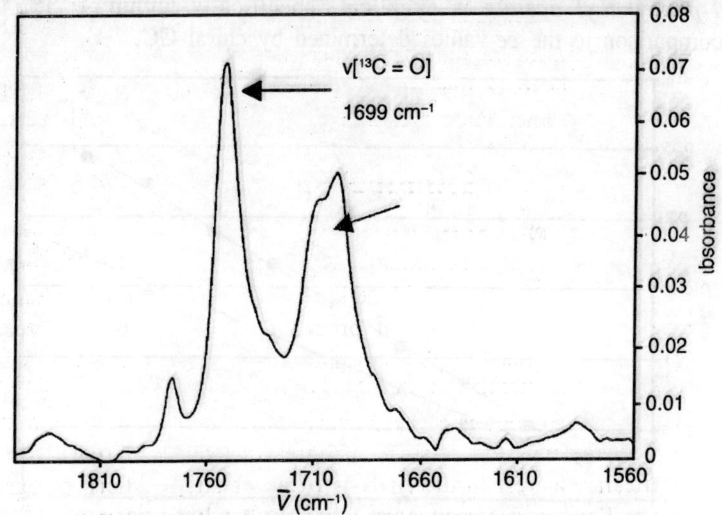

Figure 3.11 : Part of a Fourier transform infrared spectroscopy (FTIR) spectrum of a 1 : 1 mixture of (R)-18 and (S)- ^{13}C-18.

An example concerns the kinetic resolution of 1-phenylethyl acetate, previously used to illustrate the NMR-based *ee*-assay. The optimal way to proceed in the FTIR-based *ee*-assay is to apply ^{13}C-labeling in the carbonyl moiety (i.e. to prepare a *pseudo*-racemate comprising a 1 : 1 mixture of (*S*)-^{13}C-**18** and (*R*)-**18**). The FTIR spectrum 1 : 1 mixture of (*R*)-**18** and (*S*)-^{13}C-**18**, illustrating the anticipated shift of the respective carbonyl stretching vibration which allows quantification of the *pseudo*-enantiomers.

It is necessary to apply Lambert-Beer's law in the calculation of the concentrations of the *pseudo*-enantiomeric substances, and for this purpose the molar coefficients of absorbance need to be determined. Thus, solutions of (*R*)-**18** and (*S*)- ^{13}C-**18** in cyclohexane at different concentrations have to be prepared. After measuring the corresponding absorbances at the absorption maxima of the carbonyl stretching vibration, it is possible to calculate the molar coefficients of absorbance by applying Lambert–Beer's law $E = \varepsilon \times c \times d$.

It is now possible to exploit the FTIR spectra of different synthetic mixtures of the labeled and non-labeled enantiomeric compounds. After applying an automated baseline correction to the spectra and correcting the absorbance of one enantiomer in the synthetic mixtures by the absorbance of the other enantiomer at this position, the accuracy of the *pseudo*-enantiomeric system based on *1-phenylethyl acetate* is excellent, specifically within ± 3% in comparison to the *ee*-values determined by chiral GC.

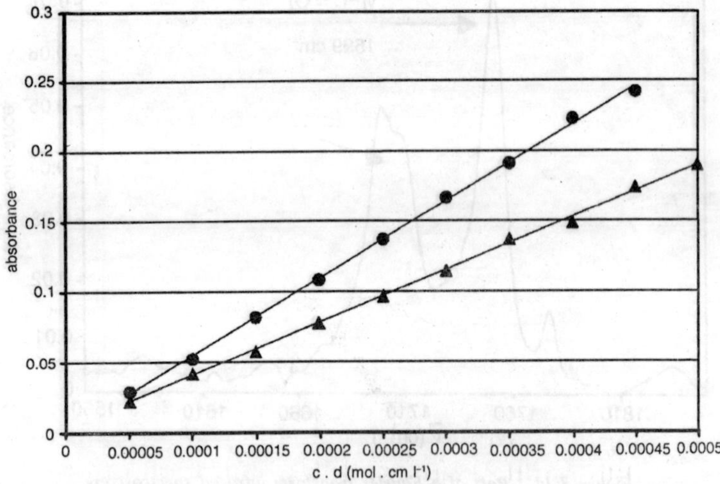

Figure 3.12 : Determination of the molar coefficients of absorbance of (R)*-18 (circles, absorption maximum: 1751 cm–1) and* (S)*-*^{13}*C-18 (triangles, absorption maximum: 1699 cm–1) by linear regression.*

Using commercially available high-throughput screening (HTS)/ FTIR systems, high-throughput measurements are easily possible. The analysis can be performed on a Tensor 27 FTIR *spectrometer* coupled to an HTS-XT system which is able to analyze the samples on 96- or

384-well *microtiter plates*. The plates are equipped with a silicon plate for IR transmittance. A great advantage of this system is the fact that the *ee-values* can be measured in culture supernatants, which is not possible in the case of the MS- or NMR-based assays. This means that the extraction step can be omitted. In order to make this simple *ee*-assay possible, Bruker has already coupled the microplate stacking device TWISTER 1 to its *microplate reader*. In this combination which is controlled by OPUS©software, 40 IR microplates can be measured automatically. In order to perform the sample loading at high throughput, the autosampler Microlab 4000 SP can be used. Both formats (96 and 384) of the Bruker silicon *microplates* are suitable for automatic loading of different types of samples (proteins, cells, culture media). At least 9000 samples can be handled per day.

High-throughput ee-Determination Based on FTIR

As a first step the molar coefficients of absorbance of the labeled and unlabeled substrate (in this case **18**) need to be determined. After preparation of a stock solution (0.200 M) of (*R*)-1-phenylethyl acetate ((*R*)-**18**) and (*S*)-(1-phenylethyl)-1- ^{13}C-acetate ((*S*)- ^{13}C-**18**) in *cyclohexane*, the solutions are diluted with cyclohexane to concentrations of 0.180, 0.160, 0.140, 0.120, 0.100, 0.080, 0.060, 0.040 and 0.020 M, respectively (total volume: 1 mL).

The absorbance of the resulting samples is measured with an FTIR spectrometer at the corresponding absorption maxima of the carbonyl stretching vibration ((*R*)-**18**: 1751 cm–1 (*S*)- ^{13}C-**18**: 1699 cm–1) with a thickness of the layers of 25.0 lm. The measurements are carried out with 32 scans and a resolution of 4 cm^{-1}. The molar coefficients of absorbance are determined by linear regression. The correlation coefficients are in both cases better than 0.995. The analysis of synthetic mixtures of the *pseudo*-enantiomers of 1-phenylethyl acetate is performed under the same conditions at a concentration of 0.10 M.

The high-throughput *ee-measurements* are performed on a Tensor 27 spectrometer in connection with a HTS-XT system provided by Bruker Optik GmbH. Aliquots of 3 lL of each supernatant mixture are transferred onto a 384-well microtiter plate equipped with a silicon plate for IR transmittance. Every sample is measured with a resolution of 8 cm–1 and a scan number of 10, so that the total time for the analysis of each sample amounted to 8.9 s, allowing a throughput of > 9000 samples per day. The resulting spectra were analysed with the software Opus and Opus Lab.

The first 14 samples are used for calibration purposes, while the remaining probes are taken as unknown samples. To evaluate the accuracy of the system, the *ee*-value of each mixture is independently determined by chiral GC analysis.

ASSAYS BASED ON GAS CHROMATOGRAPHY

It is well known that conventional chromatography (GC) using chiral stationary phases can only handle a few dozen *ee-determinations* per day. However, if medium- or high-throughput is desired, speed-up using automation is possible, although generally at the expense of precision. This disadvantage may be of no serious concern, since the purpose is to identify hits which can then be analysed by a more precise system. Indeed, in some cases GC can be modified so that in optimal cases about 700–800 exact *ee*- and *E-determinations* are possible per day. Such medium-throughput may suffice in certain applications. The first such case concerned the lipase-catalysed kinetic resolution of the chiral alcohol (*R*)- and (*S*)-**21** with formation of the acylated forms (*R*)- and (*S*)-**22**. Thousands of mutants of the lipase from *Pseudomonas aeruginosa* were created by epPCR for use as catalysts in the model reaction and were then screened for enantioselectivity.

(*R*)-21 + (*S*)-21 —[OAc (vinyl acetate), mutant lipases]→ (R)-22 + (S)_22

Although in principle it is possible to simply use several GC instruments each equipped with a sample manager and a PC, this is really not efficient because it is expensive, and at the same time data-handling becomes more difficult. The first successful construction consisted of two GC instruments (e.g. GC instruments and data bus (HP-IB) are commercially available from Hewlett Packard, Waldbronn, Germany), one prep-and-load sample manager (PAL) (commercially available from CTC, Schlieren, Switzerland) and one PC. The instruments are connected to the PC via a standardised data bus (HP-IB) (commercially available from Hewlett-Packard) which controls pressure, temperature, etc. and handles other data such as those of the detector. A wash station as well as a drawer system with a maximum of eight microtiter plates are included. The sample manager is attached to the unit in such a way as to reach both injection ports. Since the sample manager can inject samples from 96- or 384-well microtiter plates, over 3000 samples can be handled without manual intervention. The software Chemstation (commercially available from Hewlett-Packard) enables additional programs (macros) to be applied before and after each analytical run. Such a macro controls the sample manager, each position on the microtiter plate being labeled via the sequence table. Another macro ensures analysis following each sample run in a specified manner (i.e. the peaks of the chiral compound **21** are analysed quantitatively). The analytical data are transferred to an Excel spreadsheet via DDE (Dynamic Data Exchange) (commercially available from Microsoft, Unterschleissheim, Germany) in table form or in microtiter format, allowing for a rapid overview. Finally, the setup includes H_2-guards which monitor the hydrogen concentration in the ovens; at concentrations exceeding 1% (potentially explosive at > 4% H_2) the system responds and automatically switches to nitrogen as the carrier gas.

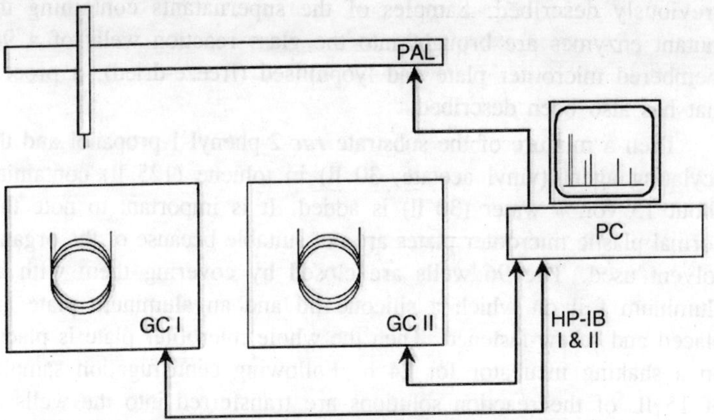

Figure 3.13 : Optimised unit for ee-determination based on two gas chromatography instruments driven by one PC.

Upon using a stationary phase (2,3-di-*O*-ethyl-6-*O*-*tert*-butyldimethylsilyl-b-cy-clodextrin) complete separation of (*R*)- and (*S*)-**21** (but not of (*R*)/(*S*)-**22**) is possible within 3.9 min. Since the unit comprises two simultaneously operating GC units, about 700 exact *ee*-determinations of (*R*)/(*S*)-**21** are possible per day. The corresponding values for the conversion and the selectivity factor *E* (or *s*) are likewise automatically provided in microtiter format which means that the *ee* of (*R*)/(*S*)-**22** is also accessible. Of course, every new substrate has to be optimised a new using commercially available chiral stationary phases.

More recently, this system has been extended to include three GC instruments operated by one PC. Experience gained in the initial GC system has shown that it is better to use one sample manager for each GC instrument, and not a single one for two or even three such instruments. The new system was applied successfully in the analysis of chiral products arising from catalysis of mutant Baeyer-Villiger monooxygenases. In this directed evolution project about 12 000 samples were screened.

Medium-throughput GC-based ee-Assay

The GC system is described here using the lipase-catalysed model reaction of *rac*-**21**. In principle mutants of any lipase can be used, but for the purpose of illustration lipase mutants from *Pseudomonas aeruginosa* were used.

Mutants from *P. aeruginosa* lipase were obtained by epPCR as previously described. Samples of the supernatants containing the mutant enzymes are brought into the glass reaction wells of a 96-membered microtiter plate and lyophilised (freeze-dried), a process that has also been described.

Then a mixture of the substrate *rac*-2-phenyl-1-propanol and the acylating agent (vinyl acetate, 30 ll) in toluene (125 ll) containing about 15 vol.% water (30 ll) is added. It is important to note that normal plastic microtiter plates are not suitable because of the organic solvent used. The 96 wells are closed by covering them with an aluminum foil on which a silicone lid and an aluminum plate are placed and screw-fastened. Then the whole microtiter plate is placed on a shaking incubator for 24 h. Following centrifugation samples of 15 lL of the reaction solutions are transferred into the wells of another glass microtiter plate together with 170 lL of toluene. The GC measurements are performed using these mixtures.

The sample manager comprises the HTS/PAL system (high-throughput-screening prep-and-load) and was incorporated in such a way that samples can be robotically taken from the wells of a collection of microtiter plates, injection into the two injection ports being possible. Accordingly, the sample manager is operated on the basis of microliter syringes, the cannulas being regularly cleaned in the wash station. The system includes drawers for storing microtiter plates. The final data are made available in the form of Excel tables or in microtiter format by the program Excel. In evaluating the GC chromatograms, calibration factors have to be determined first.

ASSAYS BASED ON HPLC

The problems arising when developing a medium-throughput *ee*-assay based on HPLC are similar to those surrounding GC assays. As before, speed is accomplished at the expense of precision, but systems can in fact be implemented which allow the identification of hits, which then need to be analysed more precisely using conventional HPLC, using longer columns and of course longer reaction times. Nevertheless, only medium throughput can be achieved, meaning the evaluation of 300–800 samples per day (which suffices in some applications).

H_3C–C_6H_4–CH_2–S–CH_3 (**23**) $\xrightarrow[\text{CHMO-mutants}]{O_2}$ (*R*)-**24** + (*S*)-**24**

The first medium-throughput HPLC-based *ee*-assay was developed to identify hits in the directed evolution of enantioselective cyclohexanone monooxygenases (CHMO) as catalysts for the asymmetric O_2-mediated oxidation of prochiral thio-ethers such as **23**.

A commercially available HPLC instrument was equipped with a sample manager and the appropriate software (Lab Solutions) to handle 96-or 384-well microtiter plates. Since enantiomeric separation of (*R*)-**24**/(*S*)-**24** has to be fast and efficient, exploratory experiments were performed using a variety of different chiral stationary phases, solvents and conditions. An efficient system turned out to be benzoylated cellulose as the stationary phase with a mixture of *n*-heptane and ethanol as the mobile phase. For rapid analysis short columns 50 mm in length and 4.5 mm in diameter were used. This system allows for at least 800 *ee*-determinations per day. Unfortunately, the *E. coli*-based expression system produces small amounts of indole, which leads to an overlap with the HPLC peak of (*S*)-**24**. Although this can be considered in the quantitative evaluation, the *ee*-values accessible under these conditions are not precise and were consequently used only to identify hits. These were subsequently analysed precisely by a conventional HPLC setup using a longer column (250 mm).

ASSAYS BASED ON CAPILLARY ARRAY ELECTROPHORESIS

The chromatographic techniques described above may well serve as practical assays in some cases, but truly high-throughput encompassing thousands of *ee*-values per day is outside of the realm of these assays. In sharp contrast, capillary array electrophoresis (CAE) has been modified to allow the high-throughput determination of enantioselectivity. It is well known that traditional capillary electrophoresis (CE) in which the electrolyte contains chiral selectors, such as b-cyclodextrin derivatives (b-CDs), can be used to determine the enantiomeric purity of a given sample. The conventional forms of this analytical technique allow for only a few dozen *ee* determinations per day. However, because of the analytical demands arising from the Human Genome Project, *inter alia*, CE has been revolutionised so that efficient techniques for instrumental miniaturisation are now available, making super-high-throughput analysis of biomolecules possible for the first time.

Two different approaches have emerged, namely CAE and CE on microchips (also called CAE on chips). Both techniques can be used to carry out DNA sequence analyses and/or to analyze oligonucleotides, DNA-restriction fragments, amino acids or PCR products. Many hundred thousand and more analytical data points can be accumulated per day.

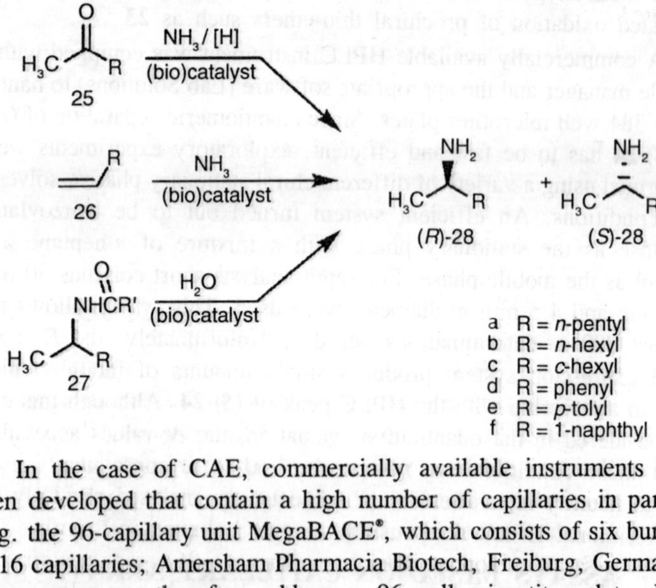

25 → NH_3 / [H], (bio)catalyst

26 → NH_3, (bio)catalyst

27 → H_2O, (bio)catalyst

(*R*)-28 + (*S*)-28

a R = *n*-pentyl
b R = *n*-hexyl
c R = *c*-hexyl
d R = phenyl
e R = *p*-tolyl
f R = 1-naphthyl

In the case of CAE, commercially available instruments have been developed that contain a high number of capillaries in parallel (e.g. the 96-capillary unit MegaBACE®, which consists of six bundles of 16 capillaries; Amersham Pharmacia Biotech, Freiburg, Germany). The system can therefore address a 96-well microtiter plate. Each capillary is about 50 cm long. This system was adapted as a super-high-throughput analytical tool for *ee* determination. In this study chiral amines of the type **28**, which are of importance in the synthesis of pharmaceutical and agrochemical products, were used as the model substrates. They are potentially accessible by enzymatic reductive amination of ketones **25**, addition of ammonia to activated olefins **26** or hydrolysis of acetamides **27** (the reverse reaction also being possible).

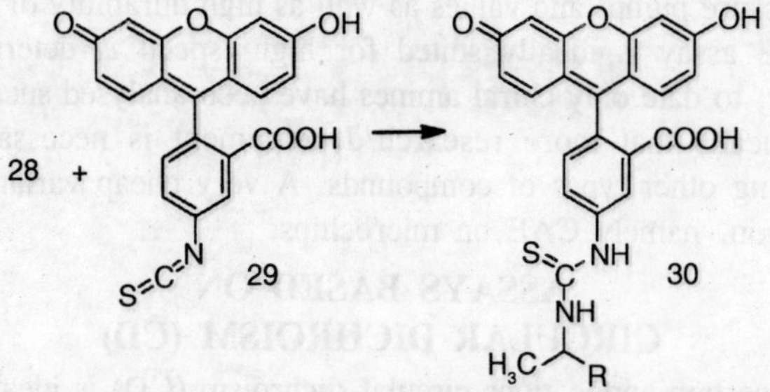

28 + 29 → 30

The conditions for conventional CE assay of the amines **28** were first optimised using various b- and b-CD derivatives as chiral selectors. In order to enable a sensitive detection system, specifically laser-induced fluorescence detection (LIF), the amines were derivatised with fluorescein-isothiocyanate (**29**) leading to fluorescence-active compounds **30**. Although extensive optimisation was not carried out (only six CD derivatives were tested), in all cases satisfactory baseline separation was accomplished. The next step involved the use of compounds **28 c**/**30 c** as the model substrates for CAE analysis using an instrument of the kind MegaBACE®. Known enantiomeric mixtures of the amine **28 c** were transformed into the fluorescence-active derivative **30 c**. The latter samples were then analysed by CAE. A special electrolyte having a high viscosity had to be developed. It is composed of 40 mM CHES pH 9.1/6.25 mM c-CD 5 : 1 diluted with a buffer containing linear polyacrylamide. The MegaBACE instrument was operated at a potential of −10 kV/8 lA and a sample injection potential of −2 kV/9 s. Under these conditions, base-line separation is excellent. The agreement between *ee*-values of (*R*)/(*S*)-mixtures of **30 c** determined by CAE and those of the corresponding (*R*)/(*S*)-mixtures of **28 c** as measured by GC turned out to be excellent.

The enantiomer separation of (*R*)/(*S*)-**30 c** on the MegaBACE instrument required about 19 min. This means that even though the conditions are far from optimised, the automated 96-array system provides more than 7000 *ee* determinations in a single day. In related cases optimisation resulted in shorter analysis times for enantiomer separation so that a daily throughput of 15 000–30 000 *ee* determinations is realistic. Such super-high-throughput screening for enantioselectivity is not readily possible by any other currently available technology. In view of the possibility of chiral selector optimisation and the fact that CAE has many advantages such as extremely small amounts of samples, essentially no solvent consumption, absence of high-pressure pumps and values as well as high durability of columns, this CAE assay is ideally suited for high- speed *ee*-determination. However, to date only chiral amines have been analysed successfully, which means that more research/development is necessary when considering other types of compounds. A very cheap variation is on the horizon, namely CAE on microchips.

ASSAYS BASED ON CIRCULAR DICHROISM (CD)

For certain applications circular dichroism (CD) is ideally suited as a high-throughput *ee*-assay, although to date it has been used in enzymatic reactions only once. An alternative to HPLC employing chiral columns which separate the enantiomers of interest is the use of normal columns that simply separate the starting materials from the enantiomeric products, the *ee* of the mixture of enantiomers then being determined by CD spectroscopy. Recently, it was shown that this classical method can be applied in high-throughput *ee*-screening. The method is based on the use of sensitive detectors for HPLC which determine in a parallel manner both the CD (De) and the UV absorption (e) of a sample at a fixed wavelength in a flow-through system. The CD signal depends only on the enantiomeric composition of the chiral products, whereas the absorption relates to their concentration. Thus, only short HPLC columns are necessary. Upon normalizing the CD value with respect to absorption, the so-called anisotropy factor *g* is obtained:

$$E = \frac{\Delta c}{\varepsilon}$$

It is thus possible to determine the *ee*-value without recourse to complicated calibration.

The fact that the method is theoretically valid only if the *g*-factor is independent of concentration and if it is linear with respect to *ee* has been emphasized repeatedly. However, it needs to be pointed out that these conditions may not hold if the chiral compounds form dimers or aggregates, because such enantiomeric or diastereomeric species would give rise to their own particular CD effects. Although such cases have yet to be reported, it is mandatory that this possibility be checked in each new system under study. In work concerning the directed evolution of enantioselective enzymes there was the need to develop fast and efficient ways to determine the enantiomeric purity of these compounds, which can be produced enzymatically either by reduction of the prochiral ketone using reductases or by kinetic resolution of *rac*-acetates by lipases. In both systems the CD approach is theoretically possible. In the former case, a liquid chromatography column would have to separate the educt **31** from the product (*S*)/(*R*)-**19**, whereas in the latter case (*S*)/(*R*)-**19** would have to be separated from (*S*)/(*R*)-**18**.

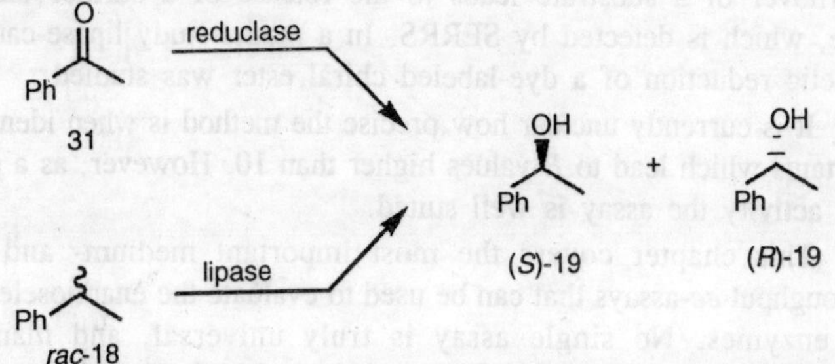

Due to the fact that acetophenone (**31**) has a considerably higher extinction coefficient than 1-phenylethanol (**19**) at a similar wavelength (near 260 nm), separation of starting material from product was absolutely necessary, which was accomplished using a relatively short HPLC column based on a reversed-phase system. In preliminary experiments using enantiomerically pure product **19** the maximum value of the CD signal was determined. Mixtures of **19** having different enantiomer ratios (and therefore *ee*-values) were prepared and analysed precisely by chiral GC in control experiments. The same samples were studied by CD, resulting in the compilation of *g*-values. Upon plotting the *g*- against the *ee*-values, a linear dependency was in fact observed with a correlation factor of $r = 0.99995$, which translates into the following simple equation for enantioselectivity:

$$ee = 3176{:}4\ g - 8{:}0$$

It was then demonstrated that the *g*-factors have no dependency on concentration in the medium used.

Although complete optimisation was not carried out, separation of **31** from (*S*)/(*R*)-**19** was in fact accomplished using reversed-phase silica as the column material and methanol/water as the eluant. In view of the results concerning the dependency of the *g*-factor on concentration, aggregation can be excluded in this protic medium. In the HPLC setup the mixture is fully separated within less than 90 s. Using the JASCO-CD-1595 instrument in conjunction with a robotic autosampler, it is possible to perform about 700–900 exact *e* determinations per day. Thus, this system is practical.

ASSAY BASED ON SURFACE-ENHANCED RESONANCE RAMAN SCATTERING

A novel method for assaying the activity and stereoselectivity of enzymes at *in vitro* concentrations is based on surface-enhanced resonance Raman scattering (SERRS) using silver nanoparticles. Turnover of a substrate leads to the release of a surface targeting dye, which is detected by SERRS. In a model study lipase-catalysed kinetic reduction of a dye-labeled chiral ester was studied.

It is currently unclear how precise the method is when identifying mutants which lead to *E*-values higher than 10. However, as a pretest for activity the assay is well suited.

This chapter covers the most important medium- and high-throughput *ee*-assays that can be used to evaluate the enantioselectivity of enzymes. No single assay is truly universal, and many are complementary. The currently most practical high-throughput systems which are potentially adaptable to many different enzymes and substrates are based on MS, NMR or CD. They also show the highest degrees of precision (better than ± 6%), which is important in the late stages of directed evolution studies. Using these assays typically 1500–8000 *ee* determinations are realistically possible per day. If in a given study medium-throughput suffices, meaning 300–800 samples per day, then adapted forms of GC or HPLC may solve the screening problem, although success will depend upon the particular nature of the product under consideration. Thus, taken together the *ee*-assays described herein can solve many analytical problems in the directed evolution of enantioselective enzymes, although not all of them have actually been tested under "operating conditions." The development of new *ee*-assays continues to be a fascinating and rewarding area of research.

4

SCREENING METHODS FOR OXIDOREDUCTASES

The interest in using enzymes as *synthetic chemistry* tools continues to grow rapidly. Among various classes of enzymes, *oxidoreductases* represent a highly versatile class of biocatalysts for specific reduction, oxidation and *oxyfunctionalisation* reactions, and are currently used for the production of a wide variety of chemical and *pharmaceutical* products. Examples of such products include l-*tert*-leucine, 6-hydroxynicotinic acid, 5-methylpyrazine-2-carboxylic acid, (*R*)-2-hydroxyphenoxypropionic acid, and steroids. *Oxidoreductases* are generally employed in whole-cell *biotransformations* or fermentation-based processes because they require expensive cofactors and coenzymes (e.g. NAD(H) or NADP(H)) to donate or accept the chemical equivalents for reduction or oxidation. To facilitate their applications *in vitro*, a large number of cofactor regeneration systems have been developed.

As with most enzyme *biocatalysts*, the discovery and engineering of *oxidoreductases* with improved stability, *catalytic* activity, and substrate specificity are highly desirable for many applications. Thanks to recent advances in recombinant DNA technology, new enzymes can now be isolated directly from *microorganisms* that are difficult to cultivate by high-throughput screening of expressed libraries of environmental DNA, whereas improved enzymes with desired functions may be engineered rapidly by directed evolution approaches. *Identifying* new or improved oxidoreductases by these routes requires the development of high-throughput screening methods that are sensitive, efficient, and simple to implement. This chapter focuses on various high-throughput screening methods recently developed for the discovery and directed evolution of *oxidoreductases*. For the purposes of this chapter, oxidoreductases have been classified into four categories: *dehydrogenases*, *oxidases*, *oxygenases*, and *laccases*. Each section begins by discussing the general uses for each oxidoreductase, then explains the details of various high-throughput screening assays, and ends with specific examples of how these methods have been used to date.

HIGH-THROUGHPUT METHODS FOR VARIOUS OXIDOREDUCTASES

Dehydrogenases

Dehydrogenases catalyse the removal and transfer of hydrogen from a substrate in an oxidation-reduction reaction and are widely used to reduce *carbonyl* groups of aldehydes or ketones and carbon/carbon double bonds. The reduction reactions carried out by *dehydrogenases* often introduce chirality into a prochiral center, which may represent a key step in *pharmaceutical* and fine chemical synthesis. The majority of dehydrogenase enzymes require *nicotinamide* cofactors (NAD(P) and NAD(P)H), while other cofactors such as *pyrroloq-uinoline quinone* (PQQ) and *flavins* (FAD, FMN) are encountered more rarely. The production or depletion of the *nicotinamide* cofactors has enabled scientists to measure the kinetics of several *dehydrogenases*. These reactions can be monitored colourimetrically by combining the redox reaction with a dye-forming reaction. High-throughput screening methods based on this approach have been developed and used to identify dehydrogenase variants with improved characteristics in various directed evolution studies.

Colourimetric Screen Based on NAD(P)H Generation

The absorbance of NAD(P)H at 340 nm is commonly used to measure the activity of dehydrogenases. Unfortunately, this approach is generally not suitable for high-throughput methods due to background noise created by cell lysates and the 96-well plate itself. Special plates designed so that they do not absorb in the UV range can be used, but tend to be very expensive when considering a large-scale screening effort. *Colourimetric assays* resolve most of these difficulties and are quite amenable to high-throughput screening methods. Thus an indirect method requiring either a synthetic compound or a secondary enzyme must be applied. *Tetrazolium* salts such as nitroblue tetrazolium (NBT) (1) are useful because they can be reduced to formazan dyes such as, which absorb light in the visible region. These reactions are essentially irreversible under biological conditions and the increase in colour can be easily monitored visually on filter discs or on a standard 96-well plate reader. A cascade reaction leading to the formation of a coloured formazan links the production of NAD(P)H to the catalytic activity of a *dehydrogenase*.

An *NBT/phenazine methosulfate* (PMS) assay has been developed that works well under several different high-throughput formats. Dehydrogenase activity on *nitrocellulose* membranes, *electrophoretic gels*, and in liquid phase has been successfully monitored using the NBT/PMS assay. The dehydrogenase cofactor NAD(P)H reacts with NBT in the presence of PMS (**3**) to produce an insoluble blue-purple *formazan*. Other transferring agents besides PMS may be used such as the enzyme diaphorase or Meldola's Blue (8-di-methylamino-2-benzophenoxazine. Compared with PMS, *Meldola's Blue* has a superior proton transfer rate and is far less sensitive to light. In addition, the cost of Meldola's Blue is approximately US$4 per gram, while the price of PMS is nearly US$16 per gram. Despite these disadvantages, PMS is used as the primary transferring agent in the NBT *colourimetric* assay. For the liquid-phase NBT/PMS assay, 0.1% gelatin is often added to the assay mix to help prevent precipitation of the insoluble blue-purple *formazan*. When scanned spectrophotometrically (range 400–700 nm), the blue-purple formazan shows a maximal absorption at 555 nm.

By measuring the increase in formazan production at 580 nm in a microplate reader, the NBT/PMS assay has been used to identify *E. coli* 6-phosphogluconate dehydrogenase (6PGDH) variants with higher activity and *thermostability*. This assay can also be used to screen dehydrogenase variants on a nitro-cellulose membrane. Briefly, a library of variants is transferred to a *nitrocellulose* filter followed by lysis of the colonies by washing in a *lysozyme* solution. A NBT/PMS cocktail is then sprayed onto the membrane to detect dehydrogenase activity. Active mutants produce a blue halo on the *nitrocellulose* filters. This approach has been successfully used by Holbrook and coworkers to produce a lactate dehydrogenase that no longer requires the expensive activator fructose-1,6-bisphosphate (FBP). The same group also used this approach to create new substrate specificities for lactate dehydrogenase.

The NBT/PMS assay has been used to identify novel alcohol dehydrogenases (ADH) in a high-throughput format without enzyme purification. This new screening approach uses *bioinformatics*, polymerase chain reaction (PCR) techniques, and the direct *in vitro* expression of enzymes to rapidly detect and characterise novel ADHs. A pool of 18 novel thermoactive ADHs with a broad substrate range was discovered and characterised.

The compound *p*-rosaniline can also be used as an alternative to NBT. No transferring agents are necessary when using *p*-rosaniline, however bisulfite must be added to help convert the dye into its rose-coloured form. *Dehydrogenase* activity can be monitored by observing the formation of a red colour on solid agar media and this allows the rapid screening of thousands of colonies.

Screens Based on NAD(P)H Depletion

NAD(P)H consumption can be monitored either directly or indirectly, similar to NAD(P)H generation in the previous section. *Enzyme kinetics* can be directly monitored by observing the decrease in absorbance at 340 nm as NAD(P)H is depleted. Coupling NAD(P)H depletion to a liquid- or solid-phase *colourimetric* assay such as NBT/PMS allows for an ideirect method for monitoring enzyme activity. Residual NAD(P)H can be titrated with NBT in the presence of PMS to produce an insoluble blue-purple formazan.

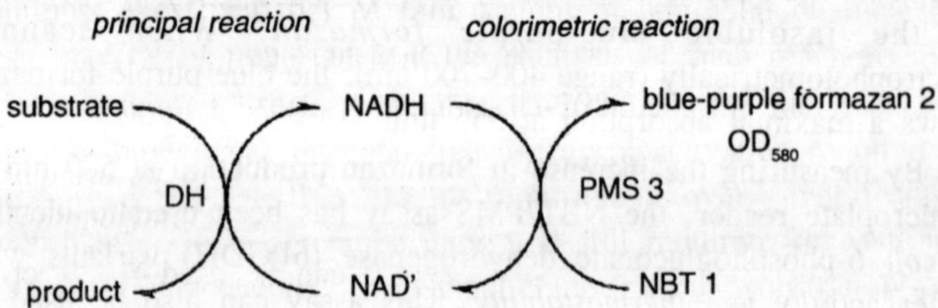

Figure 4.1 : Colourimetric assay for dehydrogenase activity. DH = dehydrogenase; PMS = phenzine methosulfate; NBT = nitroblue tetrazolium. solid-phase colourimetric assay such as NBT/PMS allows for an indirect method for monitoring enzyme activity. Residual NAD(P)H can be titrated with NBT in the presence of PMS to produce an insoluble blue-purple formazan.

High-throughput methods involve use of a *microplate* reader to measure the decrease in absorption at 580 nm or screening bacterial colonies on agar plates by visually identifying white spots on a purple background. Instead of focusing on NAD(P)H, the oxidised form $NAD(P)^+$provides an alternative. The basis for this method relies on the instability of NAD(P)H in a strong alkali environment. Under these conditions NAD(P)H breaks down to form highly *fluorescent* products. Recently this method has been adapted to a high-throughput screening method using microtiter plates. In general, any *oxidoreductase* that utilises either NADH or NADPH as a cofactor can potentially be screened using this method.

Oxidases

Oxidases catalyse the removal of hydrogen from substrates and transfer it to molecular oxygen to form either hydrogen peroxide or water. These fundamental synthetic reactions have made oxidases a prime target for directed evolution.

Some oxidases are currently being used on a large scale, such as the food antioxidant d-glucose oxidase, but unfortunately the vast majority of oxidases remain economically unattractive due to their poor *biocatalytic* performance such as low catalytic activity. Thus, several high-throughput screening methods have been developed for directed evolution studies of these oxidases with the goal of overcoming the limitations of naturally occurring enzymes.

Galactose Oxidase

Galactose oxidases catalyse the oxidation of primary alcohols to yield aldehydes and hydrogen peroxide. These *synthetically* useful chemical transformations have made galactose oxidases an attractive target of directed evolution in recent years. Standard chemical methods for oxidizing primary alcohols rely on environmentally unfriendly heavy metals and organic solvents, whereas galactose oxidase provides an environmentally friendly alternative. *Galactose oxidase* has broad substrate specificity as well as strict *regioselectivity* and *stereoselectivity* which make it widely applicable in *chemical synthesis*, *diagnostics*, and *biosensors*. For example, the enzyme can be used to detect a *disaccharide* tumor marker, Gal-GalNAc, in certain types of colon cancer. *Galactose* oxidase isolated from its natural host has insufficient expression and activity to be economically used in a large-scale industrial process. Liquid- and solid-phase high-throughput screening methods have been developed to help analyse galactose oxidase libraries to identify variants with improved expression and specific activity.

HRP/ABTS/H_2O_2 Assay

Galactose oxidase activity can be monitored by coupling the production of hydrogen peroxide (H_2O_2) to a horseradish peroxidase (HRP)/2,2'-azinobis(3-ethyl-benzthiazoline)-6-sulfonic acid (ABTS)/ H_2O_2 reaction. The enzyme HRP catalyses the polymerisation of ABTS to produce a soluble end-product that is green in colour and can be measured spectrophotometrically at 405 nm. The high solubility of

ABTS in *aqueous* solutions combined with its high extinction coefficient and reproducibility make this assay well suited for a 96-well plate format.

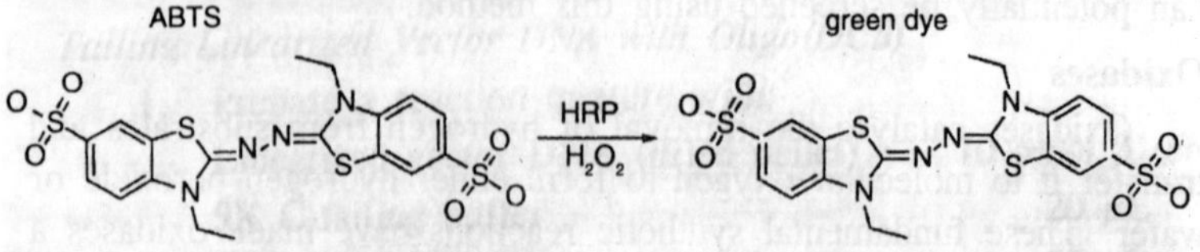

Figure 4.2 : Colourimetric assay for oxidase activity based on the HRP/ABTS/H_2O_2reaction. HRP = horseradish peroxidase; ABTS = 2,2'-azinobis(3-ethyl-benzthiazoline)-6-sulfonic acid.

Other advantages of ABTS are its low toxicity, stability at high temperature (for *sterilisation*), and the resulting green dye is relatively stable once oxidised.

However, problems such as *autooxidation*, *instability*, and a lack of sensitivity have been reported. This method has been used to create a galactose oxidase enzyme with enhanced *thermostability* and improved expression in *E. coli*. It has also been successfully used to screen fungal sugar oxidases.

HRP/4CN/H_2O_2Assay

In addition to traditional 96-well plate screening methods, a relatively new technology involving digital imaging shows great potential in screening large enzyme libraries. The efficiency of digital imaging technology provides a distinct advantage over its traditional counterpart. This system combines simple well-known *colourimetric* activity assays with single-pixel imaging *spectroscopy* on a solid-phase screening format. The assay is performed in the solid phase, and 4-chloronaphthol (4CN) is used to generate an insoluble compound which absorbs visible light maximally around 550 nm. As the colour develops on the solid support due to oxidase activity, digital imaging equipment is used to track the absorbance versus time for each pixel in the image. *Delgrave* and co-workers used this technology to isolate a galactose oxidase variant that showed a 16-fold increase in activity and a threefold lower K_m compared with the wild-type enzyme.

D-Amino Acid Oxidase

The flavoprotein d-amino acid oxidase (D-AO) catalyses the deamination of d-amino acids to their corresponding a-imino acids. This reaction has drawn significant attention recently due to its potential to deaminate cephalosporin C in a sequential enzymatic route to form 7-amino cephalosporanic acid. 7-Amino *cephalosporanic acid* is a basic building block used in the synthesis of many important *semi-synthetic cephalosporin* antibiotics. Moreover, D-AO has been used to resolve racemic amino acid mixtures and prepare a-keto acids. A high-throughput screening method based on the oxidation of the chromogen, *o*-dianisidine, has been used to detect the activity of D-AO and identify new microbial D-AO producers.

Peroxidase/o-Dianisidine Assay

The flavin cofactor $FADH_2$ is oxidised by molecular oxygen to form oxidised FAD and H_2O_2. The product H_2O_2 can then be coupled with HRP and a chromogen to form a coloured compound. ABTS was found to be unsuitable with D-AO due to problems such as *autooxidation*, instability, and a lack of sensitivity. Another chromogen *o*-dianisidine has been successfully used as an alternative to ABTS. HRP reacts with *o*-dianisidine and H_2O_2 to form a red-coloured compound under acidic conditions. The *o*-dianisidine reagent has the added benefit of being a competitive inhibitor of catalase, an enzyme that competes *in vitro* with HRP for the substrate H_2O_2. This assay might be adaptable to screening other oxidases by simply changing to the particular oxidase's substrate.

Peroxidases

Peroxidases catalyse the oxidation of a wide variety of molecules utilizing H_2O_2 as the oxidant. Many peroxidases have an iron *protoporphyrin* IX prosthetic group in the active site, while others have a vanadium or, in the case of some bacterial peroxidases, no metal cofactor. Peroxidases catalyse reactions that are potentially useful in both *industrial* and *biotechnological* applications. For example, HRP has been used as a reporter in diagnostic assays, *biosensors*, and *histochemical* staining. HRP has also been suggested as a catalyst for chemical synthesis as it catalyses polymerisation and dehydration of aromatic compounds, *heteroatom oxidations*, and *epoxidation* reactions. Furthermore, peroxidases have shown potential in *bioremediation* for removal of phenols and aromatic amines. However, many peroxidases are not stable enough under the oxidative conditions created by high peroxide concentration and there are many substrates that peroxidases do not react with. Therefore, a number of activity assays have been developed that can be used in a high-throughput fashion for the discovery and engineering of new and improved peroxidases.

ABTS and o-Dianisidine Assays

As mentioned above in the galactose oxidase and D-AO sections, ABTS and *o*-dianisidine can both be used to assay peroxidases. ABTS is oxidised by peroxidases in the presence of H_2O_2 to form a radical cation with intense green colour as shown in Figure elsewhere in this chapter, whereas the oxidation of *o*-dianisidine results in the formation of a red dye. These types of assays are carried out similarly to those described in the oxidase section, except that hydrogen peroxide is now provided as a substrate.

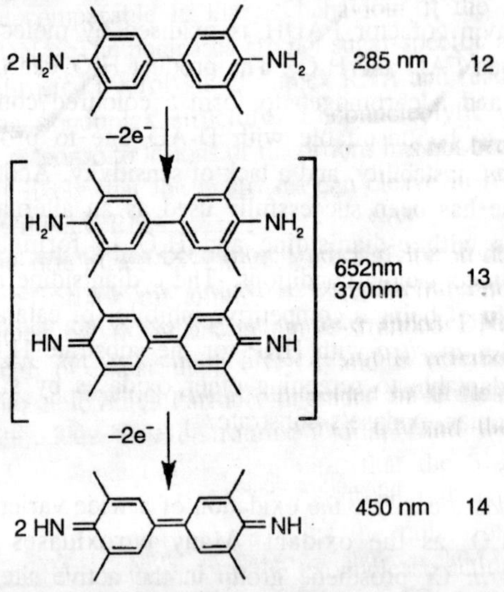

Figure 4.3 : Colourimetric assay for Peroxidase activity based on 3,3',5',5-tetramethylbenzidine (TMB).

ABTS screening has been used to find cytochrome *c* peroxidase mutants with altered specificity. Using a combination of DNA shuffling and saturation *mutagenesis*, Iffland and coworkers were able to obtain variants of the cytochrome *c* peroxidase with a 70-fold increased specificity toward ABTS compared with the natural substrate (cytochrome *c*). The ABTS assay has also been used to screen a library of HRP variants created by random *mutagenesis*, which resulted in a mutant that is more thermally stable, resistant to a variety of chemical and physical denaturants, and has a higher activity towards other organic substrates. The ABTS-based assay was also utilised to find mutants of HRP that can be functionally expressed in *E. coli* by directed evolution, to find mutants with 174-fold higher thermostability and 100-fold higher oxidative stability in *Coprinus cinereus* heme peroxidase by directed evolution, and to enhance the peroxidase activity of horse heart *myoglobin* by directed evolution. The usefulness of the *o*-dianisidine assay has been demonstrated in determining HRP mutants with expanded substrate activity. The convenience of the high-throughput *o*-dianisidine assay was further demonstrated when it was used to measure the total plasma peroxidase activity to assess the clinical severity of *sickle cell anemia* and *citrulline* therapy.

TMB Assay

Among the numerous colourimetric assays available for peroxidases, one of the most sensitive ones is based on the colourless compound 3,3',5,5'-tetramethylben-zidine (TMB). Upon oxidation of this substrate by the peroxidase, TMB forms a blue charge-transfer complex initially, followed by a stable yellow final product.

A typical cell lysate screen is performed by monitoring the production of the blue charge-transfer product at 650 nm on a microplate reader after mixing a solution of the TMB substrate, H_2O_2, and the cell lysate. This assay is commonly used in HRP-based biosensors to determine the effects of mutations on peroxidases, and would be a reasonable assay for directed evolution of a peroxidase.

Guaiacol Assay

The guaiacol peroxidase assay is another colourimetric assay that is amenable to high-throughput screening. In this case, the colourless guaiacol (2-methoxyphe-nol, is oxidised to the phenoxy radical in the presence of the peroxidase and H_2O_2. The phenoxy radical subsequently undergoes *polymerisation* to form *tetraguaiacol*, which is brown and can be measured at 470 nm. This assay can be performed in microplate format or in solid media format. Removal of background activity can be achieved by the addition of a small amount of ascorbate, which will *scavenge* the *phenoxy radicals* until all the ascorbate has reacted. The *guaiacol assay* has been successfully applied in identifying cytochrome *c* peroxidase mutants with altered substrate specificity from a library of mutants created by *combinatorial* mutagenesis. Additionally, cytochrome *c* peroxidase mutants were created by directed evolution and screened for activity with this nonnatural substrate with a final mutant displaying a 1000-fold change in substrate preference.

MNBDH Assay

4-(*N*-Methylhydrazino)-7-nitro-2,1,3-benzooxadiazole (MNBDH) has been demonstrated to be a good substrate for peroxidase screening. The hydrazine portion of MNBDH is oxidised by peroxidases in the presence of hydrogen peroxide to yield the strongly fluorescent *nitroaniline*. When measured with a fluorescent plate reader, this assay gives a very sensitive readout and is superior to *chromogenic assays* such as ABTS. However, application of this assay in directed evolution or discovery of peroxidases has yet to be demonstrated.

Oxygenases

There are several families of oxygenase enzymes whose members fall into one of two groups: those that introduce one oxygen atom into their substrate are generically referred to as *monooxygenases*, while those that introduce two oxygen atoms are referred to as *dioxygenases*. These enzymes typically catalyse the hydroxylation of nonactivated carbons or *epoxidations* of organic substrates and therefore have potential applications in synthetic chemistry, *environmental remediation*, *toxicology*, and *gene therapy*. Examples of oxygenases are the omnipresent cytochrome P450 heme monooxygenases, di-iron enzymes such as methane and toluene monooxygenases, and Rieske nonheme iron dioxygenases such as *naphthalene dioxygenase*. Molecular oxygen often serves as the oxidant, while reducing equivalents typically come from NAD(P)H-dependent reductase proteins via electron transfer.

Since reducing equivalents are transferred from a separate protein in most oxygenases (excluding those with a fused reductase domain), they typically function as large multimeric protein complexes. As such, oxygenases are often not stable enough or inadequately active towards the substrate of interest *in vitro*.

Therefore, protein engineering of mono- and dioxygenases and the discovery of new oxygenases has received much interest in the hopes of discovering or evolving oxygenases for use in organic synthesis or biotechnology. Discovery of new or improved oxygenases from large random libraries requires an efficient and sensitive high-throughput screening method for desired enzyme function. Many such screening methods have been developed and used. In this section some of these methods and their applications will be discussed. Assays based on NADPH depletion have been discussed previously. However, these assays can yield misleading results with oxygenases, since uncoupling of the reductase and oxygenase activity can occur, especially for unnatural substrates.

Assays Based on Optical Properties of Substrates and Products

Many substrates of mono- and dioxygenases are *aromatic* and upon oxidation become coloured or have a shifted absorbance *spectrum*. In other cases, the reaction products will autooxidise further to form coloured products under the proper conditions. Although this is not applicable to all oxygenase products, it is the most convenient method of assaying oxygenase activity in a high-throughput manner. 2-Hydroxybiphenyl-2-monooxygenase (HpbA), in addition to its natural activity, catalyses at low efficiency the regioselective *o*-hydroxylation of many different 2-substituted phenols to form the corresponding catechols. Random *mutagenesis* of HpbA was used to create mutants with enhanced monooxygenase activity with 2-substituted phenolic substrates. Screening was accomplished by the formation of colour when the reaction products *autooxidise*. Using this method, mutants were discovered with greater than eightfold improvement in catalytic efficiency of *hydroxylation* of 2-methoxyphenol and a fivefold improvement in turnover rate with 2-*tert*-butylphenol.

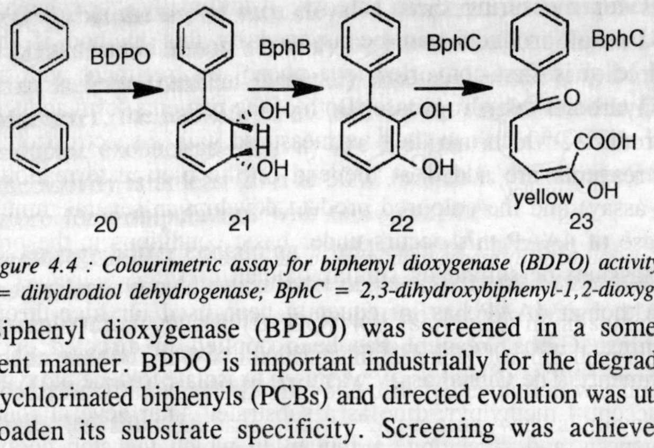

BDPO → BphB → BphC → BphC
OH OH OH OH O COOH yellow OH
20 21 22 23

Figure 4.4 : Colourimetric assay for biphenyl dioxygenase (BDPO) activity. BphB = dihydrodiol dehydrogenase; BphC = 2,3-dihydroxybiphenyl-1,2-dioxygenase.

Biphenyl dioxygenase (BPDO) was screened in a somewhat different manner. BPDO is important industrially for the degradation of polychlorinated biphenyls (PCBs) and directed evolution was utilised to broaden its substrate specificity. Screening was achieved by coexpressing *dihydrodiol dehydrogenase* (BphB) and 2,3-dihydroxybiphenyl-1,2-dioxygenase (BphC), which converted the BPDO reaction product to a diol and finally into a ring *meta* cleavage product that was yellow in colour. This method was first developed to screen a BPDO library created by DNA shuffling for enhanced PCB degradation capabilities and later used to obtain dioxygenase variants with enhanced activity towards monocyclic aromatic hydrocarbons such as toluene and bensene. In another example, hybrid genes of catechol-2,3-dioxygenase were screened for enhanced thermostability based on the colour of the product. In this case, the substrate, catechol, is colourless, while the product, 2-hydroxymuconic semialdehyde, is bright yellow. Using this method of screening, mutant dioxygenases with 26-fold enhancement in thermostability were isolated. Finally, mutants of toluene 4-monooxygenase were screened for enhanced activity on *o*-cresol and *o*-methoxyphenol based on the formation of the red-brown autooxidation products. Mutants created by saturation mutagenesis were identified that had sevenfold and twofold higher catalytic rate with *o*-meth-oxyphenol and *o*-cresol as substrates, respectively.

Assays Based on Gibbs' Reagent and 4-Aminoantipyrine

4-Aminoantipyrine (4AAP, and 2,6-dichloroquinone-4-chloroimide (Gibbs' reagent, are very similar in the products they are able to detect. They both react strongly with ortho- and meta-substituted phenolic compounds to produce strongly coloured products. Additionally, they react well with para-substituted compounds if the substituent is a halide or alkoxy moiety to form coloured products. Since many oxygenases produce such compounds, these are useful assays for measuring their activity. Additionally, dioxygenase *cis*-dihy-drodiol products can be assayed by this method if the *cis*-dihydrodiol is first converted to a phenol by treatment with acid or *cis*-dihydrodiol dehydrogenase. Both of the reagents form adducts with the products, which can then be measured with a spectrophotometer. Both reagents are added at the end of the biotransformation (end-point assay) and the coloured product develops in several minutes. In the case of 4AAP this occurs under basic conditions in the presence of potassium persulfate or similar oxidant.

Although 4AAP has infrequently been used for high-throughput screening, Gibbs' reagent has been applied in directed evolution experiments. The Gibbs' assay was used to isolate toluene dioxygenases that accept 4-methylpyri-dine as a substrate. After several rounds of mutagenesis and screening, a mutant in which the stop codon was replaced with a threonine was found to have sixfold higher activity with 4-methylpyridine and enhanced activity towards the natural substrate.

para-Nitrophenoxy Analog (pNA) Assay

The chemical methods available for linear chain hydrocarbon oxidation are energy intensive and hazardous to the environment. Several monooxygenases such as the P450s can catalyse alkane hydroxylation, but at slow rates. An ingenious assay involving the release of *para*-nitrophenolate from a *para*-nitrophenoxy analog (pNA) of a linear chain hydrocarbon has been developed and utilised. When the analogs are hydroxylated by the oxygenase, an unstable hemiacetal is formed that dissociates into an aldehyde and *para-nitrophenolate*; the latter is bright yellow. The obvious disadvantage of this P450 *mono-oxygenase* BM-3 mutants were screened using an octane pNA. Typically this mono-oxygenase only has activity on longer chain hydrocarbons (at least 12 carbons), but after two rounds of mutagenesis and screening with the analog, the activity of this enzyme on octane was improved fivefold. Another group used site-specific saturation mutagenesis and pNA screening to create mutants of the

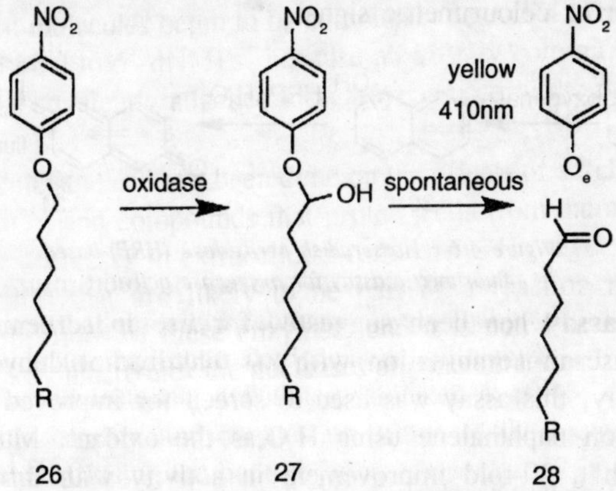

Figure 4.5 : Colourimetric assay for the activity of P450 monooxygenases. assay is that the pNA substrate is not the desired substrate and may result in the evolved enzyme only acting on pNA. However, when compared with other alternatives, this is the easiest assay for long-chain hydrocarbon hydroxylation by monooxygenases. The assay is accomplished by the addition of pNA and NADPH (or use of a regenerative system) and then monitoring the increase in absorbance at 410 nm.

same enzyme with similarly enhanced activity toward smaller substrates. Recently this P450 was optimised for activity in organic cosolvents by directed evolution using a pNA assay. Activity of the P450 was increased 10-fold in the presence of 2% *tetrahydrofuran* (THF) and sixfold in the presence of 25% DMSO. Overall, this assay has proved to be very useful for screening monooxygenases that act on long-chain hydrocarbons.

Horseradish Peroxidase-coupled Assay

The use of HRP/H_2O_2 to detect the reaction products of oxygenases has shown promise. This assay can be used when the products of the oxygenase reaction on aromatic substrates, e.g. naphthalene are or can be converted into hydroxylated aromatic compounds. These hydroxylated aromatic compounds (e.g. naphthol) are then converted by oxidation into coloured or fluorescent compounds (e.g. **30** in the presence of HRP and H_2O_2). Mono- and dihydroxylated aromatic compounds form dimers and polymers in this assay and, depending on their structure, they will display different colourimetric and fluorescent properties, allowing determination of the desired product. This assay can be carried out in cell strains expressing HRP either on solid phase or in liquid assay by adding H_2O_2 to the solid media or cell culture. The positive mutants will then display an enhanced fluorescent or colourimetric signal.

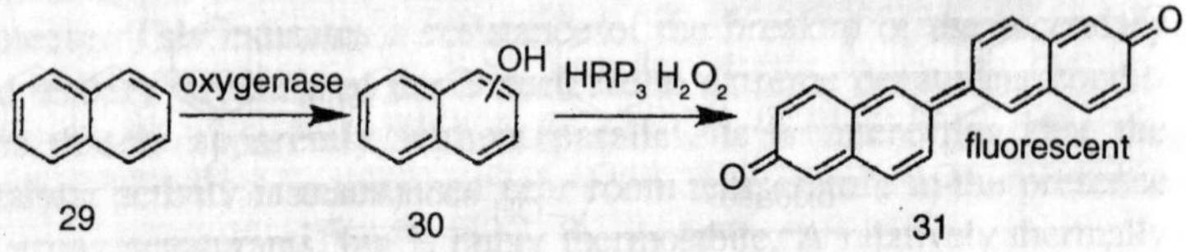

Figure 4.6 : Horseradish peroxidase (HRP)-based fluorimetric assay for oxygenase activity.

This assay has been suggested for use in screening toluene dioxygenase in combination with *cis*-dihydrodiol dehydrogenase. Additionally, this assay was used to screen for improved activity of P450cam on naphthalene using H_2O_2as the oxidant. Mutants were found with a 20-fold improvement in activity with differences in regioselectivity of hydroxylation in some cases.

Indole Assay

Indole can be oxidised by mono- and dioxygenases to various 2 and 3 position hydroxyl and epoxide indoles. Upon exposure to air, the generated compounds further oxidise and dimerise to form indigo and indirubin. Both of the formed compounds are intensely coloured and are easily monitored colourimetrically.

This assay, coupled to site-specific saturation mutagenesis, was used to enhance the activity of P450 BM-3 on indole from a level that was too low to detect up to a turnover rate of 2.7 s–1. This assay was recently used to broaden the substrate specificity of toluene dioxygenase such that it would accept *p*-xylene as a substrate. One round of *mutagenesis* and screening resulted in a 4.3-fold activity improvement on *p-xylene* as well enhancements with several other unnatural substrates. This method of assaying oxygenases is well suited for removing inactive clones from a library or for evolution of altered substrate specificity.

Laccases

Laccases (EC 1.10.3.2) belong to the blue copper oxidase family and are *multicopper enzymes* capable of oxidizing phenols, *polyphenols*, *substituted phenols*, *diamines*, *anilines*, and other similar compounds. Electrons are removed one at a time from the substrate by a type 1 blue copper ion and transferred to a tri-nuclear copper cluster and molecular oxygen is utilised as the electron acceptor. The product radical can undergo further oxidation catalysed by the laccase or it may undergo a *nonenzymatic* reaction such as hydration or polymerisation. Laccases are common enzymes in nature, but are primarily produced by fungi and plants.

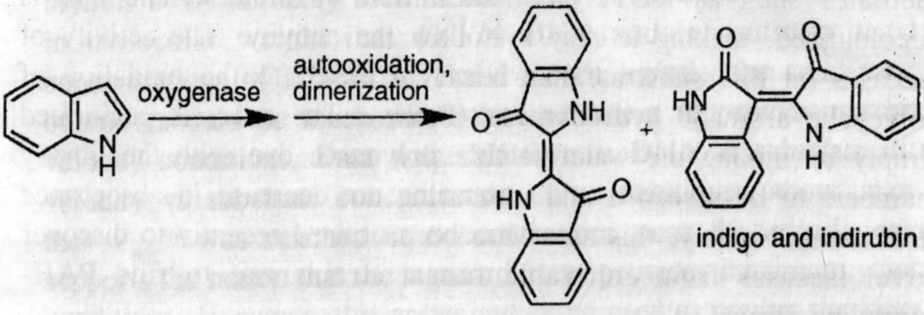

oxygenase
autooxidation, dimerization
indigo and indirubin
32
33
34

Figure 4.7 : Indole-based assay for oxygenase activity.

Laccases are currently receiving a lot of attention as industrial enzymes with proposed applications including textile dye bleaching, pulp bleaching, effluent detoxification, and bioremediation. However, a common hurdle that must be overcome with these enzymes is their low expression level in native hosts.

Additionally, laccases can catalyse the oxidation of *polycyclic* hydrocarbons (very toxic pollutants) in the presence of primary substrates such as 1-hydroxy benzothiazole (HBT). These primary substrates are often toxic, expensive, and result in side-reactions. Therefore, high-throughput screening applied to the discovery and/or directed evolution of laccases is critical to overcome these hurdles. Several screening methods have been developed and are described below.

ABTS Assay

The ABTS assay is a very flexible assay applicable to many oxidases, peroxidases as well as *laccases*. Again the production of the green radical cation is observed spectrophotometrically. This assay has been utilised to improve the functional expression of a fungal laccase in *Saccharomyces cerevisiae*. Directed evolution was utilised to improve protein expression by eightfold to the highest level yet reported with an additional increase of 22-fold in the turnover number.

The ABTS assay has also been used to screen for functional expression of another laccase from *Trametes versicolour* into the heterologous host *Yarrowia lipolytica*. This assay is well suited for development of functional selection and enhanced stability screening, and should prove useful in the future.

Poly R-478 Assay

Poly R-478 is a polymeric dye that can act as a surrogate substrate for lignin degradation by laccases in the presence of a mediator such as HBT. Upon oxidation by laccases the dye is decolourised, making it easy to follow the activity. The activity of laccases on this substrate has been correlated to the oxidation of polycyclic aromatic hydrocarbons (PAHs). The assay is performed simply by mixing the water-soluble polymeric dye with the library members to be screened and measuring the decrease in absorbance at 520 nm. Notably, this assay has been utilised recently to discover novel laccases from cultivable fungal strains and to find PAH-degrading strains.

Other Assays

Several other assays have been described for laccases that have not yet seen significant application as high-throughput screening methods. In one such assay, anthracene is converted to 9,10-*anthroquinone* by the laccase enzyme, followed by subsequent chemical reduction to 9,10-anthrahydroquinone, which is orange in colour and can be measured at 419 nm. Another assay is the production of triiodide in acidic conditions from *sodium iodide*. In this assay the activity of the laccase can be monitored at 353 nm as the formed triiodide is yellow. Both of these assays require the use of mediators like HBT or ABTS (**6**) and might be useful in the directed evolution of laccases that have the desired activity in their absence by screening with progressively less mediator.

High-throughput screening methods for oxidoreductases represent a key step in identifying variants with improved catalytic function by directed evolution as well as discovering new oxidoreductases by expression of environmental DNA.

In this chapter, we have discussed many effective high-throughput screening methods for various *oxidoreductases* and their successful applications in discovering and engineering new or improved *oxidoreductases*. The resulting oxidore-ductases are not only important for many potential applications such as biocatalysis, bioremediation, diagnostics, and gene therapy, but also are critical to increase our understanding of cellular metabolism and protein structure/function relationships. No doubt in the years to come, these high-throughput screening methods will be used increasingly to isolate oxidoreductases with new or improved characteristics.

5

Genetic Selection

The directed evolution of enzymes has rapidly become established as a powerful strategy for changing the properties of enzymes in a targeted manner. Such changes can make enzymes more suitable for use as biocatalysts for practical applications or alternatively engender new catalytic function. Directed evolution combines two independent areas of technology: (1) the generation and expression of random genetic libraries in a suitable host system and (2) methods for screening or selecting variants of interest that possess the desired characteristic (e.g. appropriate substrate specificity, enantioselectivity, catalytic activity, stability under process conditions, etc.). Several methods now exist for producing libraries of genes, some of which are in the public domain and hence can be freely used by academic and industrial laboratories around the world. Truly vast gene libraries ($> 10^{12}$variants) can now be routinely prepared, raising the prospect of increasing the possibility of identifying important mutations which can lead to beneficial improvements. In contrast, however, the poor availability of generally applicable techniques for identifying active variants of interest still constitutes a bottleneck which inevitably restricts the range of enzymes that can be successfully subjected to directed evolution approaches. The developments of novel analytical tools and protocols for finding the "needle in the haystack" is an area of active research, particularly in terms of procedures that allow very high-throughput assessment of variants.

It is important to be clear about the difference between screening and selection, and the author has chosen to adopt the nomenclature outlined by Hilvert and coworkers. Random screening involves examin-

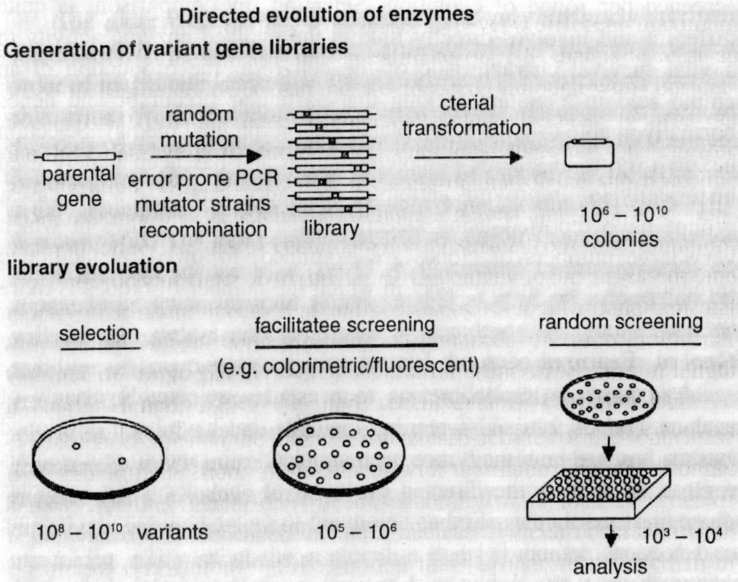

Directed evolution of enzymes

Generation of variant gene libraries

random mutation
parental gene
error prone PCR
mutator strains
recombination
library
cterial transformation
$10^6 - 10^{10}$ colonies

library evolution

selection
facilitatee screening
(e.g. colorimetric/fluorescent)
random screening
$10^8 - 10^{10}$ variants
$10^5 - 10^6$
$10^3 - 10^4$
analysis

Figure 5.1 : General strategies for the directed evolution of enzymes.

ing each variant of the library in turn to check for a particular characteristic (e.g. enantio-selectivity) without any prior evaluation of the library. For example, it may be necessary to pick individual clones from an agar plate, either manually or robotically using a colony picker, grow each clone in a 96-well microtiter plate, and then subject each well to analysis via liquid chromatography/mass spectrometry (LC/MS) or gas chromatography/mass spectrometry (GC/MS). Clearly random screening is a laborious process and severely limits the size of libraries that can be practically screened. Facilitated screening reduces significantly the numbers of variants to be screened by associating the desired characteristic with a phenotype that is easy to recognise, for example a colourimetric or fluorescent reaction. Also included here, although not discussed in detail below, would be zone-clearing assays, in which the activity of a hydrolytic enzyme is detected when secreted by a colony because of the formation of a zone of clearing in the turbid solid medium that contains the substrate. In general the size of the halo is a good indication of the enzyme activity secreted by the growing colony. This type of assay has been most effectively used when screening for lipases and other hydrolytic enzymes. Facilitated screening of colonies on agar plates is reasonably high through-put, allowing several hundred thousand clones to be screened per round of evolution. Selection, by comparison, is quite different and involves directly linking the evolved enzyme activity to a vital survival characteristic of the cell (e.g. the production of an amino acid in a mutant lacking that particular precursor and hence normally unable to grow when plated-out on minimal media). Selection has its origins in genetic complementation and has been widely used historically for improving antibiotic resistance genes. In some cases, groups have reported using selection as a primary assay, in which very large numbers of clones are examined, followed by facilitated screening as a secondary assay in which the number of clones is more manageable.

General Features of Agar Plate-based Screens

The aim of this chapter is to present a selected overview of various agar plate-based screening methods that have been developed and to illustrate how they have been coupled with random mutagenesis regimes to achieve the directed evolution of enzymes. The examples given are broadly classified as belonging to either facilitated screening or selection methods.

In general agar plate-based screens are attractive options for high-throughput screening for the following reasons:

- They can typically be used to evaluate from about 10^5–10^6(facilitated screening) up to 10^{12}(selection) colonies per round of directed evolution. A typical standard size Petri dish can accommodate about 2000–3000 colonies when plated-out at high density for facilitated screening such that individual clones can be reliably picked.
- "Hit" colonies can be rapidly picked, replated at lower density and then grown in liquid culture at small scale to confirm the activity in a complementary assay (e.g. by LC/MS or GC/MS).
- A wide range of characteristics can be screened for by adjusting the assay conditions, for example, substrate specificity, thermostability, resistance to product inhibition (see below).
- Once established they are inexpensive to operate, requiring minimal equipment. However there are still some factors that currently limit their more widespread application including:
- A requirement for general skills in molecular biology/microbiology that would be outside the expertise of some laboratories.

- A need to identify specific enzymatic transformations that result in either a coloured/fluorescent product (facilitated screening) or the production of a metabolite that leads to improved levels of growth in the host cell (screening).

Although there are some important differences between the various agar plate-based methods reported in the literature and discussed below, in general the overall procedure is based upon the protocol. First the library of transformants is plated-out onto a nitrocellulose membrane which is placed on top of an agar plate. Diffusion of nutrients through the membrane allows the colonies to grow as efficiently as if they were directly plated-out onto the agar. At this stage any reagents (e.g. IPTG isopropyl-b-d-thiogalactopyrano-side)) necessary to induce

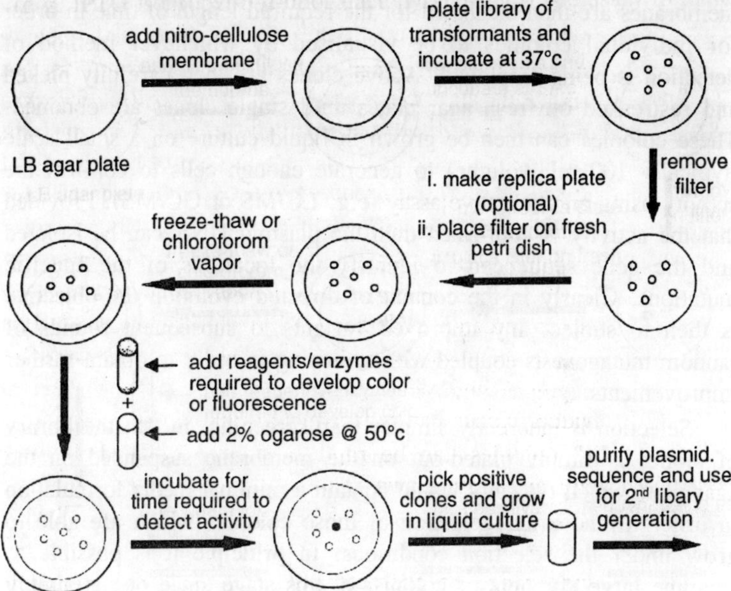

Figure 5.2 : Protocol for screening or selection of colonies on agar plates.

the enzyme activity of interest can be added to the agar. Once sufficient growth of the colonies has occurred the membrane is then lifted prior to assay. Many groups opt to make a replica plate at this point in order to ensure that there is an archive of the library of colonies. This step is especially important if the assay method is toxic to the cells resulting in no surviving population that can ultimately be picked. The advantage of having the colonies plated-out on a membrane is that it is easy to introduce additional screening criteria such as heat treatment of the membrane to try to identify variants that are more thermostable.

Prior to carrying out the assay of the colonies it is usually necessary to partially lyse the cells in order to release the intracellular enzyme activity. Lysis is typically carried by either using a carefully controlled freeze-thaw procedure or by immersing the membrane in a vapor of an organic solvent such as chloroform. A further option is to add lysozyme to ensure partial degradation of the cell wall. Thereafter the cocktail of substrates/enzymes necessary to generate the coloured/fluorescent product is added together with a solution of agarose which cools rapidly upon contact and helps to prevent diffusion of the colour generated within individual colonies. The membranes are then incubated for the required length of time in order for individual colonies to be visualised by whichever method of detection is being employed. Active clones are then carefully picked and restreaked on fresh agar plates until single clones are obtained. These colonies can then be grown in liquid culture on a small scale (typically 100 mL volume) to generate enough cells to confirm the activity using an alternative assay (e.g. LC/MS or GC/MS). Provided that the activity is confirmed then the plasmid DNA can be isolated and the gene sequenced to identify the locations of the specific mutations. Clearly in the context of directed evolution the objective is then to subject any improved variants to subsequent rounds of random mutagenesis coupled with screening in order to obtain further improvements.

Selection is inherently simpler than screening, in that the library of clones is simply plated-out on the membrane suspended on the agar plate which typically would contain a minimal media formulation in order to encourage selection of those colonies which are able to grow under the selection conditions. In principle it is possible to examine large numbers of colonies at this stage since one is simply trying to identify clones that grow relative to those that do not. Having identified vigorously growing colonies the procedure is essentially the same as above in that these individual colonies can be picked and subsequently processed in an analogous manner.

The remainder of this chapter is devoted to a presentation of selected examples of both facilitated screening and selection strategies using agar plate solid-phase assays. The organisation of the material is according to specific enzyme classes in order to encourage the reader to understand the underlying basis for the screening/selection method that has been devised and how it relates to the specific transformation that is occurring.

FACILITATED SCREENING-BASED METHODS

Amidase

Asano and coworkers have carried out the directed evolution of a d-amino acid amidase from *Ochrobactrum anthropi* SV3 in order to improve both its thermostability and catalytic activity. They employed a previously developed plate-based colourimetric screen in order to select variants of interest. Random mutagenesis of the *daaA* gene using error-prone polymerase chain reaction (epPCR), followed by transformation of the plasmid library into *E. coli* and growth on nitrocellulose filters gave a library of variant clones. The colonies were then lysed and heat treated for 15 min at a specific temperature (45–658C) to select thermostable mutants. In order to detect activity the filter was soaked in a solution containing the substrate (d-phenylalanine amide), phenol, d-amino acid oxidase, peroxidase and 4-aminoantipyrine. Colonies that developed a pink/red colour were picked and replated to give single clones. This assay exploits the capture of hydrogen peroxide generated from the d-amino acid oxidase reaction and conversion, in the presence of peroxidase, into a coloured product. Using this plate-based assay they carried out two rounds of directed evolution, initially involving 10 000 mutants, which led to the identification of variant B29, whose thermostability was increased by about 5 8C. This gene was subjected to a second round of evolution (40 000 colonies screened) at a temperature of 50 8C, resulting in the identification of clone BFB40. This variant had mutations at K278M and E303V and had an optimum temperature shifted higher by about 58C. Furthermore, the V_{max} was increased by about threefold relative to the wild-type enzyme, whereas the apparent K_m value was similar.

Esterase

Bornscheuer and coworkers have reported an effective screen for esterase activity based upon a combined growth selection/pH indicator assay. Libraries of the esterase from *Pseudomonas fluorescens* (PFE) were firstly generated using the Epicurian Coli®XL1-Red mutator strain (available from Stratagene, La Jolla, CA, USA) and subsequently transformed into the *E. coli* host expression strain. These libraries were assayed on minimal media agar plates supplemented with pH indicators (neutral red and crystal violet), thus allowing the identification of active esterase variants by formation of a red colour cased by a reduction in pH due to the released acid. An additional selection criterion was introduced by using the corresponding glycerol ester because release of glycerol as a carbon source facilitated growth on the minimal media. Using this approach, one double mutant (A209D/L181V) of PFE was identified which was able to hydrolyse the target substrate with moderate enantioselectivity (25% *ee* for the remaining ester corresponding to $E \sim 5$) but significantly enhanced activ- ity compared to the wild-type enzyme.

Glycosynthase

"Glycosynthase" is a term coined by Withers and coworkers to describe enzymes derived from glycosidases that have considerably reduced hydrolytic activity and correspondingly higher glycosyl transfer activity towards sugar acceptors.

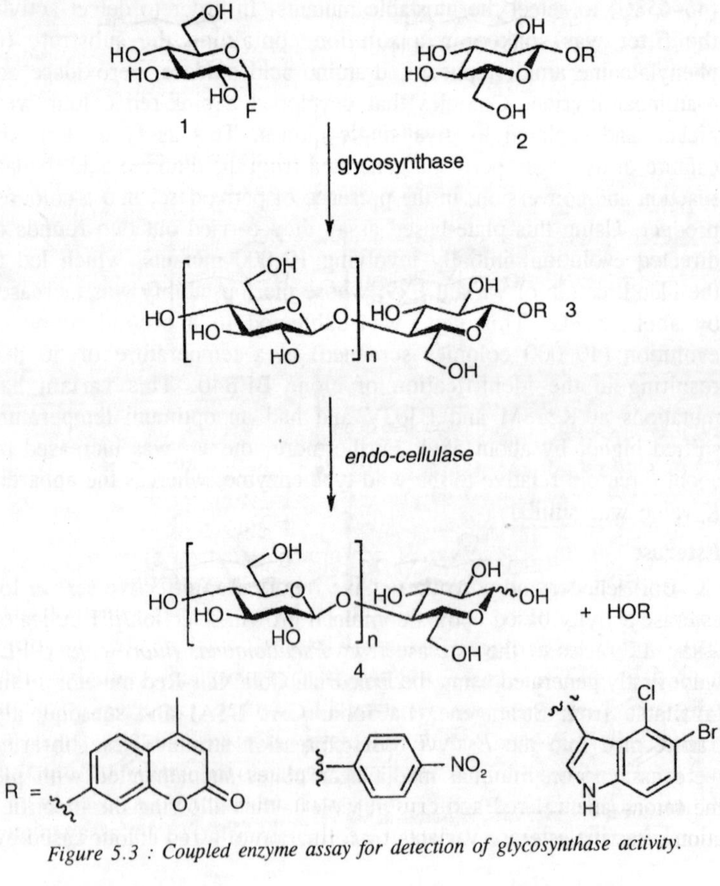

Figure 5.3 : Coupled enzyme assay for detection of glycosynthase activity.

Glycosynthases are generated by mutating the active site catalytic residue (glutamate) to an alanine. Such enzymes become hydrolytically inactive but can transfer an activated donor sugar, typically a glycosyl fluoride, to a sugar acceptor.

In order to improve the activity of the glycosynthase from *Agrobacterium* sp., Withers and coworkers have reported directed evolution of the gene using a novel two-plasmid screening system. In their system, the glycosynthase gene is located on one plasmid and the reporter gene, in this case *Cel5A*, which codes for an *endo-cellulase*, is placed under separate replication on a second plasmid. The *endo-cellulase* enzyme can only catalyse release of the fluorescent umbelliferone from **3** to yield **4** after prior transfer of a glucose residue to the substrate. In order to improve the contrast of the fluorescent colonies they introduced a replica plating step by transferring the colonies to a filter paper. A small library of variants, in which residue 385 was fully randomised, was screened (about 136 colonies), resulting in the identification of a number of mutants (E385G, E385S, E385Cys, E385Ala) that possessed comparable glycosynthase activity. Recently, in order to identify mutants with significantly enhanced activity, the same group has reported applying this screen to larger libraries of glycosynthase variants. Interestingly the group also decided to convert the "two-plasmid" system to a one-plasmid equivalent in order to facilitate some of the manipulation steps, in particular transformation and selection of reagents for induction. Through three rounds of directed evolution, involving screening of about 20 000 variants, they were able to identify a mutant (2F6) which possessed 27-fold improvement in catalytic efficiency (k_{cat}/K_m).

Galactose Oxidase

Delagrave and coworkers (Kairos Scientific) have reported a high-throughput digital imaging screen to evolve the enzyme galactose oxidase from *Fusarium sp.*. Their system, known commercially as "Kcat", combines single-pixel digital imaging spectroscopy with solid-phase screening of colonies on agar plates in order to select variants with improved activity. Transformed cells were plated-out on a polyester membrane placed on top of an agar plate. After overnight growth the membrane was transferred to another agar plate containing ampicillin in order to induce galactose oxidase. Thereafter, lysis of the microcolonies (radii < 0.4 mm) was carried out by transferring the membrane to a chloroform vapor chamber for 45 s. Membranes containing the lysed colonies were then transferred to assay plates containing a cocktail of the substrate, 1% agarose, 4-chloronaphthol, and peroxidase to generate the coloured colonies indicative of galactose oxidase activity. Using the insoluble galactose-containing polymer guar as substrate they screened about 100 000 colonies and found a number of mutants that had greater activity than the wild-type enzyme. The most active variant (Cys383Ser) showed a 16-fold improvement in activity, partly due to a threefold improvement in K_m.

The authors estimate that it should be possible to screen up to 100 000 mutants in a day using this technology. The same group has reported the use of the Kcat technology to evolve the enzyme b-glucosidase from *Agrobacterium sp.*. By simultaneously assaying against two different substrates (glucose and galactose) tagged with spectrally distinct chromogenic reporters they were able to identify enzyme variants that showed specificity for either glucosyl or galactosyl substrates.

Monoamine Oxidase

Turner and coworkers have recently developed a novel method for the efficient synthesis of enantiomerically pure a-amino acids and amines by deracemisation of the corresponding racemic mixtures. Their approach relies upon coupling enzyme-catalysed enantioselective oxidation of a chiral amine with nonselective chemical reduction of the intermediate imine **6** in a one-pot process. Provided that the enantioselectivity of the oxidase enzyme is high, then only seven catalytic cycles are required to transform a racemic mixture into a single enantiomer. In order to identify enzymes with appropriate substrate specificity and enantioselectivity towards target substrates of interest they have developed an agar plate-based screening assay to evolve the enzyme monoamine oxidase N (MAO-N) from *Aspergillus niger*. The principle of the assay is similar to that described above by Asano and coworkers and Youvan and coworkers in that it relies upon capture of the hydrogen peroxide generated from the oxidase reaction in the presence of the substrate diaminobenzidine to generate coloured colonies.

In their work a plasmid harboring the MAO-N gene was transformed into *Escherichia coli*, and used to express the wild-type enzyme which was found to have very low, but detectable activity towards a-methylbenzylamine, the amine chosen for the model studies. Interestingly there was also evidence that the wild-type enzyme was inherently (*S*)-enantioselective although the intrinsic rates were very low. Random mutagenesis of the gene, using the XL1-Red mutator strain, yielded a library of variants (about 150 000) that were screened against (*S*)-a-methylbenzylamine as substrate using the agar plate colourimetric assay.

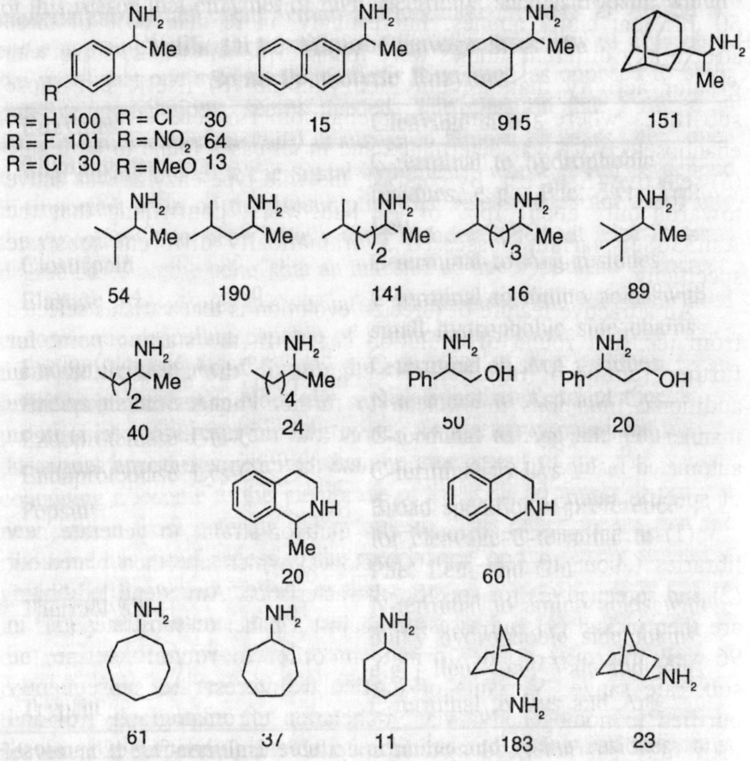

Figure 5.4 : Substrate specificity of Asn336Ser variant of MAO-N (numbers beneath structures denote relative activities).

The frequency of mutation was adjusted such that individual clones possessed about 1–2 nucleotide mutations per gene. Approximately 30 clones were selected at this stage, on the basis of their activity in the screen, and amongst these one clone in particular was found to have very high (*S*)-selectivity. Small-scale growth of this clone and partial purification of the amine oxidase revealed that this variant possessed approximately 47-fold greater activity towards a-methylbenzylamine compared with the wild-type enzyme. This variant was sufficiently catalytically active to be used in combination with ammonia-borane to deracemise a-methylbenzylamine, yielding the (*R*)-enantiomer in 77% yield and 93% *ee*.

Sequencing of this variant revealed that it possessed a single mutation (Asn336Ser). This variant amine oxidase was purified to homogeneity and examined for its reactivity towards a panel of about 50 different chiral amines. Interestingly, more than 60% of these structurally different amines were found to be substrates, some with greater activity than a-methylbenzylamine (**5**) itself. For those substrates where the individual enantiomers were available, the enantioselectivity was determined and in all cases found to be highly (*S*)-selective ($E > 10$). By contrast, the wild-type enzyme was active towards only about 15% of the substrates, confirming that the Asn336Ser variant possessed a dramatically different substrate specificity.

An attractive feature of directed evolution is that variants selected from the first round of evolution can form the starting point for further rounds of mutagenesis/selection in the anticipation than additional mutations might lead to further changes in an additive manner. Further cycles can be carried out rapidly (3–4 weeks) in an automated fashion in order to fine tune the enzyme towards substrates of specific interest.

(1) initial passage through the mutator strain, to generate new libraries (about 10 000–300 000 clones) which are then plated-out (2) and screened (3) for specific characteristics. Any identified "hits" are then picked (4) and assessed against a panel of substrates (5), in 96-well microtiter plate format, in order to roughly define the substrate range. Variants of particular interest are subsequently purified to homogeneity, via Ni-chelation chromatography (6) and then fully characterised against specific chiral amines of interest, including determination of k_{cat} and K_m values. Using this approach, Turner and coworkers recently reported the identification of a new variant (Asn336Ser/Ile246Met) that is able to deracemise racemic cyclic secondary amines on a preparative scale.

P450 MONOOXYGENASES

Arnold and coworkers have reported a fluorescent plate-based assay for the directed evolution of P450cam from *Pseudomonas putida*. In their system a library of about 200 000 P450cam variants were coexpressed with an engineered horseradish peroxidase (HRP) by cotransformation of two plasmids into *E. coli*.

After incubation for 16 h, the colonies were replicated using a nitrocellulose membrane and transferred onto a fresh agar plate containing naphthalene and hydrogen peroxide. Colonies were then screened for activity using fluorescence digital imaging which indicated formation of the oxidised and coupled product. Three clones with enhanced fluorescence relative to the wild-type enzyme were picked and grown for further study. A second round of evolution led to the identification of a variant with about 20-fold improvement in activity.

Arnold and coworkers have extended the underlying principle of this digital imaging screen, using HRP as a coupling enzyme to generate coloured/fluorescent products derived from the initial hydroxylation reaction, in order to screen for other monooxygenase activities. For example toluene dioxygenase, which catalyses the conversion of aromatic substrates (e.g. chlorobenzene) to the corresponding 3-substituted-catechol, generates products possessing a red-brown color (k = 500 nm) in the presence of HRP. They note that the colours generated by the coupling reaction are sensitive to the site of hydroxylation and hence may be used to identify variant enzymes with altered regiospecificities.

Guengerich and coworkers have devised an alternative system for screening P450 monooxygenase activity based upon production of the blue pigment indigo. This screening method is based upon their observation that during expression of recombinant cytochrome P450 2E1 some of the colonies developed a blue colouration which was subsequently shown to be indigo. They proposed that the indigo arose from dimerisation of 3-hydroxyindole, itself derived from P450-mediated hydroxylation of indole, which is a breakdown product of tryptophan in bacteria. Subsequently they exploited this colourimetric screening method for the directed evolution of human cytochrome P450 2A6.

Random libraries of the gene were expressed in *E. coli* and screened for their ability to hydroxylate indole and hence generate blue colonies. One mutant, Phe209The, showed a reduced rate of hydroxylation for indole compared with the wild-type enzyme but a k_{cat} *13-fold higher for hydroxylation of coumarin at the 7-position. Li and coworkers have also employed this indigo screening system, in their case to evolve the fatty acid hydroxylase P450 BM-3 from *Bacillus megaterium*. They identified a number of variants, in particular the triple mutant Phe87Val, Leu188Gln, Ala74Gly, which were able to hydroxylate indole.

Significantly the wild-type P450 BM-3 is unable to carry out the equivalent hydroxylation reaction.

Carotenoid Biosynthesis

Carotenoid biosynthesis in a noncarotenogenic microorganism

such as *E. coli* requires extension of the general terpenoid pathway with the genes for geranyl-geranyl diphosphate synthase (*crtB*) and phytoene synthase (*crtE*) for the production of the C-40 carotenoid phytoene. Subsequent desaturation by phytoene desaturase (*crtI*) and further modifications by, for example, cyclases, hydroxylases, and ketolases, result in the production of different carotenoids.

Arnold and coworkers have reported molecular breeding of these acyclic carotenoid pathways. By employing a two-plasmid-based screen, genes that produced the carotenoid precursors that serve as substrates for the target enzyme were cloned on a pACYC184-derived plasmid. Genes for the enzymes that were subjected to evolution *in vitro* were cloned on a pUC19-derived plasmid. A library of desaturases, generated by *in vitro* homologous recombination (DNA shuffling) of the genes from *E. herbicola* and *E. uredovora*, was then transformed into phytoene synthesizing *E. coli* JM101 harboring pAC-*crtEEU-crtBEU*. Colonies were transferred to nitrocellulose membranes, which provided a white background for visual screening of the clones based on colour. Approximately 10 000 colonies were screened with about 30% appearing white as a result of inactivation of the desaturase. Twenty colonies turned yellow, indicating the presence of carotenoids with fewer conjugated double bonds than lycopene. In addition, one pink clone was isolated suggesting the introduction of additional double bonds into lycopene by this mutant. Analysis of cell extracts by high-pressure liquid chromatography (HPLC) allowed full characterisation of the reaction products.

Liao and coworkers have reported analogous experiments on the directed evolution of carotenoid-producing pathways. Geranylgeranyl diphosphate (GGPP) synthase is an important rate-controlling enzyme in the production of carotenoids in *E. coli*. They constructed a library of GGPP synthase variants from *Archaeoglobus fulgidus*, which was expressed in *E. coli* and screened on agar plates. Mutants were picked on the basis of the orange colour generated, which is characteristic of production of the final product astaxanthin. Eight mutants were isolated and the optimal one found to increase lycopene production by 100%. Sequencing of the variants indicated that the mutations were clustered in four "hot spots".

The same group has applied directed evolution to the improvement of phytoene desaturase from *Rhodobacter sphaeroides*. In this example they screened for libraries of variants that were able to synthesise lycopene (pink colonies) rather than neurosporene (yellow colonies).

BIOTIN LIGASE

Heinis and coworkers have described a variation on the standard method for screening on agar plates which they term "colony filter screening". *E. coli* cells expressing the enzyme of interest are grown on a porous master filter that receives nutrients from an agar plate by diffusion in the normal manner. The enzyme is expressed in the cytoplasm of the cells and hence does not diffuse in the master filter. This porous master filter, containing the bacterial colonies, is then placed on top of a second filter (reaction filter). Enzymes are released from the bacterial cells by freeze-thawing and can then convert suitable substrates on the reaction filter.

As a model enzyme they studied biotin ligase (BirA), which catalyses the formation of biotinyl-5'-adenylate from biotin and ATP. The enzymatic activity of the secreted enzyme is assayed as follows and shown further below.

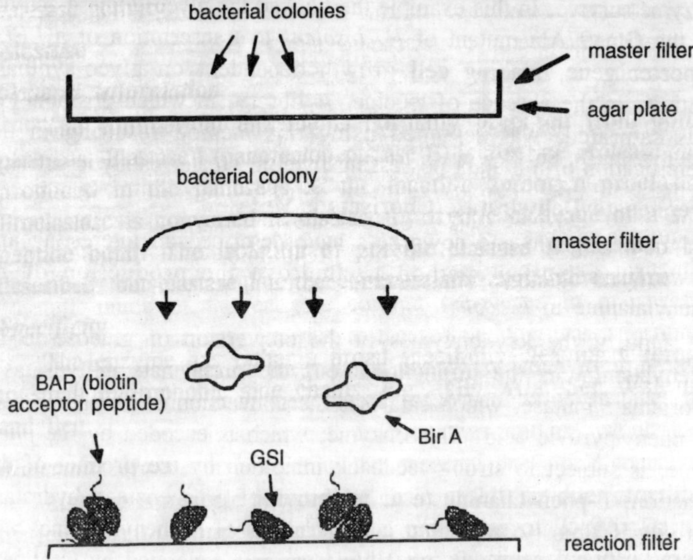

Figure 5.5 : Filter sandwich assay for biotin ligase.

The reaction filter is coated with the substrate, in this case the biotin acceptor peptide (BAP), which is biotinylated if it comes into contact with the BirA ligase secreted from the master filter. Finally the reaction filter is washed, and product formation detected with streptavidin/HRP and a chemiluminescent substrate. Importantly, the biotin ligase released by the bacterial cells does not diffuse, permitting the identification of individual colines expressing the desired enzyme activity.

IN *VIVO* SELECTION-BASED METHODS

Glycosynthase

Cornish and coworkers have devised an alternative to the Withers' assay described above, namely their yeast three-hybrid chemical complementation assay, in order to carry out directed evolution of the *Humicola insolens* glycosynthase. Chemical complementation detects enzyme-catalysed bond-forming or bond-breaking reactions based upon coupling of two small molecule ligands *in vivo*. The heterodimeric small molecule reconstitutes a transcriptional activator, turning on the transcription of a downstream reporter gene. The assay is high throughput since it can be run as a growth selection.

In this case a dexamethasone (Dex)–methotrexate (Mtx) yeast three-hybrid system is used whereby only cells containing functional enzyme survive. In this example they linked the glycosynthase activity of the Glu197Ala mutant of *H. insolens* to transcription of a *LEU2* reporter gene, making cell growth dependent on glyco-synthase acitivity in the absence of leucine. A library, in which position 197 was saturated, was screened using this method resulting in the identification of a mutant possessing fivefold greater catalytic activity.

Prephenate Dehydratase/Chorismate Mutase

Fotheringham and coworkers have described an interesting and powerful selection method to improve the production of L-phenylalanine in *E. coli*.

One of the key enzymes in the conversion of glucose to L-phenylalanine is the bifunctional enzyme prephenate dehydratase/ chorismate mutase, which catalyses the conversion of prephenic acid to phenylpyruvic acid. This enzyme, which is encoded by the *pheA* gene, is subject to strong feedback inhibition by the product of the reaction, L-phenylalanine (e.g. inhibition of prephenate dehydratase activity is almost complete at concentrations of L-phenylalanine > 1 mM). A library of about 1000 *pheA* variants generated by treatment with nitrous acid was screened on agar plates impregnated with 10–20 mM b-2-thienyl-L-alanine, which is an analog of L-phenylalanine and is toxic to the cells. A number of colonies showed improved growth, implying greater production of L-phenylalanine in order to counteract the toxic effect of b-2-thienyl-L-alanine, and four in particular proved to have greater resistance to feedback inhibition.

Sequencing of these variants revealed that all mutations occurred within the region specified by codons 304–310. Some mutations resulted in specific amino acid changes (e.g. Gln306Leu, Gly309Cys) whereas others involved excision of residues 307–310 and 304–306.

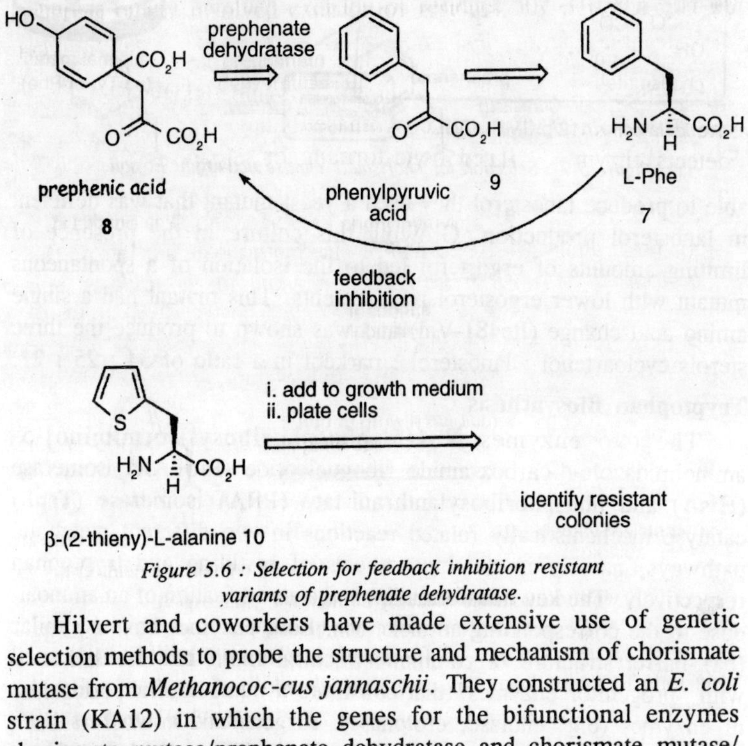

prephenate dehydratase

prephenic acid 8

phenylpyruvic acid 9

L-Phe 7

feedback inhibition

i. add to growth medium
ii. plate cells

identify resistant colonies

β-(2-thieny)-L-alanine 10

Figure 5.6 : Selection for feedback inhibition resistant variants of prephenate dehydratase.

Hilvert and coworkers have made extensive use of genetic selection methods to probe the structure and mechanism of chorismate mutase from *Methanococ-cus jannaschii*. They constructed an *E. coli* strain (KA12) in which the genes for the bifunctional enzymes chorismate mutase/prephenate dehydratase and chorismate mutase/ prephenate dehydrogenase were deleted and replaced by the monofunctional versions of the dehydratase and dehydrogenase. Random gene libraries were generated and introduced into this strain and the colonies were then grown on minimal agar media lacking added phenylalanine and tyrosine. Colonies that grew under this regime were indicative of having evolved chorismate mutase activity.

Terpene Cyclase

Hart and coworkers have employed an *in vivo* selection method to evolve the enzyme cycloartenol synthase, which catalyses the conversion of oxidosqualene to cycloartenol via the intermediate lanosteryl cation. The wild-type enzyme yields cycloartenol and parkeol in a ratio of 99 : 1. In order to select for variant enzymes

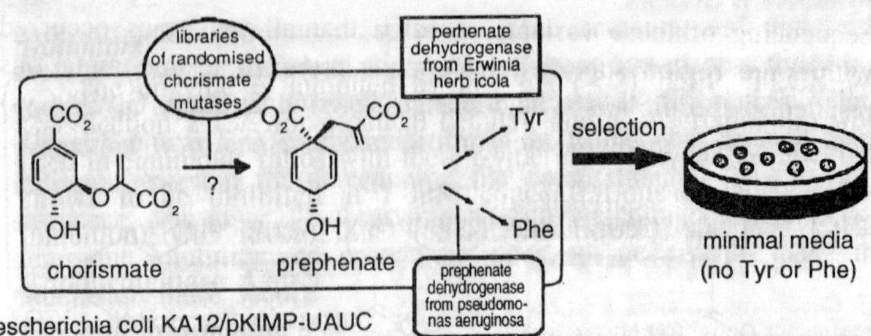

Figure 5.7 : Selection for chorismate mutase activity in E. coli.

able to produce lanosterol they used a yeast mutant that was deficient in lanosterol production. Growing this culture in the presence of limiting amounts of ergosterol led to the isolation of a spontaneous mutant with lower ergosterol requirements. This mutant had a single amino acid change (Ile481-Val) and was shown to produce the three sterols cycloartenol : lanosterol : parkeol in a ratio of 54 : 25 : 21.

Tryptophan Biosynthesis

The two enzymes *N*'-[(5'-phosphoribosyl)formimino]-5-aminoimidazole-4-carbox-amide ribonucleotide (ProFAR) isomerase (HisA) and phosphoribosylanthrani-late (PRA) isomerase (TrpF) catalyse mechanistically related reactions in two different metabolic pathways, namely in the biosynthesis of histidine and tryptophan respectively. The key transformation is the isomerisation of an aminoal-dose to the corresponding aminoketose. Both enzymes have a similar $(ba)_8$-barrel structure, a common structural motif that is associated with "progenitor enzymes" that constitute a number of superfamilies of enzymes (e.g. enolase, crotonase). Jürgens and coworkers have used random mutagenesis and selection to identify a number of HisA variants from *Thermatoga maritima* that were able to catalyse the TrpF reaction while still retaining some HisA activity. In order to select appropriate variants they used an *E. coli* strain that lacked the *trpF* gene on its chromosome and hence was unable to grow without the addition of tryptophan. This strain was transformed with a plasmid library containing randomly mutated *hisA* genes generated by DNA shuffling. After growth for 3 days on a medium lacking tryptophan, three colonies were selected. Of the mutations identified in the variants, the Asp127Val change appeared particularly important in imparting TrpF activity.

Ribitol Dehydrogenase

HomsiBrandeburgo and coworkers have reported obtaining a

mutant ribitol dehydrogenase from *Klebsiella aerogenes* that has improved xylitol dehydrogenase activity. The host strain containing the ribitol dehydrogenase gene was grown under selective pressure to improve growth on xylitol as the sole carbon source.

Inteins

Liu and coworkers have reported the directed evolution of an intein (self-splicing protein element) using an elegant selection/screening based method. The idea is based upon the development of inteins as molecular switches for activation of proteins in response to small molecules. Inteins are attractive in this respect in that their insertion into a protein blocks the target protein's function until splicing occurs. Initially they began with the *Mycobacterium tuberculosis* RecA intein using *Saccharomyces cerevisiae* as the screening host. Insertion of a natural ligand-binding domain into this minimal intein resulted in destruction of the splicing activity. In order to restore the activity they linked protein splicing to cell survival or fluorescence in the presence of the small molecule 4-hydroxytamoxifen (4-HT). Several rounds of random mutagenesis coupled with selection/screening led to efficient splicing activity in a range of different systems.

Aminoacyl-tRNA Synthetase

The Schultz group has for a number of years been pursuing their ambitious goal of developing strategies for the introduction of unnatural amino acids into proteins. Expanding the amino acid composition of proteins beyond the 20 natural building blocks could potentially lead to novel enzyme activities by incorporating unnatural functional groups into specific parts of the protein (e.g. the active site of an enzyme). Recently this group has reported a combined genetic selection and screen that allows the rapid evolution of aminoacyl-tRNA synthetase activity. The basis of their method is to use an "orthogonal" aminoacyl-tRNA synthetase and tRNA pair that cannot interact with any of the naturally occurring synthetase/tRNA pairs in *E. coli*. A chloroamphenicol-resistance gene was used as a primary selection method to select for highly active synthetase variants and an amplifiable fluorescence reporter gene was used in combination with fluorescence-activated cell sorting (FACS) as a secondary screen to identify variants with changes in amino acid specificity. Through iterative rounds of evolution three new variants of a tyrosyl-tRNA synthetase were identified which allowed selective incorporation of amino-, isopropyl-, and allyl-containing tyrosine analogs into a protein.

CONCLUSIONS AND FUTURE PROSPECTS

It is generally accepted that the major limiting factor in the application of directed evolution approaches is the availability of a suitable high-throughput screen/selection system for the enzyme activity of interest. In this respect, agar plate-based screens are not different and although the past 10 years has witnessed a substantial increase in the number of systems available, there still remains a need for a wider range of screens and genetic selection procedures to expand the range of enzyme activities that can be addressed. For example in the area of P450-mediated hydroxylation, the screening assays described above are either highly product specific or rely upon screening for a surrogate activity (e.g. hydroxylation of indole and subsequent formation of indigo) rather than directly on the substrate of interest. Thus there is a need for more general hydroxylation assays that can be carried out on agar plates which would undoubtedly increase the application of this important class of enzymes. Among the other classes of reactions that are difficult to screen for on agar plates one could also include reduction of carbon–carbon double bonds, sulfoxidation, dehalogenation, carbon–carbon bond formation (e.g. using aldolases, oxynitrilases, transketolases) and many more. Similarly, in the area of selection using genetic complementation, the systems described above are generally confined to primary metabolism (e.g. amino acids, carbohydrates, nucleotides) or involve the production of coloured compounds in secondary metabolic pathways (e.g. carotenoids).

Given the inherent power of selection in terms of throughput then undoubtedly developments in these areas will have a major impact on enzyme evolution work and in this context it is interesting to note that new systems based upon transcriptional activation are starting to be developed and have the potential to be more generally applicable to a wide range of enzyme-catalysed reactions.

6

Selections of Enzyme-encoding Genes

The availability of vast *gene repertoires* from both natural (genomic and cDNA libraries) and artificial sources (gene libraries) demands the development and application of novel technologies that enable the screening or selection of large libraries for a variety of enzymatic activities. This chapter describes recent developments in the selection of enzyme-coding genes for directed evolution and functional genomics. We focus on high-throughput screening (HTS) approaches that enable selection from large libraries (> 10^6 gene variants) with relatively modest means (i.e. *nonrobotic systems*), and on *in vitro* compartmentalisation (IVC) in particular.

The basis of all screening and selection *methodologies* is a linkage between the gene, the enzyme it encodes, and the product of the activity of that enzyme. The difference between screening and selection is that screening is performed on individual genes or clones and requires some spatial organisation of the screened variants on *agar plates*, *microtiter plates*, *arrays*, or *chips*, whereas selections act simultaneously on the entire pool of genes. The focus of this chapter is selection techniques, although formally speaking, several of the techniques described are in fact screening methods. The techniques described herein are applicable in the context of both functional genomics and directed evolution.

Directed enzyme evolution has been used in the past two decades as a powerful approach for generating enzymes with desired properties. Enzyme variants have been evolved for catalytic activity under extreme conditions such as high temperatures, *acidic* and *alkaline* environments, and organic solvents and with improved *catalytic* activities and new substrate specificities. Directed evolution experiments are based on the principles of natural *Darwinian* evolution, and therefore consist of two major steps: (1) creation of genetic diversity in the target gene in the form of gene libraries; and (2) an effective selection of the library for the desired catalytic activity. A large variety of methods for creation of genetic diversity are currently available.

However, the typical library size is still many orders of magnitude larger than the number of protein variants that can be screened. The same restriction applies to cDNA and genomic libraries derived from natural sources (e.g. environmental libraries) the diversity of which is almost unlimited. Further, while methods for creating gene diversity are generic, the screens for activity need to be tailored for each enzyme and reaction. The "*bottleneck*" for most enzyme isolation endeavors is therefore the availability of a genuinely high-throughput screening or selection for the target activity.

THE BASICS OF HIGH-THROUGHPUT SCREENS AND SELECTIONS

Screening and selection methodologies should meet the following demands: (1) They should be, if possible, directly for the property of interest – "you get what you select for" is the first rule of directed evolution. Thus, the substrate should be identical, or as close as possible to the target substrate, and product detection should be under multiple turnover conditions to ensure the selection of effective catalysts. (2) The assay should be sensitive over the desired dynamic range. The first rounds of any evolution experiments demand isolation with high recovery–all improved variants, including those that exhibit only several-fold improvement over the starting gene, should be recovered. The more advanced rounds must be performed at higher stringency to ensure the isolation of the best variants. A limited dynamic range seems to be the drawback of most selection approaches. (3) The procedure should be applicable in a high-throughput format.

Numerous assays enable the detection of enzymatic activities in agar colonies or crude cell lysates by the production of a *fluorophore* or *chromophore*. Assays on agar-plated colonies typically enable the screening of > 10^4 variants in a matter of days, but they are often limited in sensitivity: soluble products diffuse away from the colony and hence only very active variants are detected. Assays based on insoluble products have higher sensitivity, but their scope is rather limited. The range of assays that are applicable for crude cell lysates is obviously much wider, but their throughput is rather restricted: in the absence of *sophisticated robotics* that are usually unavailable to academic laboratories, only 10^3–10^4 variants are typically screened. These low-to-medium throughput screens have certainly proved effective for the isolation of enzyme variants with improved properties as described in a number of recent reviews. However, a far more efficient sampling of sequence space is required for the isolation of rare variants with dramatically altered *phenotypes*.

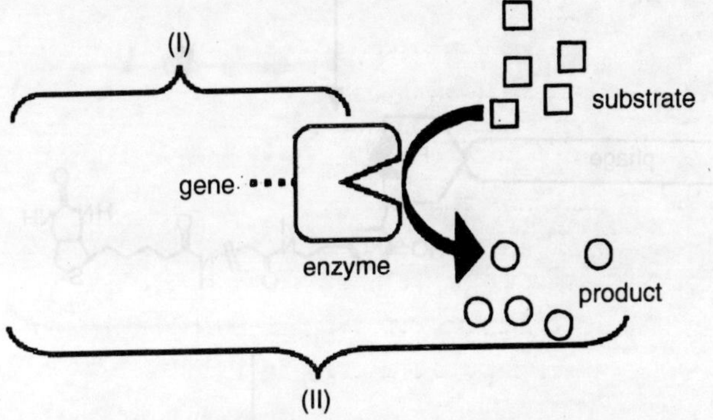

Figure 6.1 : The basis of all selection and screens for enzyme-coding genes, whether in a directed evolution or functional genomics context, is the linkage between the gene, the enzyme it encodes, and the products of the activity of that enzyme.

Whilst high-throughput selections for binding activity have become abundant, *enzymatic* selections remain a challenge. The main obstacle is that catalytic screens or selections demand a linkage between the gene, the encoded enzyme, and multiple product molecules. This chapter focuses on *methodologies* that enable the selection of large libraries, typically well over 10^6 variants, for enzymatic activities. This throughput is still beyond the reach of high-throughput technologies that are based on two-dimensional arrays and robotics (microplates, chips, etc.).

HIGH-THROUGHPUT SELECTION OF ENZYMES USING PHAGE DISPLAY

The display of the screened enzyme on the surface of *bacteriophages* has several advantages: first, the phage provides the link between the gene and the protein it encodes; second, although library size is limited by transformation, 10^7 transformants can be obtained quite easily, and 10^{11}–10^{12} are not beyond reach; third, display on the surface allows unhindered accessibility of the substrate and reaction conditions of choice (e.g. buffer, pH, metals).

The main challenge with display systems is maintaining the linkage between the enzyme and the products of its activity.

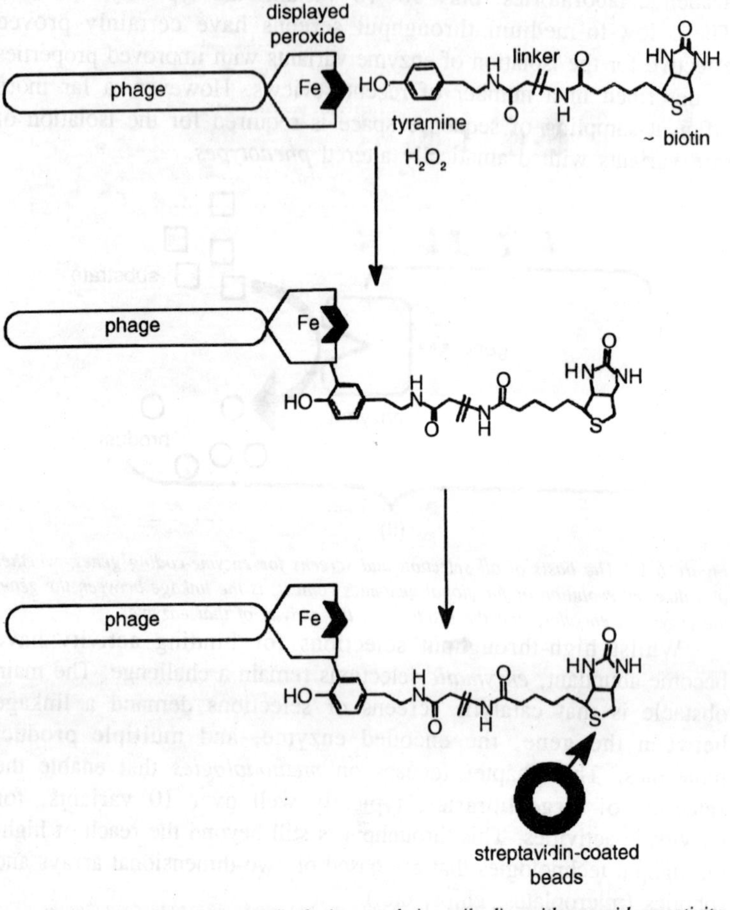

Figure 6.2 : Selection of phages displaying catalytic antibodies with peroxidase activity. The selection was based on oxidation of the peroxidase substrate tyramine that is known to undergo rapid condensation with phenolic side-chains on protein surfaces upon oxidation.

Previously, selections for enzymatic activities of either catalytic antibodies or enzymes displayed on phage were performed indirectly, by binding to transition state analogs (TSAs) or suicide inhibitors. In recent years, direct selections for product formation under single as well multiple turnovers have been achieved. A notable example is the selection of a synthetic antibody library for alkaline phosphatase activity. *Hydrolysis* of a soluble substrate generated a product which acts as an *electrophilic* reagent and couples onto the phage particle that displays the *catalytically* active antibody variant. These phage particles were then captured by affinity *chromatography* to the product. Two rounds of selection from an initial library of > 10^9antibody variants yielded an antibody with catalytic efficiency (k_{cat}/K_m) that is 1000-fold higher than other alkaline *phosphatase*-like antibodies generated by immunisation with a transition state analog. This high catalytic efficiency was mainly the result of an increase in k_{cat}value, demonstrating the importance of selection for multiple turnovers.

Another example regards the selection of catalytic antibodies with peroxidase activity. Selection was based on a biotin-linked tyramine substrate being oxidised and coupled to the antibody's binding site. Phages displaying active anti-body catalysts were isolated by binding to streptavidin.

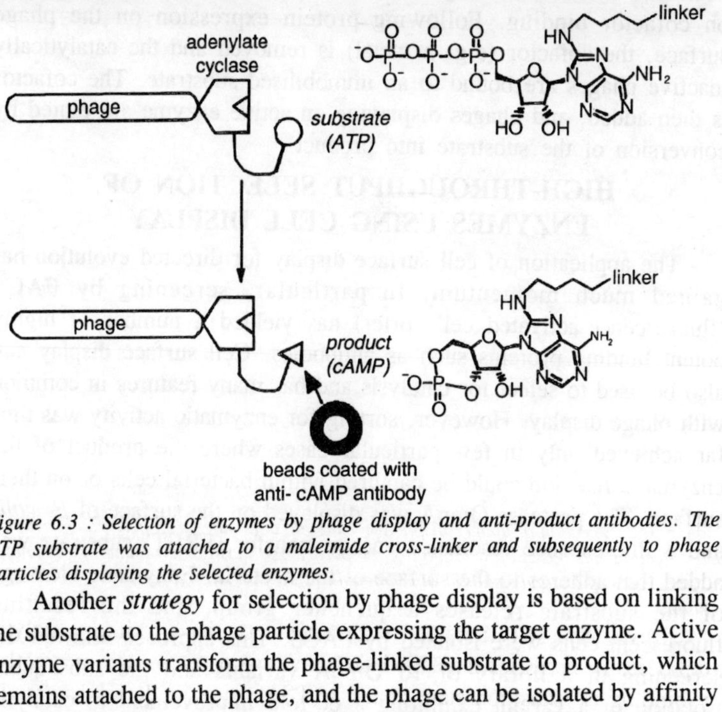

Figure 6.3 : Selection of enzymes by phage display and anti-product antibodies. The ATP substrate was attached to a maleimide cross-linker and subsequently to phage particles displaying the selected enzymes.

Another *strategy* for selection by phage display is based on linking the substrate to the phage particle expressing the target enzyme. Active enzyme variants transform the phage-linked substrate to product, which remains attached to the phage, and the phage can be isolated by affinity chromatography to the product. The strategy was recently applied for a model selection of phages displaying adenylate cyclase from a large excess of phages that do not display the enzyme. The enzymatic conversion of a chemically linked ATP into cAMP was selected using *immobilised* anti-cAMP antibodies, and an enrichment factor of about 70-fold was demonstrated. A selection for DNA polymerases was also described. In this case, the substrate (a DNA primer) was also covalently attached via a flexible tether to the phage coat protein. Active variants were selected by virtue of elongation of the linked primer, first by incorporation of the target nucleotides, and finally by a biotinylated nucleotide that was used to capture the phage onto avidin-coated beads. Using this strategy, a DNA-dependent DNA polymerase was evolved to efficiently incorporate rNTPs and thereby function as a DNA-dependent RNA polymerase.

The same selection system was used to evolve DNA polymerases that incorporate 2'-*O*-methyl ribonucleosides with high efficiency.

Several groups applied a phage selection strategy termed "*catalytic elution*" for enzymes the catalytic activity of which depends on cofactor binding. Following protein expression on the phage surface, the cofactor (e.g. a metal) is removed and the catalytically inactive phages are bound to an immobilised substrate. The cofactor is then added, and phages displaying an active enzyme are eluted by conversion of the substrate into product.

HIGH-THROUGHPUT SELECTION OF ENZYMES USING CELL DISPLAY

The application of cell surface display for directed evolution has gained much momentum. In particular, screening by FACS (fluorescence-activated cell sorter) has yielded a number of highly potent binding proteins such as antibodies. Cell surface display can also be used to select for catalysis and has many features in common with phage display. However, sorting for enzymatic activity was thus far achieved only in few particular cases where the product of the enzymatic reaction could be captured within bacterial cells or on their surface. The protease OmpA was displayed on the surface of *E. coli*, and a fluorescence resonance energy transfer (FRET) substrate was added that adheres to the surface of the bacteria. Enzymatic cleavage of the substrate releases a quencner group, and the resulting fluorescent cells were isolated by FACS. This approach allowed the screening of a library of 10^6 OmpA variants and the subsequent isolation of a variant exhibiting a 60-fold improvement in catalytic activity. A complementary approach was developed by Kolmar and co-workers, whereby the products of a surface-displayed enzyme (an esterase) was covalently linked to the cell surface using peroxidase-activated *tyramide* conjugation.

Another potentially useful screening approach is based on the immobilisation of enzyme-displaying cells on beads. The cells are adsorbed to polyacrylamide beads pre-equilibrated with growth medium to form a bead population containing, on average, a single cell per bead. The beads are immobilised on a solid glass support and cells are allowed to grow and form *microcolonies* while utilizing the medium retained within the *hydrogel* matrix. The *microbead* colonies are then equilibrated with a chromogenic or fluorogenic substrate, and the beads are screened under the *microscope*. This technique was recently demonstrated by enriching *E. coli* cells expressing -lactamase on their surface from a large excess of other cells.

IN VIVO GENETIC SCREENS AND SELECTIONS

Genetic, or *in vivo*, selections were the tools used for most early directed evolution experiments, but as the scope of targets for evolution widened, the utility of this approach became limited. After all, these selections are usually based on the evolving activity complementing an auxotroph strain in which an enzyme was knocked-out. Thus, broadly speaking, the target for evolution (substrate, reaction, etc.) needs to parallel an already existing enzyme. Nevertheless, genetic screens can still be useful, for example in the identification of *promiscuous* enzyme activities. Furthermore, the scope of genetic selections has been significantly widened by the application of the three-hybrid system to link the catalytic activity of an enzyme to the transcription of a reporter gene in yeast cells.

SCREENS FOR HETEROLOGOUS PROTEIN EXPRESSION AND STABILITY

Introduction

The high levels of expression of stable and functional proteins remain the bottleneck for many enzyme applications. Conventional approaches for production of soluble and active proteins in heterologous expression systems include low temperature expression, promoters with different strengths, modified growth media, and a variety of solubility-enhancing fusion tags.

More recently, a series of vectors and fusion partners that can be screened for high-level functional expression of a target protein have been developed.

These approaches are often successful but, at the end of the day, they cannot modify the intrinsic properties of the protein. Thus, even when a soluble fraction of the protein is obtained, the protein may still be inactive, or aggregate, during purification and storage.

An alternative to expression optimisation is protein engineering by rational design. There are now several examples of proteins that have been stabilised by the introduction of mutations with small yet cumulative stabilizing effects. It is largely unknown, however, how the stability of a protein is encoded in its sequence, and how individual amino acid changes contribute to stability, not to mention that most of the interesting targets for structural and mechanistic studies are obviously of unknown structure and are therefore not amenable to rational design.

This section focuses on the application of directed evolution to obtain efficient heterologous expression of proteins in *E. coli* and yeast, by increasing the stability and solubility of the protein in the host's environment.

Screening Methodologies for Heterologous Expression

Heterologous expression is performed primarily in *E. coli*, although yeast and insect cells are becoming increasingly popular, particularly for eukaryotic proteins that require posttranslational processing. Considerations of transformation efficiency, stability of plasmid DNA, and growth rate, make *E. coli* and *S. cerevisiae* the best-suited organisms for directed evolution experiments.

Directed evolution in the laboratory, like natural evolution, involves two key steps: generation of genetic diversity and selection for function. In the laboratory, diversity in the gene of interest is typically created by random mutagenesis or family shuffling. Whilst the methods of creating genetic diversity are generally applicable, the selection system needs to be tailored, or modified, for each target protein and aim – a selection for higher expression may not be suitable for altering substrate selectivity and vice versa.

Broadly speaking, genetic screens or selections for heterologous expression can be done in two ways: (1) screening/selecting for the activity of a reporter protein, or (2) screening/selecting for the protein's own function. Examples of both approaches are discussed below.

A variety of generic "*C-fusion*" approaches have been developed that rely on expression of the target protein as an N-terminal fusion to a reporter protein with a selectable function. An insoluble target protein leads to the aggregation of the reporter protein and loss of its function. As an example, colonies overexpressing green fluorescent protein (GFP) fused to a soluble protein can be easily distinguished owing to folding and fluorophore formation of the GFP–whereas those expressing an insoluble protein do not fluoresce.

Another generally applicable method relies on chloramphenicol acetyltransferase (CAT) fusion, whereby selection for solubility of the target protein is performed by antibiotic resistance. Using this methodology, Arnold and coworkers converted the membrane-associated, insoluble human P450 cytochrome (1A2) into a fairly soluble protein by recombination of 1A2 with a distant bacterial homolog. The resulting library of hybrid proteins was first selected by fusion to CAT, in order to isolate all in-frame and correctly folded variants. A subsequent screen for active enzyme variants was performed with a fluorogenic P450 substrate.

In the b-Gal complementation assay, a small a-fragment of b-galactosidase (LacZ) of about 50 amino acids is fused to the C-terminus of the target protein, restoring–in those cases where the target protein remains soluble – the b-galactosidase activity of a truncated LacZ (LacZ-X) by complementation in trans. Other potentially generic methodologies are based on screening by cellular stress responses to mis-folded proteins, "*proteolytic selection*" by phage display, or "protein stability increased by directed evolution" (*Proside*), the last two technologies being based on the principle that infectivity of the phage is coupled to protease resistance of the protein variants. This assumption in turn relies on the observation that *proteolysis* resistance can be used as a marker for folding.

Because the "*C-fusion*" techniques do not depend on the function of the target protein, they are generally applicable, and particularly advantageous when attempting to evolve new folds, or in cases where the target protein has no assigned function. But a screen for soluble expression can turn into a major draw-back. The blind fashion by which evolution acts ("you get what you select for" is the first rule of directed evolution) implies that selection for soluble expression may be accompanied by significant changes in activity that may even render the soluble protein nonfunctional. To drive the selection of soluble and stable variants without compromising the protein's functional properties, the selection system must be based on the protein's own activity. A large variety of assays are available for screening enzyme libraries. Indeed, functional assays have been used in a number of cases. As an example, Baik and co-workers have evolved glucose dehydrogenase (GlcDH), to increase thermal stability

400-fold by screening a library generated by family shuffling for the enzyme's activity.

Directed Evolution for Heterologous Expression

Recent examples of directed evolution for heterologous expression illustrate the power and versatility of this approach in facilitating structural and mechanistic studies. Directed evolution has thus far been the key to two crystal structures of general interest: a hyperthermophile nucleotide diphosphate kinase (NDP-K), and the first PON family member. Using a GFP "C-fusion" screen, Waldo and coworkers evolved the essentially insoluble NDP-K from *P. aerophilum*, into a 90% soluble and active variant that enabled the determination of its crystal structure.

Serum paraoxonases (PONs) provide a particularly interesting target for directed evolution. PONs comprise a family of enzymes that play a key role in *organophosphate* (OP) detoxification and in preventing atherosclerosis. PONs are widely spread in mammals and other vertebrates, as well as in *invertebrates*.

Three subfamilies of PON are known (PON1, 2 and 3) to show around 60% sequence homology, but the structure of not a single family member was known prior to the directed evolution of PON recombinant variants. Attempts to crystallise serum-purified PON1 resulted in the crystallisation of a contaminant protein, instead of PON1. Moreover, although PONs exhibit a range of hydrolytic activities towards *esters*, *phosphotriesters*, and *lactones*, their native or physiological substrate is unknown.

Family shuffling and screening for esterase activity led to the first PON1 and PON3 variants that express in a soluble and active form in *E. coli*. However, in the absence of knowledge of the native substrate, the screen was based not on PON's native function, but on its promiscuous esterase and *phosphotriesterase* activities. Interestingly, whilst evolution of PON1 led to a variant with wild type-like enzymatic properties and high *E. coli* expression, evolution of PON3 by the very same method led to a mutant with increased esterase activity relative to wild-type PON3, in addition to higher expression.

The *promiscuous* esterase activity was also used for evolving a bacterial phosphotriesterase (PTE). In this case, libraries created by random mutagenesis of the PTE gene were screened for higher esterase activity.

However, the attempt to evolve this promiscuous activity led to a variant with a 20-fold increase in functional expression, and enzymatic properties that are essentially identical to wild-type PTE. From a methodological point of view, these works highlight the advantages and drawbacks of using screens based on a promiscuous activity, particularly when the native substrate is unknown (e.g. in PON's case), when a facile screen for the native substrate is not available, or when the detection of the native activity falls out of the *dynamic* range of the screen (e.g. in PTE's case). The snag is that this approach can obviously lead to variants with increased ***promiscuous*** activity as well as, or instead of, increased expression.

Some proteins, especially eukaryotic ones, are notoriously difficult to express in heterologous systems. Arnold and coworkers have succeeded in obtaining high functional expression levels for fungal MtL laccase – a glycoprotein–along with higher activity, by directed evolution. For this purpose, they used a yeast host, *Saccharomyces cerevisiae*, and a high-throughput screen based on the enzyme's native activity. Both *in vitro* and *in vivo* shuffling were used to achieve random recombination of the isolated variants between rounds. In addition, polymerase chain reaction (PCR) and gap repair tools were used to ensure the recombination of neighboring mutations in a site-directed manner.

Directed evolution also provided some general insights into the factors that dictate protein solubility, stability, and assembly in non-native environments. It is widely assumed that protein expression levels are determined by the "*solubility*" or "*stability*" of the expressed protein. This rationale is reflected in most strategies that have been designed to improve overexpression yields. However, recent works point to a key factor that is often overlooked. The main determinant of (active) over-expression for many proteins is defined by the thermodynamic and kinetic properties of an intermediate, whether an apo-enzyme, pro-protein, monomeric state, or a folding (on- or off-pathway) intermediate, rather than by the stability and solubility of the final state. Because these expression bottlenecks are largely unidentified, the design of stabilizing mutations may prove a daunting task. Evolution, which is directed only by the phenotypic outcome of the process and requires no knowledge of the protein's structure and folding pathway, seems like an effective solution to this problem. Care must be taken, however, that selection is performed in a controlled manner that improves functional expression but does not alter the protein's function and structure.

IN VITRO COMPARTMENTALISATION

In vitro compartmentalisation (IVC) is based on water-in-oil emulsions, where the water phase is dispersed in the oil phase to form microscopic aqueous compartments. Each droplet contains, on average, a single gene, and serves as an artificial cell in allowing for transcription, translation, and the activity of the resulting proteins, to take place within the compartment. The oil phase remains largely inert and restricts the diffusion of genes and proteins between compartments. The droplet volume (~5 fL) enables a single DNA molecule to be transcribed and translated, as well as the detection of single enzyme molecules. The high capacity of the system ($> 10^{10}$ in 1 mL of emulsion), the ease of preparing emulsions, and their high stability over a broad range of temperatures, render IVC an attractive system for enzyme high-throughput screening.

IVC makes it easy to co-compartmentalise genes and the proteins they encode, but the selection of an enzymatic activity requires a link between the desired reaction product and the gene. One possible selection format is to have the substrate, and subsequently the product, of the desired enzymatic activity physically linked to the gene. Enzyme-encoding genes can then be isolated by virtue of their attachment to the product while other genes, which encode an inactive protein, carry the unmodified substrate. The simplest applications of this strategy lie in the selection of DNA-modifying enzymes where the gene and substrate comprise the same molecule. Indeed, IVC was first applied for the selection of DNA methyltransferases (MTases). Selection was performed by extracting the genes from the emulsion and subjecting them to digestion by a cognate restriction enzyme that cleaves the nonmethylated DNA. Other applications include the selection of restriction endonucleases and DNA polymerases. The selection of DNA polymerases was based on the fact that inactive variants failed to amplify their own genes and therefore disappeared from the library pool.

These modes of selection are obviously restricted not only in the scope of the selected enzymatic activities but also by the stoichiometry – typically one gene, and hence one substrate molecule (or several substrate sites), is present per droplet, together with 10–100 enzyme molecules. Despite these restrictions, several new and interesting enzyme variants have been evolved by IVC. A variant of *Hae*III MTase was evolved with up to 670-fold improvement in catalytic efficiency for a nonpalindromic target sequence (AGCC) and ninefold improvement for the original recognition site (GGCC). Active *Fok*I restriction nuclease variants were also selected by IVC from a large library of mutants with low residual activity. And, DNA polymerase variants with increased thermal stability, and the ability to incorporate a diverse range of bases including fluorescent dye-labeled nucleotides have been evolved.

The first application of IVC beyond DNA-modifying enzymes was demonstrated by a selection of bacterial phosphotriesterase variants. The selection strategy was based on two emulsification steps: in the first step, microbeads, each displaying a single gene and multiple copies of the encoded protein variant, were formed by translating genes immobilised to microbeads in emulsion droplets and capturing the resulting protein via an affinity tag. The microbeads were isolated

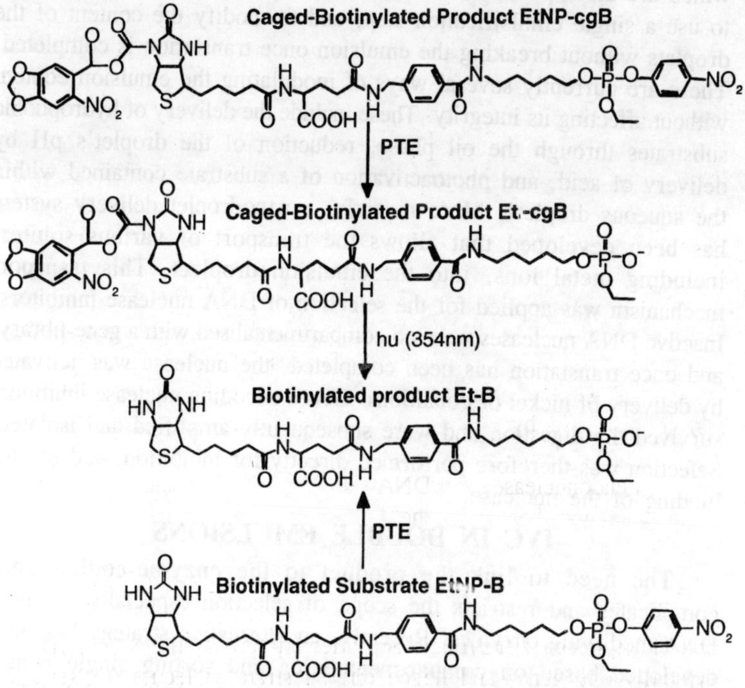

Caged-Biotinylated Product EtNP-cgB

PTE

Caged-Biotinylated Product Et-cgB

hu (354nm)

Biotinylated product Et-B

PTE

Biotinylated Substrate EtNP-B

Figure 6.4 : The application of caged substrates for selection by in vitro compartmentalisation (IVC). Paraoxon (the substrate of the target enzyme) was modified by substituting an ethyl group with a linker connected to caged biotin.

and re-emulsified in the presence of a modified phosphotriester substrate. The product and any unreacted substrate were subsequently coupled to the beads. Product-coated beads, displaying active enzymes and the genes that encode them, were detected with fluorescently-labeled anti- product antibodies and selected by FACS. Selection from a library of > 10^7 different variants led to the isolation of a variant with a very high k_{cat} value (> 10^5s–1). Microbead display libraries formed by IVC can also be selected for binding activity.

Some of the IVC selection modes take advantage of the fact that this system is purely *in vitro*, and can allow selection for substrates, products, and reaction conditions that are incompatible with *in vivo* systems. However, the cell-free translation must be performed under defined pH, buffer, ionic strength, and metal ion composition. In the selection for phosphotriesterase using IVC described above, translation is completely separated from catalytic selection by using two sequential emulsification steps, allowing selection for catalysis under conditions which are incompatible with translation. However, it is also possible to use a single emulsification step, and to modify the content of the droplets without breaking the emulsion once translation is completed. There are currently several ways of modulating the emulsion content without affecting its integrity. These include the delivery of hydrophobic substrates through the oil phase, reduction of the droplet's pH by delivery of acid, and photoactivation of a substrate contained within the aqueous droplets. More recently, a nanodroplet delivery system has been developed that allows the transport of various solutes, including metal ions, into the emulsion droplets. This transport mechanism was applied for the selection of DNA nuclease inhibitors. Inactive DNA nucleases were co-compartmentalised with a gene-library, and once translation has been completed, the nuclease was activated by delivery of nickel or cobaltions. Genes encoding nuclease inhibitors survived the digestion and were subsequently amplified and isolated. Selection was therefore performed directly for inhibition, and not for binding of the nuclease.

IVC IN DOUBLE EMULSIONS

The need to link the product to the enzyme-coding gene complicates and restricts the scope of selection especially for non-DNA-modifying enzymes. Recently, an alternative strategy has been developed based on compartmentalizing and sorting single genes, together with the fluorescent product molecules generated by their encoded enzymes. The technology makes use of double, water-in-oil-in-water (w/o/w) emulsions that are amenable to sorting by FACS. This circumvents the need to tailor the selection for each substrate and reaction, and allows the use of a wide variety of existing fluorogenic substrates. Recently it has been shown that the making and sorting of w/o/w emulsion droplets does not disrupt the content of the aqueous droplets of the primary w/o emulsions. Further, sorting

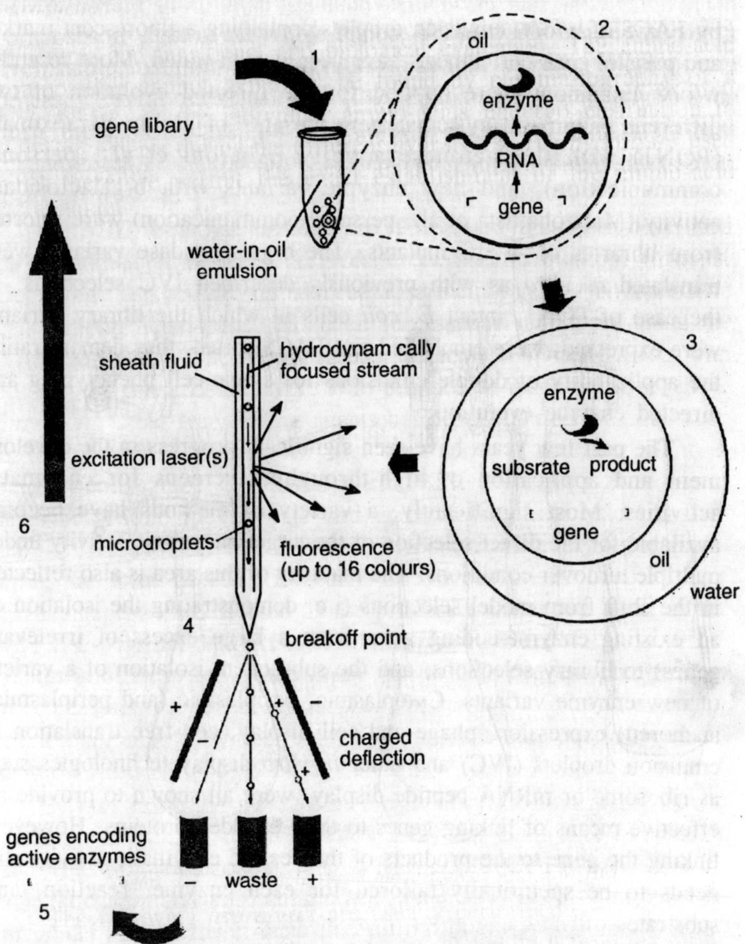

Figure 6.5 : Selections by flow sorting of double emulsion microdroplets in using a fluorescence-activated cell sorter (FACS). A library of genes, each encoding a different enzyme variant, is dispersed to form a water-in-oil (w/o) emulsion with typically one gene per aqueous microdroplet (1). The genes are transcribed and translated within their microdroplets (2), using either in vitro *(cell-free) transcription/translation, or by compartmentalizing single cells (e.g. bacteria into which the gene library is cloned) in the microdroplets. Proteins with enzymatic activity convert the nonfluorescent substrate into a fluorescent product and the w/o emulsion is converted into a water-in-oil-in-water (w/o/w) emulsion (3). Fluorescent microdroplets are separated from nonfluorescent microdroplets (or microdroplets containing differently coloured fluorochromes) using FACS (4). Genes from fluorescent microdroplets, which encode active enzymes, are recovered and amplified (5). These genes can be recompartmentalised for further rounds of selection (6).*

by FACS of w/o/w emulsion droplets containing a fluorescent marker and parallel gene enrichments have been demonstrated. More recently, w/o/w emulsions were applied for the directed evolution of two different enzymatic systems: new variants of serum paraoxonase (PON1) with thiolactonase activity (Aharoni et al., personal communication), and new enzyme variants with b-galactosidase activity (Mastrobattista et al., personal communication) were selected from libraries of > 10^7 mutants. The b-galactosidase variants were translated *in vitro* as with previously described IVC selections. In the case of PON1, intact *E. coli* cells in which the library variants were expressed, were emulsified and FACS sorted, thus demonstrating the applicability of double emulsions for single-cell phenotyping and directed enzyme evolution.

The past few years have seen significant progress in the development and application of high-throughput screens for enzymatic activities. Most significantly, a variety of methods have become available for the direct selection of the target enzymatic activity under multiple turnover conditions. The maturity of this area is also reflected in the shift from model selections (i.e. demonstrating the isolation of an existing enzyme-coding gene from a large excess of irrelevant genes) to library selections, and the subsequent isolation of a variety of new enzyme variants. Cytoplasmic, periplasmic (and periplasmic-anchored) expression, phage and cell display, cell-free translation in emulsion droplets (IVC) and other *in vitro* display technologies such as ribosome or mRNA peptide display, were all shown to provide an effective means of linking genes to their encoded proteins. However, linking the gene to the products of the desired enzymatic activity still needs to be specifically tailored for each enzyme, reaction, and substrate.

The use of fluorogenic substrates and sorting by FACS appears to be the most promising avenue. Modern FACS machines routinely analyse and sort > 10^7events per hour. Fluorescence is a sensitive, versatile, and general signal that can detect a huge range of enzymatic reactions. And, unlike selections performed in bulk, FACS allows the stringency and recovery from the selection to be fine-tuned. Thus, the combination of FACS, display technologies, and compartmentalisation in double emulsions holds much potential in the areas of directed enzyme evolution and functional *genomics*.

7

Movement Across Membrane

In 1970, Liebow and I began to explore directly the possibility that proteins move individually through both the plasma membrane of the acinar cell and the membrane of the *zymogen granule*. Our reasons for undertaking these studies were at the time threefold:

1. Several lines of evidence pointed to the existence of a cytoplasmic pool of digestive enzyme from which secretion appeared to be derived. The existence of such a pool suggested that both the granule and cell membrane might be permeable to these proteins.
2. Isolated zymogen granules released their protein contents in environments in which one would have expected them to be stable; that is, in the same environment in which they presumably exist within the cell, an isotonic medium at neutral pH. Although it was not clear that release under these circumstances was due to membrane transport, as opposed to granule lysis, beyond the observation that different proteins were not released in parallel, the fact that granules released their contents in what was presumably their "*natural*" environment raised the possibility that they might be permeable to proteins; more so in the ionic, neutral, high-temperature medium of the cell than in nonionic media, at slightly acid pH, and low temperature in which environment they are generally extracted from tissue homogenates as "*stable*" granules.

3. The digestive enzymes were contained within the granules in a bound state, and were not, as had been thought, merely suspended in an aqueous medium held in place by a limiting lipid membrane. They were bound in high-affinity, enzyme specific arrays, indicating significant intragranular organisation. This internal organisation would appear to be without purpose if secretion were merely the result of the mass transport of mixed bags of protein by *exocytosis*. On the other hand, such organisation might account, at least in part, for the selective release of different enzymes from granules that had been observed, as well as for the ability of granules to retain their protein contents if the membranes were indeed permeable to them.

In any event, we began a series of experiments to determine whether or not digestive enzymes could be released from isolated *zymogen granules*. Our approach was simple. If enzyme was released from the granule in a concentration-dependent fashion across its membrane, then if we reduced the concentration of enzyme in the medium in which the granules were suspended, more enzyme would be released. We chose to study the potential for release in what in a sense was the most unfavorable situation; that is, in nonionic media at slightly acidic pH, conditions under which granules were known to be most "*stable*." Release would be expected to be least and slowest under these circumstances; but it also might be easier to follow over time, and furthermore there would be fewer concerns about the nature of the starting material, since granules were thought to be "*stable*" or "*normal*" in this environment.

When isolated granules were diluted in larger and larger suspension volumes, thereby progressively reducing the external concentration of enzyme, release into the medium was observed. *Amylase*, *chymotrypsinogen,* and protein in general were all redistributed, as a result of dilution, in a concentration-dependent fashion, although each to a different extent. Release varied in a continuous, as opposed to a discontinuous manner (as might be expected if lysis were the cause of release) as a function of suspension volume. The larger the volume, the greater the amount released, and the ratio of enzyme contained in the medium to that in the granule could be varied continuously by over an order of magnitude in this fashion.

At the more "*labile*" pH of 7.0, the distribution of protein

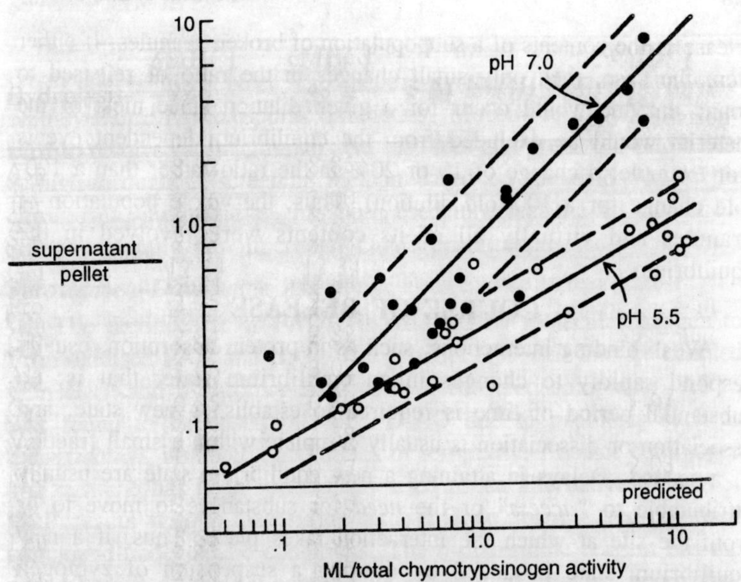

Figure 7.1 : Release as a function of the relative dilution of granule suspensions.

between granule and medium was altered to an even greater extent (by about two orders of magnitude) merely as a function of dilution; from relatively concentrated suspensions in which almost 90% of the protein was retained within the granule, to those in which 90% was recovered in the suspending medium. It is important to point out here that even our "*concentrated*" suspensions were rather dilute in terms of the solubility of the digestive enzyme proteins themselves. They ranged downward more than two orders of magnitude from a total protein concentration of approximately 1 mg/ml, at which fractional concentrations virtually all of the enzymes are soluble in water, as is the whole mixture of proteins in secreted fluid.

This 100-fold change in distribution occurred over an approximately 100-fold range of suspension volume. That is, the ratio of released to bound enzyme varied in an almost linear fashion with suspension volume, from wholly bound to fully released. This meant that virtually all of the enzyme in the population of granules was in equilibrium with enzyme in the medium. Release could not merely be attributed to an equilibrium between some small percentage of the total granule content and the medium, such as would be the case for the release or desorption of protein from the granule surface or the release of the contents of a subpopulation of broken granules. If either were the case, then only small changes in the ratio of released to bound enzyme would occur for a given dilution since most of the material would be excluded from the equilibrium-dependent events (for example, a change of 10 or 20% in the ratio rather than a 100-fold change for a 100-fold dilution). Thus, the whole population of granules and virtually all of its contents were involved in the equilibrium.

COURSE OF RELEASE

Weak binding interactions, such as in protein adsorption, usually respond rapidly to changes in an equilibrium state; that is, no substantial period of time is required to establish a new state, and association or dissociation is usually complete within a small fraction of a second. Delays in attaining a new equilibrium state are usually attributable to "*access*" or the *need* for substances to move to or from the site at which the interaction takes place. Thus, if a new equilibrium state produced by diluting a suspension of zymogen granules were merely a function of weak interactions (*ionic* or *hydrogen bonding*) without a significant intervening permeability barrier, then it would in all likelihood be rapidly attained without a macroscopically observable time course.

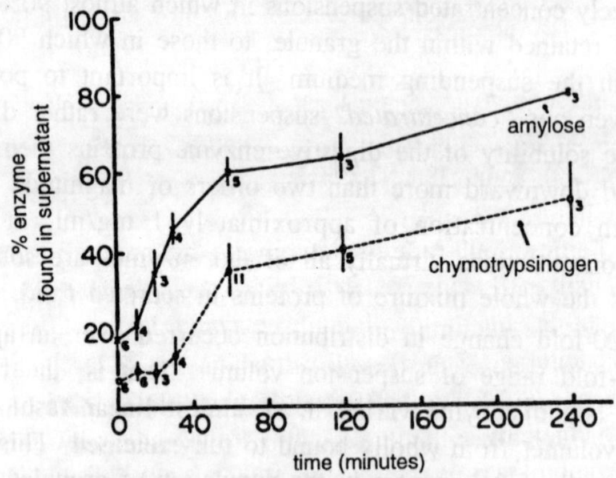

Figure 7.2 : The time course of enzyme redistribution after dilution. The percentage amylase (-x-) and chymotrypsinogen (— ° —) activity found in the medium (supernatant) of a pH 5.5 suspension of granules at various times after dilution (time 0) of granule suspensions from -1 ml /mg protein to - 10 ml /mg protein. Vertical bars represent the standard error of the mean and the number of experiments is noted by their side. No loss in enzyme activity was observed in samples incubated up to 120 min.

With this in mind, suspensions of granules were diluted and the approach to a new equilibrium state followed over time. Reequilibration was not instantaneous, but required a considerable period, suggesting the presence of a significant permeability barrier. For example, at a dilution that altered the distribution of *chymotrypsinogen* from about 90% in the granules initially, to about 50% outside, reequilibration took more than 4 hr at 37°C.

Furthermore, release occurred at different rates and to different extents for different enzymes. *Amylase* was more "*labile*," than chymotrypsinogen; that is, it was released at a more rapid rate and to a greater extent for a given dilution. For the example just noted, ty= was 30 min for amylase as compared with 60 min for *chymotrypsinogen*, and 80% of the total was released, or a medium to granule ratio of 4:1 was established for amylase as compared with 50% release or a 1:1 ratio for chymotrypsinogen. This difference argued against the possibility that a slow, time-dependent granule lysis accounted for enzyme release, because in this case, both the rate and extent of release would have been expected to be essentially the same for the different enzymes.

We also considered whether a membrane barrier was responsible for the delay in equilibration in the following manner. It is possible to solubilize the contents of zymogen granules and to then reaggregate them in non-ionic media at slightly acid pH in a form that displays similar binding characteristics to enzyme contained within the native granule. However, this reaggregated material lacks an enclosing membrane barrier. The question was whether or not a delay in reequilibration would still be seen after dilution in the absence of a membrane barrier? The answer was no. In this case, reequilibration occurred within the time required to separate the aggregate from its suspending medium. No time dependence for release could be demonstrated, although the state function, that is, the equilibrium distribution, itself was much the same as seen with the native granule.

CONCENTRATION-DEPENDENT RELEASE

As I have said, it was possible that the slow time course of enzyme release was due to the slow lysis of labile granules rather than the release of enzyme from intact granules. Although the different release characteristics for different enzymes, as well as the evidence just discussed, argued against this possibility, we attempted to distinguish between such time versus concentration-dependent release in another fashion as well. Granules were suspended in a filtration

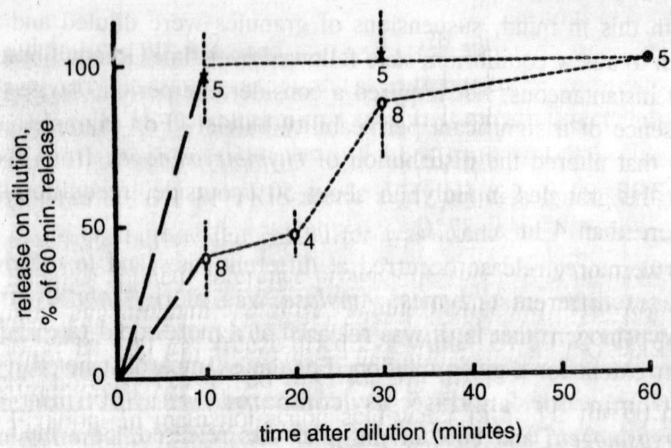

Figure 7.3 : The time course of enzyme redistribution after dilution for native granules and "reaggregated" granule material.

chamber through which a flow of fluid could be maintained and the released enzyme, although not the granules, filtered. If release was due to time-dependent granule lability, then we would expect to see much the same phenomenon that was seen when granules were merely incubated in medium in which the released contents accumulated. That is, the removal of released material should have no effect on the rate of release (rate of lysis), and the same time course and pattern

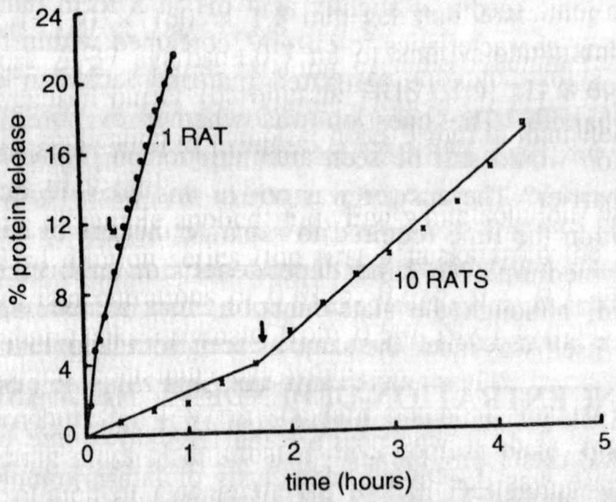

Figure 7.4 : Protein release from granules in response to continuous washing. Granules were placed in a mixing chamber and washed continuously with 0.3 M sucrose (pH 5.5).

seen for simple incubation would have been expected to be reproduced. This was not the case. The filtration of released enzyme led to the continuous release of enzyme from granules. That is, the rate of release did not fall off over time as in the simple incubation study, but continued at a relatively constant rate. Thus, external concentration and not time appeared to be determining the rate of release.

That release was dependent upon the rate of removal of already released enzyme, and hence the concentration of enzyme in the medium, was further evaluated by varying the rate of removal of released material either by increasing filtration rate or decreasing the absolute amount released from the granules for a given rate of filtration (by decreasing the number of granules in the suspending medium). In the first case, it should be possible to increase the amount of material released from the granules by increasing filtration rate, if release is not already occurring at a maximal rate. This would be most likely when relatively large numbers of granules are added to the filtration chamber, so that the amount released from the granules would be relatively great compared with its rate of removal by filtration. This would provide the opportunity for the flux of material back into the granules (that is, backflux or influx). If release were concentrationdependent, then under these circumstances we should be able to increase it merely by increasing the rate of filtration, and thereby reducing backflux Unet is increased by decreasing J; since Jnet = Je - Jt). When this was done, an increase in the release of enzyme from the granules was *seen*.

If this interpretation of the cause of increased release were correct, then we should also be able to set the system up so that release would occur at a greater rate, as a percentage of the amount of material in the granules, merely by suspending fewer granules in the chamber. In this case, for a given filtration rate, the concentration of released enzyme in the medium would be lower; that is, less material would be released (from a smaller number of granules) but into the same volume (a constant filtration rate). If release were due to some type of lytic event, independent of external concentration, then we would expect the same amount of material to be released, in proportionate terms, regardless of the number of granules in suspension. On the other hand, if release were a concentration-dependent phenomenon, then a lower concentration in the medium would reduce the backflux into the granules and hence the net flux out would be greater (that is, we would see an increase in the amount released per unit time per granule). When this was done and the number of granules suspended was reduced by about an order of magnitude ("one rat"), the rate of release per granule increased greatly (that is, m_{et} was increased because

J; was reduced due to the reduction in the enzyme concentration in the external medium). When filtration rate was increased in this case, no further increase in enzyme release occurred, which supported the interpretation that under these circumstances release was already occurring at the maximal rate, namely, at the initial rate of efflux, so that Jnet = J_e because J_i approaches zero and therefore J_{net} could not be further increased by varying filtration rate.

Thus, it was demonstrated in a variety of ways that the release of enzyme from isolated zymogen granules was the result of their concentrationdependent release from the granules into the medium, and that this flux of protein was being expressed across the granule membrane.

TEMPRATURE IN ENZYME RELEASE

The release of enzyme from granules was found to be highly temperature sensitive. For example, when a suspension of granules was diluted and release followed over a 20-min period, either at 20°C or at 30°C, the temperature dependence of release, namely, its apparent Q_{10}, was about 5.0 for protein overall and for chymotrypsinogen, and about 2 for the more "labile" amylase. That is, the rate of release increased by two to five times over a 10° temperature range.

These observations offered further support for the view that a membrane transport process was involved in the release of enzyme from granules. Such a substantial temperature dependence would not be expected if simple diffusion in an aqueous medium out of broken granules or away from their surface, or dissociation from weak bonding at either site, was responsible for release. In these situations, Q_{10} values of little more than 1.0 would have been expected. The observed values were as high or higher than those usually reported for reaction-dependent transport or for enzymatically catalyzed reactions in general (Q_{10} = 2-3).

The view that the temperature sensitivity of release was due to a membrane transport process, as opposed to a temperature-sensitive granule lysis, was supported by the fact that the temperature dependence of release was substantially different for *amylase* and *chymotrypsinogen.* A differential temperature sensitivity for different proteins would not be expected if either lysis, simple dissociation, or a combination of the two were the cause of release.

One additional characteristic of the temperature dependence of release also points to a transport, and not a lytic, process. Although the rate of release is highly temperature sensitive, the steady-state

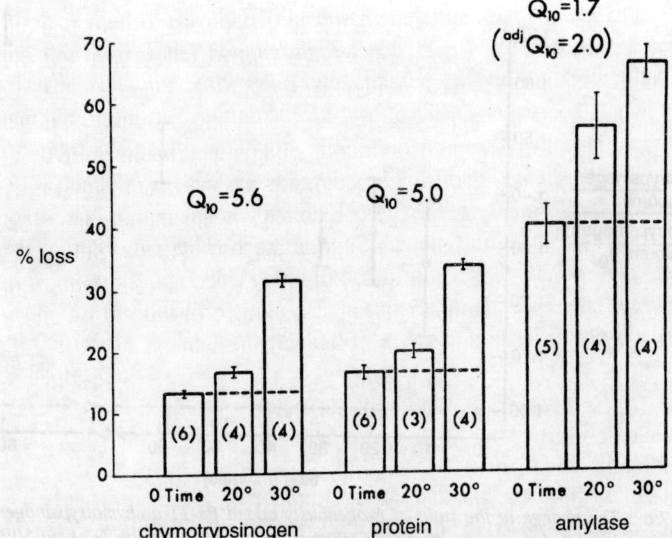

Figure 7.5 : Q_{10} determination for chymotrypsinogen, protein, and amylase release. 0 Time refers to release prior to incubation; 20° and 30° refer to release after 20 min incubation in 20° or 30°C baths.

distribution of these proteins between granule and medium is essentially temperature independent. This behavior is that expected for transport by facilitated diffusion. If a lytic event were involved, both kinetic parameters would show approximately the same temperature sensitivity.

UPTAKE OF ENZYME

Could release be reversed? That is, could the uptake of enzymes into partially depleted granules be demonstrated? Both the uptake of radiolabeled enzyme and the bulk reuptake of previously released enzyme was shown. Uptake had a substantial time dependence and was wholly reversible, at least when endogenous material was used.

That is, an initial state could be reestablished after the dilution of a suspension of granules merely by reconcentrating the enzyme in the medium (by dialysis). Therefore, enzyme could be taken up into partially depleted granules, and this uptake had the characteristics of a timedependent equilibrium, as did enzyme release from granules.

ROLE OF ISOLAND GRANULES

What do granules look like after they have released a portion, or even all of their contents? Is their visual appearance consistent with the hypothesis that we have been discussing? Ermak extracted granules

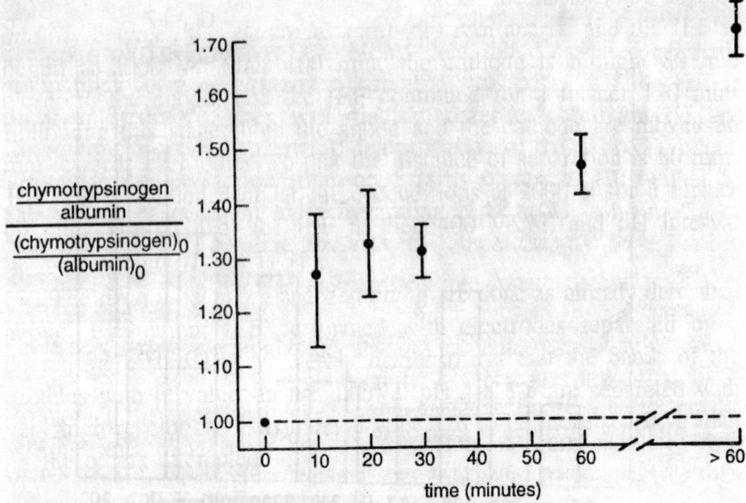

Figure 7.6 : The change in the ratio of exogenously added [31-11 a-chymotrypsinogen (bovine) to p [131]1-labeled albumin (human) with time, in granules previously about 50% depleted of chymotrypsinogen.

from tissue homogenates and compared undiluted sediments of granules to granules from the same starting material subjected to varying degrees of additional dilution. He found that, as a function of dilution, the usual electron-dense or opaque *zymogen granule* gave way to apparently intact but internally honeycombed or "*reticulated*" granules of the same approximate size. That is, granule appearance changed as a result of enzyme release. Granules did not merely decrease in number in an all-or-none fashion, with all those that remained having the

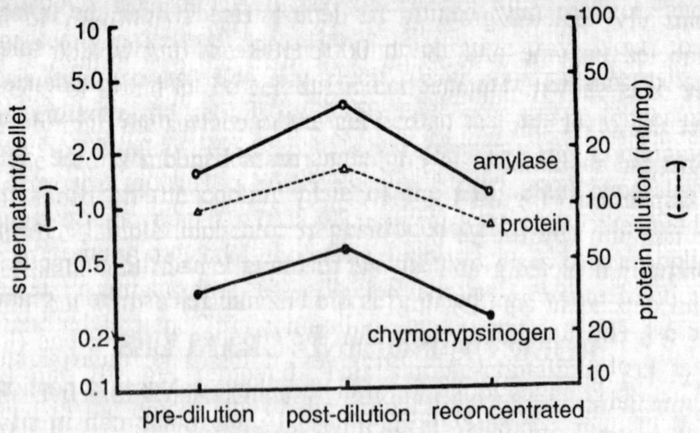

Figure 7.7 : Redistribution of enzyme upon dilution and reconcentration.

"*normal*" opaque appearance, as a lytic view of enzyme release would predict. The visual, as well as the kinetic evidence, indicated that the graded release of enzyme occurred from individual, seemingly intact, granules. Ermak also demonstrated directly that at least large segments of the membrane surface of individual granules were without apparent tears or schisms in their fabric by looking at three-dimensional images of thick sections through granules under the *transmission microscope*. Furthermore, the honeycombed appearance of the granule indicated that material was lost from throughout the whole of its interior and not from particular localised sites, or even generally from the surface towards the center. This reticulated form is consistent with the requirement of the kinetic data that all or almost all of the enzyme within the granule be involved in the equilibrium with the external medium.

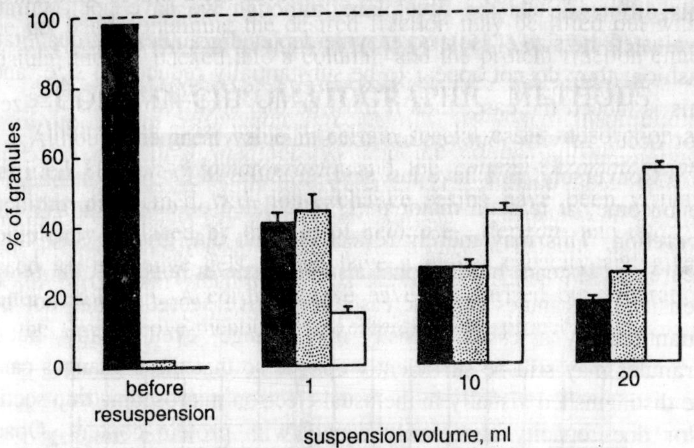

Figure 7.8 : Change in the percentage of solid (black), reticulated (stippled), and empty (white) granule profiles before resuspension and as a function of suspension volume.

Moreover, the changes in appearance were a function of dilution, both in regard to the types of granules seen and their relative frequency. As granule suspensions were made more dilute, the number of dense granules decreased, at first being primarily replaced by reticulated granules, which upon further dilution were replaced by increasing numbers of virtually empty granules of the same approximate size and general appearance. Empty granules accounted for about 50% of the total population in the most dilute suspension studied. It should be noted that the concentrations of granule suspensions used in these morphologic studies were chosen to match those employed in the kinetic studies discussed above. Thus, the microscopic exploration of the granule gave "form" to kinetic and thermodynamic observations of enzyme release. Granules appear internally structured by interlaced strands of variable thickness containing bound enzyme that border honeycombed lacunae, into which enzyme is released and from whence enzyme leaves the granule across its membrane. In this way the protein contents of the granule appear to have access to the medium.

Although these observations are reassuring and offer important support for the kinetic studies with isolated granules, they provide a very different visual picture for changes in granule structure in situ as a result of the release of enzyme during active secretion. In that case, the size of granules decreased, but their appearance otherwise remained essentially unchanged; that is, opaque granule interiors were still predominantly observed. There are several things to be said about this difference. In the first place, although we have not accurately evaluated the size of isolated enzyme-depleted granules in a quantitative fashion, they do not appear to be substantially reduced in size, and if this is indeed the case, then it must be that such reductions in size do not occur in vitro but require a cellular environment.

Conversely, as I have just said, granules in the cell do not appear to become, at least in major part, reticulated or empty during active secretion. This may merely reflect the fact that granule size can be decreased instead, but it should also be borne in mind that the relative density of granules in pancreas after active secretion has not been quantitatively assessed. Density may change, even greatly, but the granules may still be sufficiently opaque so that these changes cannot be distinguished visually in the usual electron microscopic thin section. Nor does opacity necessarily equate with protein content. Opacity may be retained in situ but reflect some other aspect of *granule morphology*.

Furthermore, it is not really correct to say that reticulated forms are not seen in situ. They most certainly are, but are thought of as "*forming*" zymogen granules (that is, *condensing vacuoles*) and not "*emptying*" granules, although the two obviously are not distinguishable on a visual basis. The appearance of some, though not all condensing vacuoles, is essentially identical to the reticulated granules that Ermak produced in vitro by causing enzyme release. The evidence that is usually offered in support of the view that condensing vacuoles are forming zymogen granules is two fold. In pulse-label experiments, labeled protein appears to accumulate in these vesicles before any substantial accumulation occurs in the mature granule population (as estimated autoradiographically); thus the hypothesis that one gives rise to the other. However, if enzyme has access to these granules from the cytoplasm (that is, if we do not assume a priori that it can only arrive there by means of a vesicular mechanism as the paradigm proposes), then partially empty granules (condensing vacuoles) might well equilibrate more readily with labeled enzyme than full or mature granules and give the same kinetic pattern whether they are emptying or filling.

The other argument in favor of the condensing vacuole as a forming zymogen granule is its geographic location within the cell. Condensing vacuoles tend to be found in greatest abundance near the *Golgi stacks* that are thought to give rise to them. They are, however, also found, although in lesser abundance, in other areas of the cell, even near the apical cell membrane. Of course, both filling and emptying may occur; some partially filled zymogen granules may be newly formed and filling for the first time, whereas others may be in the process of depletion. A quantitative evaluation of condensing vacuole number, size and form, and their location in the cell during the secretion cycle might provide additional enlightenment about the types and nature of these structures.

RELEASE OF SPECIFIC ENZYMES

From the time of our earliest studies, in the late 1960s, on the release of digestive enzymes from zymogen granules, it was clear that release tended to be different for different molecular species and varied among them in regard to rate, amount, and sensitivity to particular release effectors. Such differences argued against the idea that release was merely caused by granule lysis, in which case one would have at least expected it to be similar for different enzymes in regard to rate, if not necessarily in regard to sensitivity to particular release effectors or to amount if intragranular or surface binding produced differential distributions. For some release phenomena, essentially the whole population of granules and their contents was shown to be involved, and for this reason, differential surface desorption or the differential release of enzyme due to differences in intragranular binding affinity from a subpopulation of broken granules did not seem to provide a sufficient explanation.

Some observations of the differential release of different enzyme types are more convincing as evidence for membrane transport than others, but perhaps the most convincing case, and one that is difficult to explain as being other than the result of the membrane transport of the released enzymes, comes from relatively recent work by Grendell. We were interested in determining how end-products of digestive reactions, such as glucose and amino acids, produced the selective secretion of different digestive enzymes, a phenomenon for which we had evidence in situ. We thought that the effect might be expressed most simply and directly at the zymogen granule level, with end-products mobilizing specific enzymes from granule stores.

We initially studied two end-products, glucose and lysine, for their effect on the release of two enzymes, amylase and trypsinogen, that are involved in their respective liberation from food polymers in the gut. The results were striking. The major end-product of carbohydrate digestion, glucose, led to the release of amylase from isolated zymogen granules at glucose concentrations in the range normally found in plasma, but did not cause the release of trypsinogen. Conversely, the amino acid, lysine, led to trypsinogen, but not amylase, release and was also effective at levels of this amino acid found in plasma. Thus, a clear, indeed, an absolutely, selective distinction was observed that had potential physiological meaning. Moreover, release could be shown to involve a large percentage of the contained enzyme stores; that is, the effect was also quantitatively substantial.

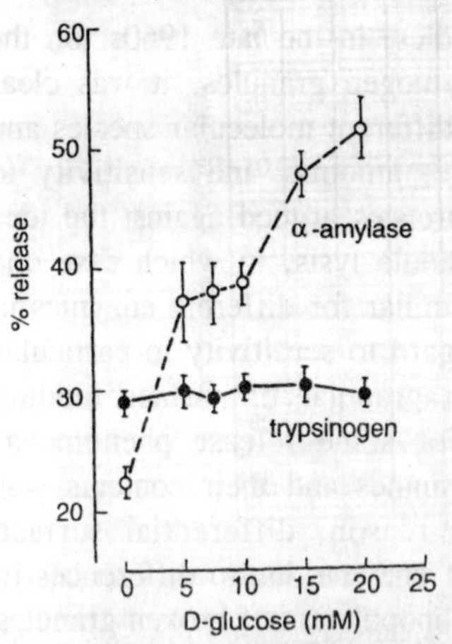

Figure 7.9 : Percent of total α-amylase (---o---) and trypsinogen released into soluble phase fromhomogenates of pancreatic tissue as a function of the concentration of D-glucose.

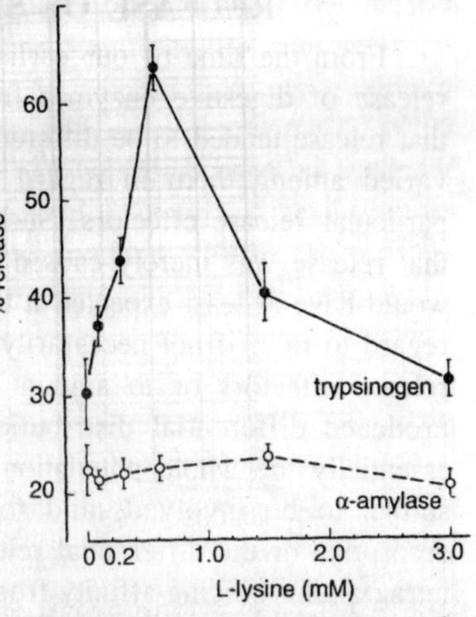

Figure 7.10 : Percent of total α-amylase (---o---) and trypsinogen released into the soluble phase from homogenates of pancreatic tissue as a function of the concentration of L-lysine..

Can such an effect be explained without invoking membrane permeability; that is, can it be explained in terms of granule lysis, and the differential surface or interior binding of different enzymes? In the first place, we would have to propose that all release effectors at millimolar, and lower, concentrations; concentrations normally found in plasma and cells, are lytic agents. Moreover, for lysine, since the release effect is lost at concentrations above 1.0 mM (and is gone at 3.0 mm), one would have to propose additionally that although low concentrations of this amino acid produce lysis, higher concentrations do not. Of course, there is no reason to think that these substances produce granule lysis, no less that they all do, no less differentially; nor, moreover, is there reason to expect them to have such effects in the range of concentrations that are normallyy found in plasma. But most significantly, there is no reason to expect, for example that glucose (or *glucose-1-phosphate*) would cause lysis that leads only to the release of amylase, whereas lysine causes lysis that leads only to the release of trypsinogen, and merely by chance at that.

PERMEABILITY IN SITU

Thus, the evidence indicates that the zymogen granule membrane is permeable to its contained secretory proteins. Can such a permeability be corroborated in situ, and can a similar permeability be shown for the plasma membrane of the secretory cell? With these

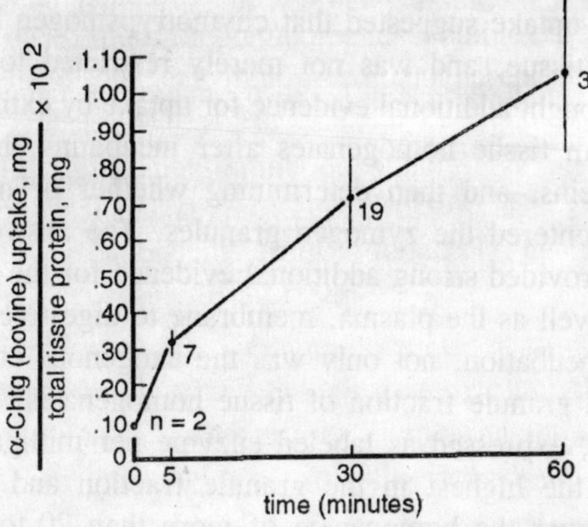

Figure 7.11 : Uptake of exogenous bovine [³H] chymotrypsinogen (Chtg) into pancreatic tissue versus time. Chymotrypsinogen uptake is corrected for albumin space. Uptake is normalised to quantity of tissue dividing by total tissue protein. Incubation was at 37°C. Numbers refer to number of experiments and error bars represent standard error of the mean.

questions in mind, Liebow performed a series of experiments in which we added radioactively labeled chymotrypsinogen to slices of pancreatic tissue in vitro, and followed enzyme uptake by the tissue. Time-dependent uptake of chymotrypsinogen was observed, and was far greater than that seen for a control protein, radiolabeled albumin. Washing the tissue to remove labeled protein from the interstitial spaces accentuated the difference between chymotrypsinogen and albumin uptake. The ratio of chymotrypsinogen to albumin uptake was 3:1 in unwashed tissues, but rose to 20:1 after a simple washing procedure that removed most of the albumin.

Table 7.1 : Effect of Number of Rinses on Relative Amounts of [^{3}H] Chymotrypsinogen and [^{131}I] Albumin in Tissue Slices.

Number of Rinses	(3H/131I) Tissue / (^{3}H/^{131}I) Bath
0	3.0 ± 0.3(9)
2	7.8 ± 0.6(26)
4	17.0 ± 0.9 (4)
2 Krebs-Ringer's and then two additional Krebs-Ringer's rinses containing 1 mg/ml cold bovine chymotrypsinogen	20.0 ± 1.6 (4)

Although the relatively large difference between chymotrypsinogen and albumin uptake suggested that chymotrypsinogen had entered the cells of the tissue, and was not merely restricted to the interstitial spaces, we sought additional evidence for uptake by extracting zymogen granules from tissue homogenates after incubating the tissue in the labeled proteins, and then determining whether or not radiolabeled protein had entered the zymogen granules. The answer was yes and the results provided strong additional evidence for the permeability of granule, as well as the plasma, membrane to digestive enzyme. After 60 min of incubation, not only was the exogenous enzyme found in the zymogen granule fraction of tissue homogenates, but the specific radioactivity, expressed as labeled enzyme per milligram of protein, was by far the highest in the granule fraction and represented an enrichment over the homogenate of more than 20-fold. (N.B.: The specific activity in the zymogen granule fraction also rose from approximately 0.2 to 2.0 from 5 to 60 min and was enriched relative to the homogenate overall as a function of time.) The ratio of labeled chymotrypsinogen to albumin in the granule fraction was 150:1; seven times the ratio for the homogenate overall. Thus, the evidence indicated that chymotrypsinogen was selectively taken up by the tissue and was accumulated in *zymogen granules*.

Furthermore, when the data were expressed in terms of chymotrypsinogen specific radioactivity; that is, the amount of labeled chymotrypsinogen per unit of total chymotrypsinogen, specific activity in the zymogen granules was about the same as in the bathing medium by 60 min. That is to say, isotopic equilibrium had been attained. Not only was exogenous *chymotrypsinogen* taken up by the cells, but it had fully equilibrated with the endogenous granule enzyme pool. This meant that complete exchange of this protein between granule and medium pools had occurred over the 60-min period of incubation.

A vesicle or endocytic uptake system of some sort cannot readily account for this observation. In such a case, isotopic equilibration would not be possible. Material taken up by granules in this way would always be diluted by the endogenous pool, and even to approach isotopic equilibrium such an uptake system would have to have moved a volume of material in an hour that is greater than the volume of the whole granule pool and even greater than that of the cell itself. This observation also demonstrates that it was indeed chymotrypsinogen that had equilibrated and not merely the label, either dissociated from the protein or reincorporated into the mixture of pancreatic enzymes. If either had been the case, then isotopic equilibration of the *chymotrypsinogen* pool would not have been possible, since much of the label would be in other substances (N.B., over 95% of the label was precipitated in trichloroacetic acid). Only a membrane transport process in which the enzyme presumably moved from bathing medium, via the cytoplasm, into the granule could seemingly account for what was observed.

This view was reinforced by the observation that the postmicrosomal supernatant fraction of the cell also essentially attained isotopic equilibrium with exogenous chymotrypsinogen (unlike mitochondrial and microsomal subfractions), and indeed for short periods of incubation (for example, 5 min) had a higher chymotrypsinogen specific activity than the zymogen granule fraction itself.

PERMEABILITY IN OTHER MEMBRANES

There are a variety of other observations, of both a direct and indirect nature, that support the notion of the membrane transport of these proteins, not only across granule and apical acinar cell membrane,

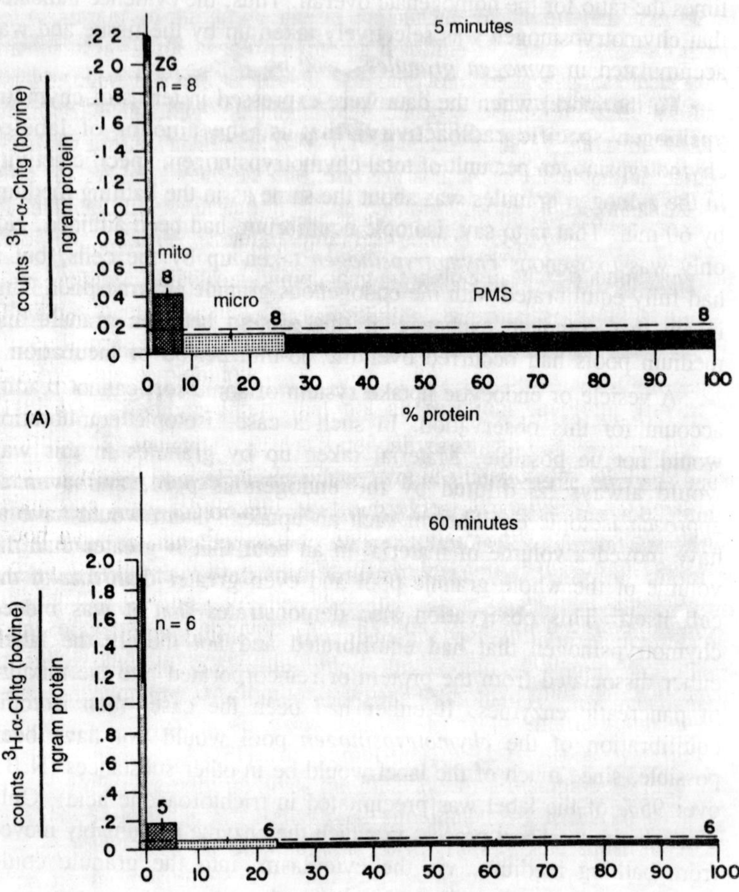

Figure 7.12 : Specific radioactivity of [^{3}H]chymotrypsinogen (Chtg) (cpm/ng protein) in different cell fractions. Fractions are: ZG (zymogen granule), Mito (mitochondria), Micro (microsomes), and PMS (postmicrosomal supernatant). Specific radioactivity is plotted against percent of total protein recovered (after washes) in these fractions. Dashed line is sum of counts collected in fractions divided by sum of protein collected. Numbers are number of individual experiments from which means were determined. Error bars are standard error of the mean. (A) Fractionation after 5 min of incubation; (B) after 60 min of incubation.

but across other biological membranes as well. Although will not consider these observations in any depth here, and must leave it to the reader to ferret them out in the primary literature, like to touch on two, both as examples and because that they are noteworthy.

The basolateral membrane of the acinar cell is permeable to at

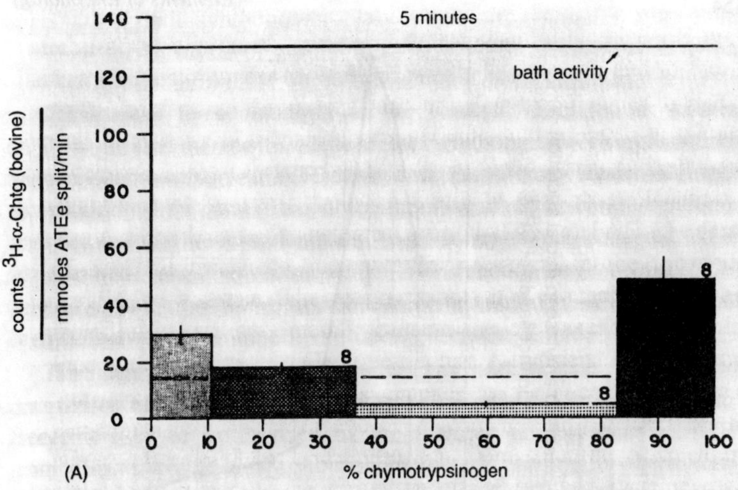

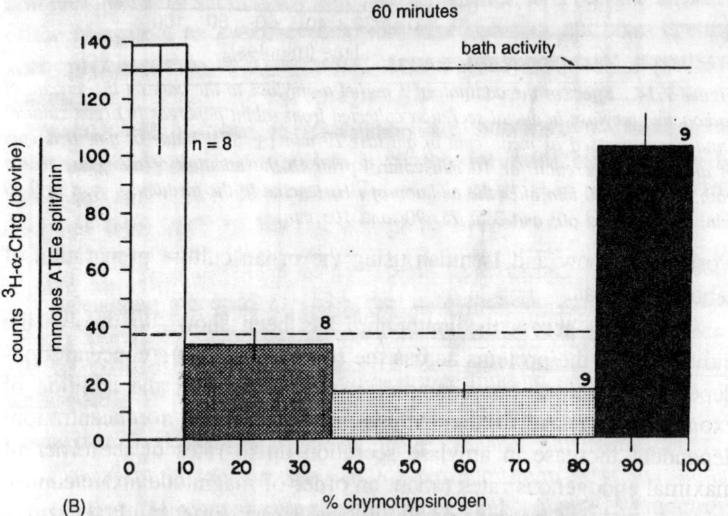

Figure 7.13 : Specific radioactivity of [^{3}H] chymotrypsinogen (Chtg) (cpm in terms of chymotrypsin activity in fractions). Fractions are from left to right: zymogen granule, mitochondrial, microsomal, and postmicrosomal supernatant.

least some digestive enzymes, by an equilibrating process suggestive of membrane transport, and not vesicular processes. In addition, digestive enzyme added to the interstitial fluid bathing the tissue can cross the epithelium and be collected in pancreatic juice. This process of transpancreatic transport has been characterised to some degree in

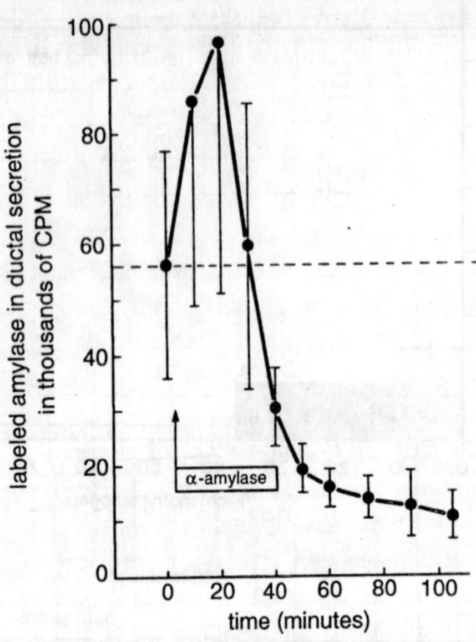

Figure 7.14 : Effect of the addition of 1 mg/ml a-amylase to the bath on the amount of radioactive amylase in ductal secretion collected from rabbit pancreas in organ culture. Data are expressed as mean cpm in amylase/10 min (± SE). Value at time 0 is cpm in amylase in secretion for 10 min control prior to the addition of a-amylase to the bath and 230-240 min after the addition of [^{3}H] leucine to the medium. n = 8 at 0-40 min; 7 at 50 and 60; and 3 at 75, 90, and 105 min.

work by Liebow and Isenman using the organ culture preparation of rabbit pancreas.

Transport across the epithelium has been shown to involve the movement of the proteins across the secretory cell in a concentration-dependent fashion. For example, it was found that the addition of exogenous amylase to the bathing medium led to a concentration-dependent increase in amylase secretion up to rates of the order of maximal endogenous rates (about an order of magnitude above control or unstimulated rates), while at the same time, and also in a concentration-dependent manner, inhibiting the secretion of previously labeled endogenous amylase (by up to 70-90% for exogenous amylase concentrations that produced magnitudinal increases in overall amylase secretion). These, as well as other similar, observations provided evidence that exogenous enzymes can be taken up by secretory cells, mix with endogenous enzyme pools, and compete with them for exit from the cell. In another experiment, Isenman incubated the gland in a medium containing a substantial concentration of chymotrypsinogen

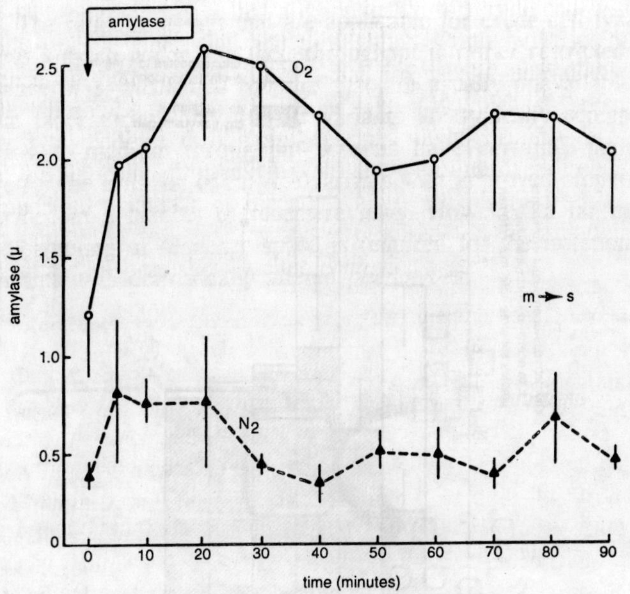

Figure 7.15 : The effect of metabolic inhibition on mucosal to serosal Heal amylase transport in vitro (Ussing chamber).

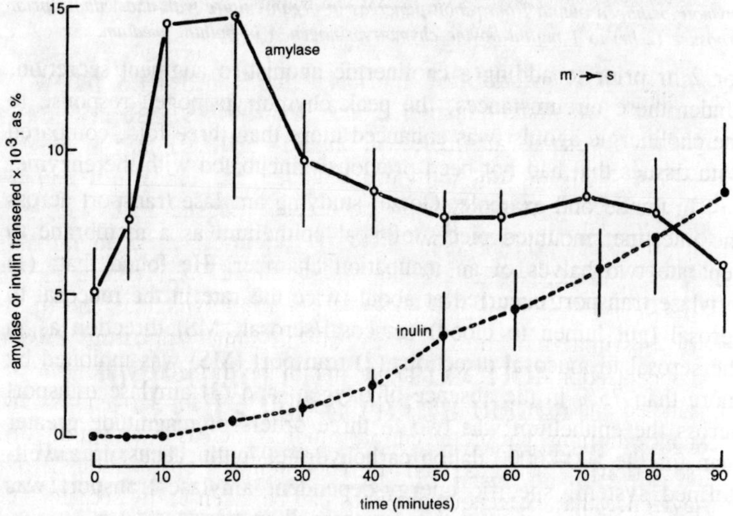

Figure 7.16 : Relative amounts of amylase (0) and [14C] inulin (e) appearing in the serosal medium following their simultaneous addition to the mucosal surface. Amylase data were corrected for endogenous release (n = 6).

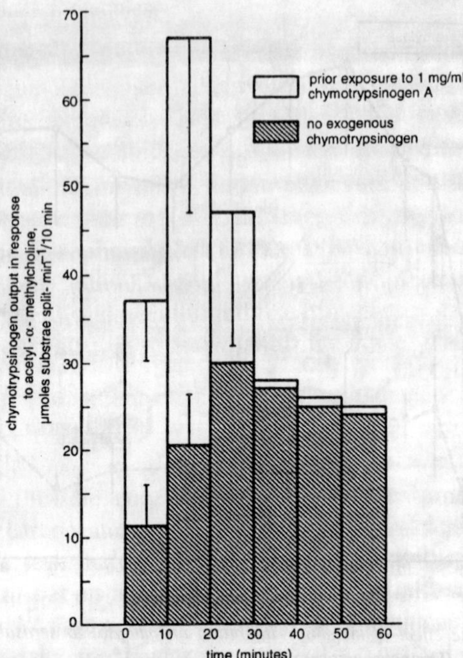

Figure 7.17 : Effect of acetyl-α-methylcholine chloride (1 mg/100 ml bath volume) on chymotrypsinogen output from rabbit pancreas in organ culture with and without prior exposure (2 hr) to 1 mg/ml bovine chymotrypsinogen A in bathing medium.

for 2 hr prior to adding a cholinergic agonist to augment secretion. Under these circumstances, the peak chymotrypsinogen response to the cholinergic agonist was enhanced more than three fold, compared with tissues that had not been previously incubated with the enzyme.

In the second example, Gotze, studying amylase transport across the intestine, mounted pieces of ileal epithelium as a membrane to separate two halves of an incubation chamber. He found that: (1) amylase transport occurred at about twice the rate in the mucosal to serosal (gut lumen to blood) (mucosal/serosal, MS) direction as in the serosal to mucosal direction; (2) transport (MS) was inhibited by more than 75% in the absence of oxygen; and (3) amylase transport across the epithelium was two to three orders of magnitude greater than for the 5000-6000 dalton carbohydrate, inulin. Thus, in a well-defined system, specific energy-dependent amylase transport was demonstrated across other than acinar cell membranes.

8

THE SIGNAL HYPOTHESES

The raison d'etre of the cisternal packaging–exocytosis theory, as well as the assumption underlying the interpretation of many experiments offered as evidence for it, is the belief that biological membranes are impermeable to protein molecules as a class. Thus, it is argued, vesicular processes of one sort or another are necessary to account for the known transport of proteins across membranes. Yet, although vesicular transport models have been constructed and evidence adduced in their behalf mainly on the belief that biomembranes are impermeable to proteins, the central model, the cisternal packaging-exocytosis paradigm for secretion, itself has had to propose paradoxically that individual protein molecules cross membranes. And an even more remarkable paradox has developed. Evidence that newly forming proteins cross the membranes of the endoplasmic reticulum individually is viewed by many as the most convincing of the available evidence in support of the vesicular model for secretion. That is, the transport of individual protein molecules across membranes stands as the most imposing and substantial argument in support of the idea that proteins are moved through and out of the cell by because they cannot cross membranes individually. The reasoning is as follows: If proteins enter the cisternal or interior spaces of the endoplasmic reticulum individually, passing through its membrane to do so, but then having done so, are unable to cross membranes again, and are, to use the common terminology, “” therein, then how else could they be moved through and out of the cell except by vesicular processes?

This peculiar state of affairs has at its origin the fact that ribosomes, on which proteins are synthesized, are either found free in the cytoplasm of the cell or bound to membranes of the (RER) facing the cytoplasm. Because proteins are thought to enter membrane-enclosed spaces, in particular the cisternae of the RER for secretory proteins, after, or actually during, their synthesis, the individual protein molecule must cross an intracellular membrane as it leaves the cytoplasmically oriented ribosome.

Thus, even the paradigm must embrace, however grudgingly, the concept that proteins are capable of crossing membranes individually in some fashion or another. The "fashion" that has been proposed to explain this event within the context of the paradigm. In addition, it will take a step back and ask how good is the evidence that proteins are transported across the membranes of the endoplasmic reticulum in the first place.

The hypothesis that has been proposed to account for the movement of protein molecules through the membranes of the endoplasmic reticulum, even though proposing that membranes are permeable to individual proteins, views such transport as being of an extremely limited nature. Indeed, it views it in such limited terms that it permits one, while admitting to the "de facto" occurrence of protein transport through membranes, to continue to assert its "de jure" absence. This is the signal hypothesis. As originally conceived, it proposes that:

(1) proteins can only cross one membrane, in the case of secretory proteins, that of the RER;
(2) this only occurs once in the life history of each protein molecule;
(3) transport only occurs in one direction across the membrane from the cytoplasmic face to a transcompartment (for example, the cisternae of the RER) and not in the opposite direction;
(4) it only occurs for the extended peptide chain, and not after the development of internal secondary and tertiary structure;
(5) transport occurs directly across the membrane from the ribosome, a two-compartment system, which excludes the cytoplasm from the transport route;
(6) transport only occurs during chain elongation, that is, during protein synthesis or cotranslationally, that is, as the genetic message is being translated;
(7) a special aminoterminal sequence of amino acids is required on the extending peptide chain for transport to occur; and

(8) this special aminoterminal sequence is cleaved during or immediately pursuant to transport to prevent its recurrence.

This series of hypotheses is called the "signal hypothesis" because the aminoterminal sequence of the transported protein is thought to act as a signal for its transport. The idea appears to be derived from the observation that in a cell-free system that synthesizes proteins, many proteins can at least in part be produced with an additional series of amino acids at the amino terminus. The originators of the signal hypothesis thought that this sequence might provide a special modus operandi for secretory proteins to cross membranes and, as it was later proposed, for the insertion of membrane proteins as well.

The idea that the aminoterminal sequence was cleaved as a result of transport seems to have been derived from the observation that the additional sequence was generally not found on secretory proteins in situ and was in a sense an artifact of the cell-free synthesizing system. Its absence in situ was proposed to be the result of the immediate cleavage of the signal sequence as it crossed the membrane, even as the peptide chain was still being formed on the ribosome. This led to another curious situation. The transport of proteins across membranes was attributed to a sequence of amino acids not generally found on the transported proteins in the cell, but seen only under special in vitro circumstances.

It is hard to imagine a more limited view of protein transport across membranes than that provided by the signal hypothesis. Transport occurs only once in the life history of a protein, in one direction only, requires a special chemical modification of the protein, occurs only when the peptide chain is elongating and before any internal structure has developed, and a covalent bond is broken to prevent its recurrence.

The signal hypothesis is also an extremely complex hypothesis, and its complexity must be viewed not only in its own right, but as adding to the overall complexity of the whole of the cisternal packaging—. This may indeed reflect the complexity of the real world, but the scientific method requires that such complex hypotheses only be accepted as being descriptive of real events after the most demanding scrutiny, and most importantly, the exclusion of simpler explanations and interpretations for experimental observations. This has not been the case. But for the moment, how complex is the signal hypothesis? Although its eight postulates listed above, might seem sufficiently complex in themselves, they only reflect the surface complexity.

Underlying each postulate are necessarily other assumptions and predictions about the system, many of which remain purely hypothetical.

ATTACHMENT PHASE

Let us begin with the themselves. There was an attempt over a period of many years to demonstrate that free ribosomes synthesized intracellular proteins, whereas those bound to the RER were given the task of synthesizing proteins for "" or secretion. This attempt failed in the sense that no differences between the two groups of ribosomes could be established, and the mRNA for cellular or secretory proteins does not prefer ribosomes of one type or another as far as we can tell. Thus, it seems reasonably clear at this point that free ribosomes are capable of synthesizing proteins that are normally secreted by cells, and there is no reason to think that in the normal course of events that a meeting between a mRNA molecule for a secretory protein and a free ribosome would not occur and lead to the formation of a new secretory protein molecule that would be released into the cytoplasm. In the paradigm's terms, such an occurrence would leave the protein to languish in the cytoplasm since it would have to, but be unable to, cross membranes in order to leave. The signal hypothesis deals with this unacceptable state of affairs by the addition of a corollary hypothesis: All protein synthesis begins on free ribosomes whatever the ultimate fate of the protein may be. As the amino terminus is revealed, extending beyond the confines of the ribosome itself, if the polypeptide chain contains the aminoterminal signal sequence, then the ribosome attaches to the membrane of the RER and transport ensues. Indeed, it is this attachment that the amino terminus is supposed to signal.

This subsidiary hypothesis, that synthesis begins on free ribosomes, leads to further predictions and requirements. For example, since the signal hypothesis insists that transport only occurs for extended peptide chains, and since the formation of secondary and tertiary structure occurs spontaneously as the chain elongates, we must propose that a mechanism exists to assure the attachment of the ribosome containing the elongating chain to the membrane prior to the formation of secondary and tertiary structure. If too much time elapses, then higher-order structure would develop, protein transport would be prevented, and the new protein would nevertheless end up in the cytoplasm. If an estimated rate of formation of the peptide bond is milliseconds, then not more than a few seconds overall is likely to be available for attachment before secondary and tertiary structure develops. It is not clear whether or not the distances, number of putative binding sites, and number of ribosomes involved would permit one to account, as one must, for a high percentage of attachments on purely statistical grounds involving only the of the ribosome, or whether a special transport mechanism for ribosomes containing signal sequences would have to be postulated in addition. In lieu of such a modus vivendi, it has been proposed that translation is temporarily halted at the end of the signal sequence to allow the ribosome the time necessary to attach to the membrane. Moreover, the ribosome cannot inadvertently associate with other membranes in the cell, that is, inappropriate membranes, to any appreciable degree, and therefore there must be specific binding sites for the signal sequence on the membranes of the RER that are absent from other membranes. Thus, the signal hypothesis specifies the timing and location of the attachment of ribosomes to membranes.

We can summarize the various subsidiary and deductive hypotheses of the attachment phase of the signal hypothesis as follows:

(1) the initiation of the synthesis of all proteins begins on free, not bound, ribosomes;

(2) only ribosomes whose nascent polypeptide chains contain the signal sequence of amino acids at the amino terminus can attach to the membrane of the RER;

(3) a mechanism exists to assure that these ribosomes attach to the membrane in a timely fashion and that all signal sequence-containing ribosomes do so;

(4) the signal sequence is only able to associate effectively with RER membrane, and not other membranes in the cell;

(5) there are specific binding sites on RER membranes for the signal sequence;

(6) proteins that are not destined for "export" or insertion into membranes do not contain the signal sequence of amino acids at the amino terminal; and

(7) such proteins are not synthesized on attached ribosomes.

We should add items 8 and 9 to the list.

(8) There probably has to be an additional receptor site on the RER membrane that is not for the signal sequence itself, but for the ribosome proper. The reason for this can be

understood if we consider the fate of the signal sequence for a moment. According to the signal hypothesis, although the signal sequence binds to the membrane of the RER that interaction is only transient, and the aminoterminal sequence moves through the membrane and as it does is cleaved by a proteolytic enzyme. Once this transport event begins, the signal sequence can no longer act to maintain the ribosome closely apposed to the membrane, and without another binding site, the ribosome might well diffuse or otherwise move away from the membrane even as protein synthesis continues. Therefore, there must be an additional receptor site on the membrane to hold the ribosome in place and of course a specific binding element on the ribosome that acts as the ligand at this site. However, since every ribosome would have to have the capability of binding to this receptor, whether or not it was synthesizing a protein with a signal sequence at that particular time, or indeed, whether or not it was synthesizing a protein at all, we have to further propose that.

(9) binding only occurs pursuant to the interaction between the signal sequence and the membrane. That is, in the absence of an association between the signal sequence and the membrane, the ribosome cannot bind to its membrane site; otherwise ribosomal sites on membranes would not be exclusively occupied by ribosomes engaged in synthesizing secretory and membrane proteins, and would include those synthesizing cytoplasmic and other cellular proteins, or those not engaged in protein synthesis at all.

Thus, the attachment of ribosomes to the membranes of the RER, or the "" of the signal hypothesis, requires a considerable array of subsidiary hypotheses. What evidence is there for these ideas? We must begin with evidence for the occurrence of such a process in the first place. Unfortunately, it is at this level, a description of the process itself, that the concept lacks satisfactory experimental verification. It is not clear that a ribosome cycle, free to bound to free, occurs during the synthesis of each protein molecule, no less that such a cycle quantitatively accounts for events. It has not been demonstrated that protein synthesis cannot be initiated on attached ribosomes, as well as free ribosomes. Nor is it clear that the timeliness of such a cycle can be ensured. Thus, essential elements of the attachment phase of the hypothesis (certainly in eukaryotes) remain without adequate experimental verification, and the signal sequence is proposed to carry out a function in a process whose existence itself has not been satisfactorily and independently established. Of course, it is necessary first to establish the existence as well as the characteristics of a process before one attempts to determine it occurs. Unfortunately, things have developed the other way round, and the signal sequence was proposed to account for a process whose existence had not been independently established.

Even if we accept the existence of such a cycle and its transport sequelae the signal sequence into question. That is, there is falsifying evidence for its role in the signal hypothesis' own terms.

As I have said, the central postulate of the "" portion of the signal hypothesis is that a leading amino acid sequence exists that is required for the attachment of the ribosome to the membrane of the RER and that, as a result of this attachment, protein transport across the membrane occurs. If there were proteins that did not contain this sequence of amino acids, but that nevertheless were secreted, then we could conclude that such a sequence was not required for secretion, nor for the attachment of the ribosome to the membrane, assuming that this occurs, nor for the transport of the protein across it, assuming that this occurs as well. Thus, if we assume that the process occurs as envisioned and that the general vesicular theory is correct, if such signal-free proteins were found, then the particular explanation that the aminoterminal amino acid sequence is required for ribosomal attachment and protein transport would seem to be incorrect.

Indeed, there is at least one well-established example of such a protein. Ovalbumin, a secreted protein which synthesized in the usual cell-free system does not contain the additional sequence of amino acids at the amino terminal, and is manufactured as seen, at least in gross structural terms, in situ and in secreted material. That is, ovalbumin does not contain the signal sequence, and yet is of course secreted. There has been an attempt to incorporate this discrepant observation into the signal hypothesis by proposing that ovalbumin does indeed contain a signal sequence, not at the amino terminus, but further along the peptide chain.

This raises the question of just what structural characteristics of a segment of a peptide chain identify it as a signal sequence if it is not its location at the amino terminus? The common structural characteristic shared by most of the numerous aminoterminal segments that have been sequenced thus far is that they contain a relatively high proportion of socalled "" residues, that is, this sequence tends to contain a higher percentage of amino acids with purely aliphatic side-chains than found in the polypeptide chain overall. Although this characteristic is hardly invariant or unambiguous, it does represent the best, perhaps the only, successful structural generalisation made to date concerning these peptide sequences.

Returning to ovalbumin then, the internal signal sequence that has been proposed is a relatively hydrophobic region of amino acids in the middle of the chain; that is, this sequence is proposed to act as the signal for the attachment of the ribosome to the RER and for the subsequent transport of the polypeptide chain across the membrane. Of course, most proteins, and globular proteins in particular, contain such internal hydrophobic regions, which account at least in part for the folding of the protein, with the more hydrophobic portions tending to face inward away from the polar suspending medium. It is not clear how this particular segment in ovalbumin, or any particular signal segment for that matter, differs from any other hydrophobic region on the polypeptide chain, beyond placement at the amino terminus. If this internal region in ovalbumin is a signal, then why wouldn't internal hydrophobic regions of other proteins similarly promote the binding of the ribosome to the membrane and lead to the transport of inappropriate proteins into the cisternae of the endoplasmic reticulum? Perhaps there is something special about this sequence in ovalbumin, either in its primary or higher-order structure, that could account for such special behaviour, but this has yet to be satisfactorily demonstrated. Even so, we cannot retain the topological special place of the aminoterminal sequence in the signal hypothesis.

This is not to say that this sequence is unrelated to the ability of ovalbumin molecules to cross membranes. Of course, the movement of individual protein molecules across membranes is likely to be a function of their particular structural characteristics, and, in this sense, such specific characteristics can be considered signals. However, this is quite a different matter from the signal envisioned by the signal hypothesis that leads to the attachment of the ribosome to the membrane and the formation of a special pore within the membrane to accommodate a passive extending peptide chain.

It should also be noted that secondary and tertiary structure will have developed in the elongating chain by the time that this internal signal is exposed, and therefore, even in the model's own context, transport must be envisioned as occurring after the establishment of internal structure and not solely as extended peptide chains. Thus, the postulate that transport only occurs for extended peptide chains must be incorrect, and there are many contrary examples at this point. Therefore, even in the paradigm's own terms, the transport of proteins containing internal structure must be entertained.

In regard to the absence of the signal sequence on secreted proteins, even for those protein molecules for which signal sequences have been identified, not all, nor even necessarily most, molecules recovered from the cell-free synthesizing system contain the additional series of amino acids. This discrepancy often goes unnoticed, but when it is noticed, it is usually explained by the proposal (ad hoc hypothesis) that the system is contaminated by the enzyme that cleaves the signal sequence normally and that it acts to form the postsignal protein. To evaluate the adequacy of this explanation or other similar problems, it is necessary to recover the putative signal peptide fragments and to demonstrate, not only their presence, but also that they are present in quantities that are equivalent to the number of signal-less proteins recovered. To my knowledge, such a demonstration has yet to be made, and in its absence we must consider the possibility that proteins lacking the signal sequence are synthesized in just that form.

Now let us turn the "" of it around. Are there proteins that contain or are thought to contain signal sequences that are not secretory or membrane proteins? If so, what is their fate? Shouldn't all ribosomes that contain such signals attach to the membranes of the RER and deposit their contents therein? If this is not the case, then doesn't this mean that the signal hypothesis is incorrect, and that the signal sequence does not lead to the attachment of the ribosome to the RER? Alternatively, if there are nonsecreted, nonmembrane proteins that contain the signal sequence and are sequestered in the paradigm's terms, then the idea that the transport of proteins across the membranes of the RER is a means of separating or segregating secretory proteins from intracellular proteins must be incorrect. Examples of both situations have been reported.

Signal sequence-containing mitochondrial proteins are released as completed proteins from the ribosome into the cytoplasm of the cell, and then cross the outer mitochondrial membrane to enter the mitochondrion. Thus, in this case, the signal sequence does not lead to the attachment of the ribosome to the membranes of the RER. There is no obvious structural reason why the sequence of amino acids on these mitochondrial proteins should be ineffective in promoting the attachment of ribosomes to RER membrane and the transport of the peptide chain across it, whereas those of other proteins, seemingly as similar to the mitochondrial sequences as they are dissimilar from each other, should be effective. This raises the question of whether or not ribosomal attachment has anything to do with the signal sequence. Of course, as with ovalbumin, it is possible that there is some special structural characteristic of these signal sequences that has yet to be discovered and that will explain why attachment does not occur. But until such a discovery is made, we cannot assume that it is there to be discovered, and that the signal hypothesis is saved and the mitochondrial story is merely one additional exception to a general rule.

Also, as with , these proteins cross membranes (mitochondria,) with the protein's tertiary structure already having been established. Thus, once again, even accepting the signal hypothesis' explanation for the behaviour of the signal sequence, it is clear that proteins in their full globular form cross membranes.

There are also examples of proteins that are neither secreted nor membrane proteins, but that nevertheless are thought to be sequestered; that is, they contain the signal sequence that leads to the attachment of the ribosome to the RER and their transport into its cisternal space. One example is the lysosomal enzymes; that is, enzymes that function within the membrane-bound organelle, the lysosome. The sequestration of such proteins tells us that the purpose of the transport of proteins across the membranes of the RER into its internal spaces (assuming that this is what occurs) cannot be to segregate secretory protein from cellular proteins, because this is simply not the case.

Moreover, assuming that these proteins are sequestered within the RER, that is, are transported into the of the RER without the ability of leaving across its membrane in the absence of vesicular processes, then specific vesicular processes for these particular proteins must be proposed in addition to those that lead to the en masse secretion of protein. A vesiclesorting system similar to the small vesicle system that has been proposed for the filling of the condensing vacuole, has been proposed for "filling" the lysosome with its enzymes. Such a system requires that special vesicles be formed from the RER that would exclude nonlysosomal proteins. The formed vesicle would, of course, have to find its way to the lysosome, deposit its contents therein, and return home to the RER. In the absence of convincing evidence that such a system exists and accounts for the transport of these proteins, they cannot be assumed to be sequestered; that is, either they do not enter the cisternal spaces of the RER, or if they do, they are able to leave across its membrane without the interposition of such a complex vesicular system.

Therefore, in terms of the “” of the matter, some proteins are secreted that do not contain signal sequences, and yet others that are not secreted contain such sequences. Among the latter group are those that are released from the ribosome into the cytoplasm to cross intracellular membranes after they are fully formed, whereas others appear to be “” in the cisternae of the RER and sorted out from secretory proteins later. The , as it was originally envisioned, appears to have many exceptions to its rules. And most significantly, there is no convincing independent evidence for the existence of a ribosome attachment cycle of the sort that has been proposed, or if such a cycle exists, that the signal sequence is necessarily the responsible party. We cannot say with any certainty that ribosomes only bind to the membranes of the RER if they contain a signal peptide, nor do we know that “” ribosomes are necessarily engaged in protein synthesis at all times.

TRANSPORT PHASE

The has been proposed to signal two events. The first is, as we have discussed, the attachment of the ribosome to the membrane. The second is the formation of a pore within the membrane of the RER through which the nascent protein travels to enter the cisternal spaces. This is the transport phase of the . As was the case for ribosomal attachment, there are two questions to be asked. Do these proteins indeed cross the membranes of the endoplasmic reticulum? That is, does a transport process exist? And, if it does and they do cross the membrane, then by what means? The transport phase of the signal hypothesis is a proposal for how transport occurs, based on the assumption that it does. There seems to be evidence for the existence of the process that is being proposed and proteins may well be transported across the membranes of the RER, although, this evidence is uncertain in important respects. For the moment, we will assume that transport occurs and consider briefly the nature of the mechanism that the signal hypothesis proposes.

We can describe the subsidiary and deductive hypotheses of the transport phase of the signal hypothesis as follows:

(1) the attachment of the signal sequence to a receptor leads, either by itself or in concert with other molecules, to a molecular reorganisation of the membrane which produces a pore that traverses its width;

(2) it is the hydrophobic nature of the signal sequence that is instrumental in forming or helping form the pore;

(3) this having occurred, the signal sequence is released from its binding site and passes through the pore;

(4) the pore's patency is maintained after the signal sequence has passed through, that is, its presence is not required for its maintenance;

(5) the pore disassembles or otherwise closes after the carboxyl terminal of the peptide chain passes into the cisternal spaces; and

(6) this leads to a decreased affinity of the ribosome for its membrane site and it is released into the cytoplasm.

Such a mechanism in the following terms: "Such pores could … be fugacious in nature opening up to let the enzyme in and then closing; … A fugacious pore hypothesis is more appealing (than a permanent pore) and might be achieved by the action of a third molecule or group of molecules acting on or near the ribosomal attachment to the membrane in a way that would change the membrane structure to permit the entry of the particular enzyme, but not 'other' protein molecules dissolved in the cytoplasm. After the entry of the enzyme, the third molecule, perhaps a type of permease, would dissociate from the membrane and the membrane would again be ''. If such a system were operative then only ribosomes making enzyme for export could have this 'third party', otherwise intracellular proteins would also enter the cisternae."

Thus, the signal hypothesis is a specific proposal for just such a system of fugacious pores. What the details of such a system might be is fertile soil for the imagination. How would the pore be formed? How would receptor binding be time-limited? What kind of pore would be able to permit the free transfer of the whole range of some 20 amino acids in almost any conceivable order simultaneously, accepting the polar and the nonpolar, the aliphatic and aromatic, binding neither, and yet being of such minimum width as to keep the extending peptide chain on the straight and narrow, that is, preventing “” in the chain and the formation of higher-order structure within the pore? Despite these interesting questions, when all is said and done, there is not much to say about this phase of the signal hypothesis beyond describing what it proposes. It is not a hypothesis that is supported by direct experimental evidence; other than evidence for the existence of the transport process itself and the existence of the aminoterminal sequence. That is, it is not known that pores exist in these membranes, no less fugacious pores with characteristics of the type proposed by the signal hypothesis. There is no direct evidence for the particular transport mechanism proposed by the signal hypothesis as opposed to other particular potential mechanisms and explanations for protein transport across this membrane that might be considered, assuming that it does occur. Unfortunately, there is nothing more that can be said about the transport phase of the signal hypothesis.

CLEAVAGE PHASE

This brings us to the third, final, and perhaps most problematic phase of the signal hypothesis: that of the cleavage of the signal sequence. This refers to the removal or “” of the aminoterminal peptide sequence from the protein as it passes out through the cisternal face of the membrane. Cleavage is usually envisioned as occurring pari passu with passage, that is, cotranslationally as the particular peptide bond passes the active site of a specific membrane enzyme whose purpose is to cleave the bond. In this way, each and every molecule passing through the membrane would have its signal sequence “” and this would explain why such sequences are not found on secretory proteins in whole cells, in secreted material, or even in cell fractions of the RER. If cleavage took place subsequent to the transfer of the protein into the cisternal spaces, rather than as it exited the pore, then at any given time it might contain some proteins with the additional sequence. Moreover, if the former were the case, then the hydrophobic amino terminus would have to be prevented from being buried within the various protein molecules, as might be expected as higher-order structure develops, or otherwise made inaccessible to the hydrolase.

In any event, wherever or however cleavage might occur, the first question is of course, does it occur? I have heard it argued that since the peptide segment is not seen in situ, then it must be cleaved. The inference being that it if had not been, it would be seen. Of course, this is merely circular reasoning. Simply because the added sequence can be demonstrated in cell-free systems, does not mean that this same translation takes place in situ. This may or may not be the case, and it requires independent evidence. The crucial evidence that must be forthcoming is for the existence of the other end product of the putative reaction, namely, the small signalcontaining peptide, in situ. For some proteins, there is qualitative proof that small peptides can be recovered from a cell-free system under the right circumstances. It has not however been demonstrated that the presence of the short-chain protein (signal sequence-free protein) and the aminoterminal peptide are quantitatively related, that is, that a stoichiometric relationship exists between the two—mole for mole, in the cell-free system, no less in situ. Of course, such a relationship is to be expected if cleavage accounts for the presence of both products (the small peptide and the signal sequence-free protein). This must be demonstrated in vivo. Although the larger signal-containing proteins are usually absent in eukaryotes, there is no evidence that I am aware of for the presence of the signal sequencecontaining small peptide. It is central to the validation of the signal hypothesis, not only that the small peptide remnants of the proposed cleavage be demonstrated in vivo, but that their fate be determined quantitatively. In the absence of this evidence, we cannot conclude that the larger protein is synthesized in situ, no less exclusively; nor that it is cleaved to form the observed smaller molecule, no less that the signal sequence accounts for the transfer of protein into the cisternal spaces of the RER, assuming that that occurs.

But even though we may not yet be able to say that these short aminoterminal peptides are produced in situ, or in what quantity, we nevertheless can consider the cleavage hypothesis from the point of view of the chemistry of the signal sequence itself. At this time there is a substantial catalog of sequences, as well as presumed points of cleavage, and therefore comparisons can be made between them. Unfortunately, there does not appear to be a unique point of hydrolysis common to all of the different proteins. A variety of different bonds seem to be involved and one cannot, as of yet, generalize satisfactorily about the putative cleavage site. Cleavage appears to occur at different points in the chain in different molecules and involves different cleavage bonds. If this is correct, then either there are a variety of different cleavage hydrolases for the different bonds (even in the same cell), or there is a rather general hydrolase that can cleave them all, and both have been suggested.

Let us attempt to put such a system into place in the membrane. Since we cannot know in advance which particular protein will pass through which formed pore, each pore must have residing at its cisternal face all of the hydrolases necessary for the range of proteins that might pass through this membrane in a particular cell. If we envision the simpler idea of a rather general hydrolase capable of breaking a variety of peptide bonds, then this problem would be simplified, but we would have to propose some fashion in which a relatively nonspecific enzyme would be prevented from splitting bonds that are not at the signal sequence cleavage point willy-nilly. Although this problem might appear less severe if a variety of specific hydrolases were available, it would really remain the same: merely many enzymes rather than one being responsible. That is, our system must be designed so that hydrolysis only occurs at one site, and not at other potential loci along not only a single peptide chain but also along all other peptide chains that are able to pass through such a pore. In the absence of such restrictions, we would have a series of uncontrolled chaotic reactions producing a wide range of different peptides of different length, which apparently does not happen. One could propose that the signal sequence "signals" or denotes the appropriate site for cleavage. Perhaps other regions on the peptide chain could be made inaccessible by protein folding, if the enzyme acts lumenally rather than in the pore. But can we account for the fact that not all signal sequences are split? In any event, a very restrictive, perhaps multienzymic, system is required that would allow for the cleavage of a range of peptide bonds at the amino terminus, while preventing the cleavage of similar bonds elsewhere on the same and other proteins. These enzymes cannot merely be capable of cleaving certain kinds of bonds, but must cleave only one particular bond of a particular type. This is certainly possible and remarkably specific enzymes do exist; however, it is uncertain that they do in this case.

The proposed cleavage must be shown to occur in situ, and be stoichiometrically established by the quantitative recovery of end product under these circumstances. In the absence of such evidence, the signal hypothesis cannot explain the absence of the signal sequence on secretory proteins in cells and in secretion, and it must be considered that its presence in cell-free systems may at least in these cases be an interesting artifact that may have some significance concerning the translation of genetic information, but that does not occur normally in situ. It cannot be assumed as it seems to have been, that the absence of the longer, signalcontaining protein in situ is proof of the occurrence of cleavage in situ.

ESCAPE INTO THE CYTOPLASM

As mentioned earlier that the signal hypothesis is based on the twin assumptions that transport into the cisternal spaces occurs, and that the transferred proteins are unable to leave these spaces without the interposition of a vesicle system to accomplish the task. There is also a third assumption that underlies the signal hypothesis-each and every newly manufactured secretory protein is released from the ribosome into the cisternal spaces of the RER. That is, none end up in the cytoplasm. This assumption is reflected in the signal hypothesis' concepts that transfer occurs by means of a two-compartment system, ribosome to cisternal spaces, excluding the cytoplasm from the transport route, and that the attachment of the ribosome to the membrane occurs in a timely fashion to prevent the inadvertent release of a new protein with a signal peptide into the cytoplasm.

Let us consider the basis of this assumption. Unfortunately, it is not direct experimental evidence that demonstrates that newly manufactured secretory protein is not released from the ribosome into the cytoplasm. Rather, the view that this does not occur appears to be based primarily on the belief that a cytoplasmic pool of secretory protein does not exist, and hence that release could certainly not occur into a nonexistent pool, as well as equating events such as proteolytic processing, glycosylation, protection from proteolysis, and so forth, with the cotranslational transport of the protein through the reticular membrane (see below for further discussion of these issues).

There is substantial evidence for a cytoplasmic pool of digestive enzyme in the pancreas, and this subject has been considered elsewhere in this volume. However, whatever one may think of the evidence for a cytoplasmic pool of digestive enzyme, there is certainly no experimental justification for drawing the strong negative conclusion that one can assume that such a pool does not exist and hence that it is not necessary to deal directly with the compartmental fate of new proteins released from the ribosome. This is not to say that there is no experimental evidence that bears on the question of the release of digestive enzyme from the ribosome. However, what evidence there is does not lead one to the conclusion that release into the cytoplasm does not occur. Quite to the contrary, evidence consistent with the view that release occurs into the cytoplasm can even be found in the work of proponents of the cisternal packaging model.

In a review of the subject of protein secretion in 1975, data from a cell-free system was found in which the compartmental disposition of proteins synthesized by vesicular fragments of the RER, called microsomes, was evaluated. In these experiments important forerunners of those that later led to the proposal of the signal hypothesis, newly (labeled) manufactured protein was recovered in:

(1) the ribosomes themselves (separated from the membrane of the microsomal vesicle by a detergent);

(2) in composite, the internal contents of the vesicle, the membranes of the vesicle, and the proteins associated with the vesicle at its membrane's surface (elements of the microsome that were soluble in the detergent); and

(3) in the medium in which the microsomes were suspended. My purpose in discussing these experiments was to note and evaluate the fact that labeled protein was recovered in the suspending medium-the in vitro equivalent of the cytoplasm. How did it get there?

What could we infer from its presence? The scientists who performed these experiments, in accordance with the belief that the protein could not end up in the cytoplasm in situ, assumed that its presence in the medium must have of necessity been artifactual, and proposed accordingly that it reflected the release of new protein from broken microsomal vesicles into the suspending medium. They did not test this ad hoc hypothesis, but thought it sufficient to note that such an explanation was possible. Certainly, it would not have been surprising if some vesicles had been broken and that it might be an artifact of this sort, although today sealed lipid vesicles are often thought to form spontaneously to minimize free energy.

In any event, it certainly was possible to imagine that some vesicles were damaged or not fully formed and that perhaps their contents were being released into the medium. However, the observed kinetics of the appearance of the new protein in the various compartments of the system raised questions about this interpretation. The appearance of labeled protein in the detergent-soluble material, assumed to be the contents of the vesicle (we will consider this assumption below), increased at a different, slower, rate than the labeled protein content of the medium. If broken vesicles had been the source of this material, then one would have expected the two time-courses, vesicle (detergent-soluble material) and medium content of labeled protein to increase roughly in parallel, which they did not. Moreover, labeled enzyme in the medium reached a steady-state concentration, even as that presumably entering intravesicular spaces continued to increase. This suggested that the labeled protein in the medium did not come from broken vesicles at all, or at least not in great part, and if this was not the case, then they must have come directly from the ribosomes; that is, they were released from the ribosome into the medium. Therefore, this experiment provided indirect evidence for protein release from the ribosome into the cytoplasm in situ. This interpretation was reinforced by the fact that the function describing the release of material into the medium and that for the accretion of labeled protein on the ribosomes themselves were roughly parallel; that is, they had the same form and approximately the same time course, as would be expected if the medium's labeled protein content were directly derived from the ribosomes.

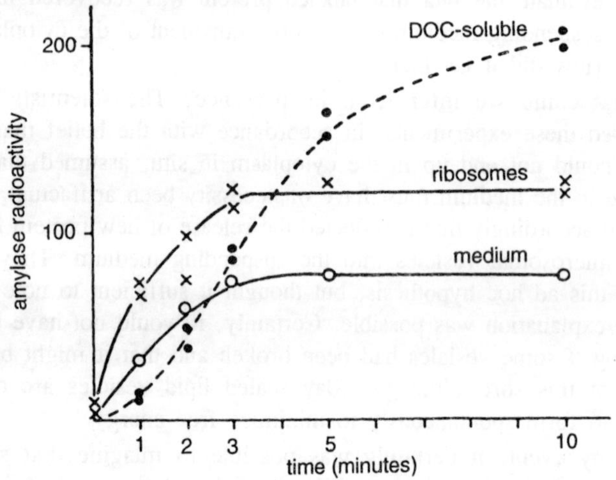

Several years later, an alternative "" explanation was suggested, in which it was proposed that a time-dependent lysis of a subpopulation of weak vesicles might account for the observed kinetics. The idea was that there were weak vesicles among the microsomes that broke open over time, and, as they did, released their contents. Breakage would be greatest for a short period just after the addition of the labeled amino acid precursor to the medium, and within a few minutes only sturdy vesicles would remain. Why such a time-dependent pattern of breakage should be produced reproducibly as a result of the addition of a labeled amino acid to the medium was not explained. But even though I do not understand why such a possibility is thought credible, credible or not, it is nevertheless certainly conceivable. And that is my point here. It seems that merely being able to imagine a conceivable explanation for an event within paradigmatic borders is viewed by some as being sufficient to retain the paradigmatic view of a process. No proof that one's ad hoc hypothesis is correct, as opposed to other possible explanations, seems needed. The question that I raised in 1975 remains experimentally unanswered almost 10 years later, even though during this time the signal hypothesis, whose premises depend in important ways on the presumption that release does not occur into the cytoplasm, has been widely used and developed, at least to some degree based on the belief that an untested ad hoc hypothesis is correct.

Thus, microsomes are vesicular fragments of the rough-surfaced endoplasmic reticulum capable of synthesizing protein in vitro under the right circumstances. Simply the existence of such a system raises interesting questions about the signal hypothesis, in particular its proposal of a ribosome-binding cycle. As already pointed out above, the signal hypothesis requires that the attachment of the free signal-containing ribosome to the membrane of the RER be timely to prevent new protein from entering the cytoplasm. This either means that the ribosome has rapid access to the membranes on the purely statistical grounds of Brownian motion, or that a special process exists to ensure its timely attachment. If we can imagine such a cycle occurring in the microsome-synthesizing system, as it must if the hypothesis is correct, then we realize that there can be no special cytoplasmic transport mechanism for ribosomes to assure timely attachment, because the suspending medium is not ordered. Brownian motion is the only means available to accomplish the cycling of ribosomes, if such occurs. Although to my knowledge this system of microsomes has not been studied from this point of view, there are some interesting questions to be asked. In the acinar cell, about 1 volume of RER is suspended in 2.5 volumes of cytoplasm, based on stereological estimates. Although one cannot make accurate estimates from the literature, the microsomes in synthesis experiments probably represent less than 10% of the volume of suspending medium (and perhaps as little as 1% or even less). If the cycle concept is correct, then one should be able to replace the short, signal-free proteins recovered from the usual microsomal system, by signal-containing proteins merely by diluting microsomes in larger and larger volumes of medium, and thereby decreasing the efficiency or timeliness of ribosome binding to membranes. Moreover, this increase in signal-containing proteins should be strictly proportional to a decrease in the RNA (ribosomes) content of the microsomes and an increase in the number of free ribosomes that should also be produced by such dilution. Alternatively, if synthesis is halted while awaiting ribosome attachment, then the rate of synthesis should be reduced in proportion to such dilution in the absence of mixing. Such quantitative results need to be demonstrated if we are to prove the hypothesis of a ribosome-binding cycle and the role of the signal sequence in such an event.

PROTEIN TRANSPORT

It may seem somewhat odd to question the existence of the transport of individual protein molecules across the membrane of the endoplasmic reticulum given the perspective of the author, but that is just what I would like to do now. My view about the existence of this process has changed over the years from being quite certain that it occurred in 1968 ["it seems clear that newly manufactured enzyme can enter the cisternae of the endoplasmic reticulum"], including a consideration of how it happened, to some hesitancy in 1975 ["Do nascent peptide chains pass into the cisternal space of the endoplasmic reticulum? The (answer is) a cautious 'yes']. Although the evidence in favor of such a process has not actually decreased over the years, it has not in my opinion been substantially bolstered by what new evidence has been brought to bear on the subject, and many of the problems with the old evidence remain unresolved even after many years. This is not to say that I believe that such a transport process does not exist. I merely remain unconvinced and would like to consider the basis of my doubts in what follows.

There are two related types of observation that provide much of the evidence that supports the existence of such a transport process. The first is the attempt to demonstrate directly that transport occurs; and the second is an attempt to measure the contents of the endoplasmic reticulum and to demonstrate the presence of the substance of interest in this compartment. In the exocrine pancreas, there is both evidence that transport occurs and that the digestive enzymes are normally found in the cisternae of the RER, but I believe that this evidence is rather weak, particularly when one considers that it is this particular evidence that forms the experimental foundation of the signal hypothesis and the large body of work derived from it. That is, the signal hypothesis is necessarily based on the belief that, if nothing else, the existence of this transport process has been established; otherwise, it proposes a mechanism for a process whose existence is not known; and that would be an unacceptable state of affairs. In addition, the cisternal packaging theory proposes a complex system of small vesicles to carry material out of the RER, a process whose existence is almost completely unsubstantiated even after many years of experimental work, based on these observations and the conclusion drawn from them that transport occurs. It is in this rather imposing context that one must evaluate the evidence for such a transport process.

DOES TRANSPORT OCCUR?

A great many experiments have been performed using microsomal vesicle preparations of the sort, either to study protein transport in vitro, or to demonstrate that protein had entered the cisternal spaces in situ (by extracting microsomes from whole tissue after its exposure to radioactive amino acids). Both types of study have generally used one of two methods for evaluating whether or not a particular protein was transported into the cisternal or vesicular space. In the first, the contents of the space are extracted as part of a detergent-soluble fraction of the whole microsome; that is, this detergent-soluble fraction contains the contents of the intravesicular space. Unfortunately, however, this material also contains the vesicle membrane and its constituents, and proteins that had been absorbed or otherwise bound to the surface of the membrane. Thus, this technique does not distinguish between these three potential sources, and therefore to the degree that all three are credible possibilities, fails as a means of determining the contents of the intravesicular compartment.

The second technique involves the addition of proteolytic enzymes to a suspension of microsomes and then determining if the protein is protected from the destructive action of the enzyme. If protection is afforded, then the inference is made that the membrane is a geometrical barrier and that the protected proteins are inside of the vesicle. But of course protection is merely protection, and it can be provided in other ways-in particular, as a result of the association of the protein with the membrane of the vesicle, or as the result of protein-protein interactions in the medium. This uncertainty is particularly relevant when protection is not complete, namely, when some of the protein is damaged by the , as is usually the case. That is, not all or even most of the given enzyme is protected, only a fraction, and sometimes a small and unquantified fraction at that. Moreover, two proteins that have structural differences, such as signal-containing and signal-free proteins, may be differentially protected from hydrolase activity in a variety of ways as a result of these structural differences. Thus, protection from enzymic destruction is an ambiguous, as well as an indirect, criterion for determining the compartmental location of a substance. Moreover, "" observations using two ambiguous techniques, detergent solubility and protection, does not necessarily make the conclusion more secure.

One example of this problem of compartmental analysis related to the use of a detergent-soluble fraction to determine the compartmental source of labeled protein. In the classical experiment that demonstrates the "" of protein, a tissue is pulse-labeled with radioactive amino acids, and then chased by a short incubation in a medium containing high concentrations of the same, although unlabeled, amino acid. According to the cisternal packaging concept, the "" in cold media should transfer essentially all of the newly synthesized (labeled) protein into the cisternal spaces from the ribosomes. Under these circumstances, a pulse followed by a brief period of chase, virtually all of the new or labeled protein in the cell should be within the cisternae of the RER. We performed this type of experiment and obtained a microsomal fraction of tissue homogenates.

Was all or most of the labeled protein associated with the microsomes within its intravesicular space as was to be expected, or was a substantial portion external to it? We tested the latter possibility simply by washing our microsomes in the usual suspension medium, resedimenting them, and then removing the ribosomes from the washed vesicles with a detergent. We found that about 10% of the labeled protein was still associated with the ribosomes despite the "," but, more significantly, 75% was removed by a simple wash, and thus was only "" associated with the microsome, either bound at its surface, or capable of crossing the membrane of the vesicle from within to outside. Thus, only 15% of the labeled protein at the most (i.e., assuming that none was tightly associated with granule membrane) was clearly attributable to an intravesicular location. The important point is that it is just such microsomal preparations that provide the evidence for concluding that all of the enzyme is sequestered in accordance with the cisternal packaging theory. In fact, no more than about 15% can be attributed to an intravesicular site. "" accounts for almost 90% of the ""-not the most auspicious of experimental situations.

Table 8.1 : The Distribution of Labeled Protein in the Microsomal Sediment after a 3-Min [^{3}H] Leucine Pulse Followed by a 5-Min Chase

	dpm/g Tissue		% of dpm in First Microsomal Sediment
First microsomal sediment	610,803 ± 81,841	6	(100)
Washed microsomes (3.0 ml resuspension volume)	196,726 ± 46,833	6	32.2
Washed microsomes (30.0 ml resuspension volume)	146,086 ± 13,596	5	23.9
(a) ribosomes	57,704 ± 5,370	5	9.4
(b) DOC-soluble fraction	88,382 ± 8,225	5	14.5

Scheele et al. have drawn a similar conclusion concerning the existence of substantial protein adsorption to microsomes based on studies in which they exogenously added enzyme to microsomal preparations. In accordance with paradigmatic views, they did not entertain the possibility that the membranes might be permeable to the added proteins and explained the association with the microsomal fraction completely in terms of adsorption. Moreover, the conclusion they draw from their observations, in part based on their assumption that no cytoplasmic pool of digestive enzymes exists, was that although much of the enzyme content recovered in microsomal fractions was apparently external to the vesicle, this was a fortuitous artifact. This was because the material initially released from the RER during fractionation was then merely adsorbed to its surface, and the interpretation of earlier cell fractionation studies, which were important elements in generating the cisternal packaging concept in the first place, were not jeopardised by the fact that a major portion of the material was not in the vesicle, but presumably on its surface [Jamieson and Palade reported that 71 to 85% of the labeled protein of collected microsomes is released into solution in an ionic buffer at pH 8.4].

Of course, one might argue that this seems reasonable. It would not be surprising if a large percentage of the RER contents were released during its fragmentation in the homogenizer. However, we must be careful to distinguish between evidence and hypothesis. That is, even if our hypothesis that the protein was in the cisternae in the first place is correct, we cannot use the evidence derived from such experiments as proof that the protein is indeed in the vesicle. That would, of course, be fallacious reasoning. We assume our hypothesis that the material was present in the cisternal spaces in the first instance and released during cell fractionation to be correct. Then we argue that its apparent location on the surface of the vesicle is fortuitous and that our interpretation holds as if they were inside.

The amount of material that we were able to attribute to the intravesicular space at the most accounted for 1.5% of the total labeled protein recovered from the homogenate (in a microsomal fraction that contained 50% of the tissue's RNA) at a time when virtually 100% should have been within these spaces according to the cisternal packaging model. Perhaps in other hands, the number might be 2%, 5%, or even an order of magnitude higher at 10%, but there is no question that the great majority of material that is recovered is not where it should be according to our predictions. Whatever our justifications may be for its assumed natural origin within the cell, this fact cannot be avoided and it highlights the vagaries of this type of cell fractionation study.

We must also turn the question around and ask if it is possible that the material we were able to attribute to intravesicular spaces might not be the remnants of a lost pool at all, but may have entered the vesicles during their formation when the tissue was being homogenised. Certainly, if enzyme is released from the ribosome into the cytoplasm, then some new enzyme would be likely to be trapped within the forming vesicle. We tested this possibility by homogenizing our tissue in a labeled compound that we added, the polysaccharide inulin, and by determining the percentage of the labeled inulin recovered in the portion of the microsomal fraction attributable to an intravesicular pool (the detergent-soluble phase of washed microsomes). If all of the protein had been released from the ribosomes into the cytoplasm, and if they were merely trapped inside the forming vesicle, then the percentage of the tissue's labeled protein and added inulin recovered in this fraction should be approximately the same. When this experiment was performed, the two values were indistinguishable, or about 1.8% of the inulin in the homogenisation medium and 1.5% of the labeled protein of the tissue was recovered as "" content. This is not to say that none of the labeled protein found in the microsomal fraction was present in the cisternal spaces of the intact tissue. It merely points out that we cannot make the statement that it was, based on its mere presence within microsomal vesicles.

Table 8.2 : Distribution of Inulin and Labeled Protein in DOC-Soluble Fraction of Washed Microsomes

	[^{14}C]Inulin, dpm/g Tissue	**^{3}H-labeled Protein, dpm/g Tissue**
DOC-soluble fraction of washed microsomes	113,980±51,590°(3)	88,382±8,225(5)
Homogenate	6,350,360±202,750(3)	5,930,126±794,572 (5)
DOC-soluble fraction as % in homogenate	1.8%	1.5%

Thus, the hypothesis that all or even a substantial percentage of new (labeled) protein enters the cisternal spaces of the endoplasmic reticulum immediately upon its synthesis, cannot be said to be affirmed by this type of evidence. And we certainly cannot use the hypothesis that much of the material was present in the cisternal spaces in the first place and merely lost during homogenisation as evidence for our thesis that the contents of the microsomal fraction reflect its initial presence. Much of what we measure in the microsome is either outside of the vesicle, or, as we shall discuss below, indicates that these vesicles are permeable to proteins contained within them. Since the hypothesis that cisternal packaging occurs is in great part based on the presence of labeled protein in microsomes recovered after labeling tissue in this manner, this hypothesis cannot be said to rest on very firm ground. Finally, the means of determining the compartmental localisation of the proteins-namely, detergent-soluble fractions and protection from proteolysis-as they have been used, are ambiguous, indirect, and uncertain means of determining whether a molecule has moved from one compartment to another.

Secretary Proteins in the Cisternae of the Endoplasmic Reticulum

“”-oriented studies, at least as performed to date, have presented numerous interpretative difficulties, and they do not permit one to draw the conclusion unambiguously that secretory proteins are transported from ribosome to cisternal spaces, no less to conclude how such transport occurs, if it does. Perhaps the second line of evidence concerning the "" of the cisternal spaces of the RER is more compelling. Do they contain secretory products in situ? In considering this subject of once again use the digestive enzymes of the exocrine pancreas as the example and will discuss one chemical and one anatomical approach to the question.

The concept of specific activity enrichment, borrowed from protein chemistry, is often used to evaluate whether or not a substance is a normal constituent of a particular intracellular organelle or compartment. Briefly, the idea is that if a subcellular fraction derived from tissue homogenates comprised primarily of a particular organellar structure is enriched in a specific protein (usually an enzyme and usually expressed in terms of the activity of the enzyme in the fraction relative to the overall protein content of the fraction compared to that in the cell overall), then such enrichment provides supportive evidence for the presence of the protein as a natural constituent in the organelle or compartment. On the other hand, if a cell fraction is impoverished relative to the whole cell, then the converse inference is drawn; that the substance does not appear to be a natural constituent of the cellular structure. Of course, things are not this simple, and there are many problems and uncertainties associated with this technique.

Nevertheless, if we apply this criterion to the presence of the digestive enzymes in the subcellular fraction of the cell derived from the RER, the microsomes, we certainly cannot properly draw the conclusion from such evidence that it is likely that these substances are normally in this part of the cell. The microsomal fraction is not enriched in these proteins, but either contains them at approximately the same concentration as found in the cell overall, or is impoverished. For example, in the rat the concentration of digestive enzymes in the microsomal fraction (per milligram of microsomal protein) is between one-third and two-thirds of that in the homogenate overall depending upon the enzyme and circumstances, and no more than one-fourth of their concentration in the zymogen granule fraction of the tissue homogenate. This value the -90% of the digestive enzyme content of such fractions that is either ribosomal or apparently external to the vesicle, and that cannot be attributed to the original contents of the RER. An objective view of these data requires one to draw the conclusion that they provide no basis for the position that the digestive enzymes are normal constituents of the cisternal spaces of the RER.

Although it has not been demonstrated that these products are concentrated in this location based on cell fractionation studies, particularly those who are active in this field, that even though they might agree about the objective facts of data, they would feel that such a negative conclusion is not warranted. "," they would say, "much of the contents of the RER may have been released during ." And so it may, but how is one to know that the digestive enzymes were part of these contents in the first place, or if so, to what extent? Certainly, if we found a low concentration of (LDH) (to choose a cytoplasmic enzyme) in the microsomal fraction, we would not conclude that it had been lost from the RER due to fragmentation during homogenisation. We would not draw this conclusion, because we do not believe that LDH is a component of this part of the cell in the first place. If we are comfortable with the view that the digestive enzymes were in the cisternae of the RER in the first instance, then our comfort cannot be based on the evidence, but must either be derived from a belief in the correctness of our hypothesis regardless of the evidence, that is, based on our assumptions about the process (for example, membrane impermeability to proteins), or on other evidence, most notably that derived from the anatomical approach to the question of the contents of the RER.

The anatomical approach involves attempts to label secretory proteins in one way or another so that they can be identified in electron microscopic sections through cells. It is commonly thought that autoradiography, in which isotopic emissions from radioactively labeled compounds in a tissue section expose a photographic emulsion immediately overlying the section, has demonstrated the presence of secretory proteins in the cisternae of the RER. Unfortunately, this technique is not capable of making spatial distinctions of the necessary resolution, and autoradiographs, at least as applied to this point, are only capable of showing that labeled enzyme is in the basolateral region of the cell. It cannot distinguish between its presence within the cisternae, on the reticular membrane, or between cisternae in the cytoplasm.

More recently, antibodies to secretory proteins, labeled with agents such as ferritin or gold that produce small particles on the surface of tissue sections where they bind to the antigen, have been used in attempts to identify proteins within the cisternae of the RER. This technique has better spatial resolution than autoradiography, but it has many potential technical problems of its own, most notably the loss or redistribution of material from the surface of the section during preparation for viewing.

Results using this technique have thus far been no more reassuring than cell fractionation data regarding the digestive enzyme contents of the cistemal spaces. Neither of two major published studies present a convincing case for their presence in these spaces, no less their presence in substantial quantities. Kraehenbiihl et al. failed to find a significant number of particles over the cisternal spaces, but believing that they should be found, the authors concluded that the protein must have been lost during tissue treatment, although they provided no evidence for this view. In Bendayan et al. grains were found over the endoplasmic reticulum at between 10 and 50%, depending upon the enzyme, of the concentration seen over the zymogen granules. However, the values were for particles over the "" cytoplasm, an area that not only includes the cisternal spaces, but ribosomes, reticular membranes, and cytoplasm as well. The authors of this study present one high-magnification micrograph of the endoplasmic reticulum from which we can estimate the number of particles over the cisternal spaces, relative to those associated with ribosomes or reticular membrane, as well as in the intercisternal cytoplasm. By far, the great preponderance of particles appears to be associated with the ribosomes and membrane, and no more than about 25 to 33% can be relatively unambiguously located over the cisternal spaces, at least if we assume that no particles that appear over such spaces are associated with membranes sectioned "." If we correct the measured particle density so that only those particles clearly associated with cisternal spaces are included, then for the nine different enzymes measured, it was either the same as or less than that found over the nucleus or mitochondria.

This is not to say that the digestive enzymes are not in the cisternal spaces of the endoplasmic reticulum, nor even to say that they are not concentrated there, but rather to point out that the anatomical evidence for such a pool is as weak and uncertain as the chemical evidence. Thus, neither the ""-oriented studies, nor the ""-oriented studies provide convincing evidence for the existence of a cisternal pool, no less for the accumulation of digestive enzyme at this location. At best, we can say that it may be so and that our hypothesis has not been conclusively shown to be false; that is, we cannot say that it is clear that there is no digestive enzyme stored within the cisternal spaces. In any event, in the end, at least for the pancreas, the tissue for which the cisternal packaging theory was developed, the evidence does not allow us to assert that there is a natural and substantial pool of digestive enzyme in the cisternal spaces of the endoplasmic reticulum, no less that all secretory proteins are sequestered in its spaces upon their synthesis.

Therefore, the central underpinning of the signal hypothesis, the existence of the transport process for which it proposes a mechanism, cannot be said to have been established with a reasonable degree of certainty in the exocrine pancreas. It is not clear beyond a substantial doubt that these secretory products are within the cisternal spaces, nor that they cross the membranes of the RER, nor that if they do, that all secretory proteins are to be found in these spaces soon after their synthesis.

PERMEABILITY TO DIGESTIVE ENZYMES

Beyond its central assumption that the membranes of the endoplasmic reticulum are permeable to secretory proteins, the signal hypothesis assumes, that this permeability is expressed across the membrane between a ribosomal compartment of protein and the cisternal spaces of the RER, and does not involve the passage of proteins through the cytoplasm en route. In this way, only the extending or elongating peptide chain crosses the membrane; that is, only it displays permeability to the membrane, and not the formed protein. In addition, the signal hypothesis assumes that the permeability of the membrane to proteins, if such it is, is unidirectional, and that once the protein is within the cisternal spaces it cannot escape. Neither of these assumptions, that is, no cytoplasmic passage and unidirectional transport, are supported by direct experimental observation.

Assuming that the reticular membrane is permeable to secretory proteins, determining their route of transport requires a kinetic compartmental analysis of that transport. Do new protein molecules pass into the cytoplasm or the medium bathing isolated microsomes prior to entering the cisternal spaces? To answer this question, one must measure the timecourse of the passage of the proteins through the putative compartments. We cannot infer the route of passage by measuring the end-state distribution of material, nor can we accept as appropriate discounting a priori the presence of new protein in the medium bathing suspensions of microsomes as merely being due to an artifact of the system, without requiring that this ad hoc hypothesis be demonstrated to be correct and complete. Indeed, as I have discussed, the evidence seemingly does not permit one to draw that conclusion.

Moreover, there is evidence that cytoplasmic passage occurs during the insertion of some membrane and other intracellular proteins. That is, the formed protein with signal sequence is released from the ribosome into the cytoplasm, and then enters the membrane from this pool spontaneously as the result of a thermodynamically favorable partitioning between cytoplasmic and membrane phases. That is, insertion does not occur, as the signal hypothesis envisions it, in the absence of intervening cytoplasmic passage and the development of higher-order structure within the protein.

This suggests that the converse event also occurs; that is, proteins leave the membrane spontaneously. This concept of the insertion and removal of membrane proteins is appealing in its simplicity and can account for differences in the turnover rate of different membrane proteins, which is otherwise difficult to account for without complex ad hoc hypotheses, namely, if one views the turnover of membrane proteins as being part of a general turnover of membrane segments rather than of individual protein molecules.

This brings us to the second assumption: that the protein, once having entered the cisternae, is trapped in that space and cannot leave of its own accord. This is an unsubstantiated hypothesis. Just as the appearance of a newly manufactured protein in the medium bathing synthesizing microsomes cannot be properly dismissed as an artifact without the necessary experimental proof, so the release of new protein from microsomes as a function of washes, and so on, cannot be assumed to be due to desorption from its surface alone, assuming that occurs, and not transmembrane movement. This may or may not be the case, and the hypothesis that these membranes are bidirectionally permeable to their contained proteins, much as proteins may be inserted and removed from membranes spontaneously, is a hypothesis that requires experimental test and not prejudgment. In fact only one group has attempted to distinguish between the two possibilities, and this group concluded that the membranes were bidirectionally permeable to secretory proteins.

Thus, granting the permeability of reticular membrane to proteins, it has not been established that such transport does not involve , nor that it is irreversible. These are not only unsubstantiated hypotheses, but there are experimental observations consistent with both cytoplasmic passage and reversible transport, although such observations have generally been discounted, often by untested ad hoc hypotheses that might explain them, at least to some extent, in ways that are consonant with paradigmatic precepts.

TESTS OF THE PARADIGM

The vesicle or cisternal packaging–model was perhaps the first paradigm to be spawned by the relatively new discipline of cell biology. As a paradigm it is more than just a popular model for a particular process. It is a complex, although incomplete, system of knowledge and belief, which is only in part a specific theory that makes explicit predictions about a process. It is associated with certain kinds of experimental and technical approaches to the study of cells that includes electron microscopy, cell fractionation, and electron microscopic autoradiography. Furthermore, and perhaps most important for our discussion, it makes certain assumptions about the nature of cells, some of which we have already considered. The cisternal packaging-exocytosis model is itself a model for other models; and as such it embodies a gestalt of the cell, how it functions, and how one goes about learning its secrets. It views the cell as an extremely complicated factory crammed full of machines of various sorts that account for the movement of the cell's chemical products hither and yon. For this reason we should not only ask whether it provides us with a relatively satisfactory explanation for the secretion of protein by the pancreas and other protein secreting cells, but whether or not it is an adequate for the range of cell functions and their study that it touches upon. This is the broader question, and no doubt the more important one.

The work on secretion and protein transport has gradually led us to question the cisternal packaging-exocytosis paradigm and to formulate an alternative general view. This alternative, the , has two primary sources. The first is a series of negative tests of the exocytosis model in the mammalian pancreas. A discussion of some of these will be the major focus of the chapters that follow. These negative tests obliged us to develop, if we could, an alternative model that could more successfully explain the behaviour of this system.

The other source of the equilibrium theory was the belief that another paradigm, a for the transport of molecules across membranes, had deeper theoretical roots and needed to be experimentally tested for goodness of fit to the observed kinetics and thermodynamics of protein transport. A diffusion-based model has precedence because it embodies a more fundamental conceptualisation of transport events, and for this reason has been the major experimental and theoretical tool guiding the study of biological transport for more than a century. It is founded on two central beliefs. The first is that simple physical explanations for the movement of molecules must be favored over the more complex alternative of chemical or reaction-mediated movement, unless and until the evidence requires us to consider such processes. The second is that observed molecular movements should be assumed to result from the intrinsic kinetic energy of the moving molecule itself, and not from an external energy source of some kind, unless and until the evidence requires that we consider such complex processes.

When a diffusion-based hypothesis is applied to a process such as the transport of protein by the pancreatic acinar cell, it becomes apparent that in addition to its more general character, such an hypothesis is superior in at least one important additional respect. Although unambiguous experimental proof may be a chimera, the equilibrium hypothesis presents us with a crystal-clear predictive situation when compared with the murky case of the cisternal packaging—. This is so because all diffusionbased models stem from a single, well-established mathematical relationship, the Fick diffusion and similar equations, which correlate the observed flux of a molecule to the variable relevant to that flux, namely, concentration gradient, as well as to the phenomenological constraints of the system such as surface area and diffusivity. These established relationships make predictions clear and permit us to test the hypothesis in a fashion that allows for the falsification of its premises, leads to quantitative predictions for various specific models derived from it, and permits an orderly development of hypotheses. Of course, a in these terms has no greater hold on truth or reality, but it does offer a rational path to follow. It offers the opportunity of starting with the simplest possible case, and being drawn to the more complex only by gained knowledge of the system.

The equilibrium hypothesis has yet another value. It can be viewed as antithetical to the cisternal packaging-exocytosis model. This permits us to ask whether or not a particular observation, consistent with the cisternal —, is not also perfectly consistent with its logical antithesis. Without the foil of an antithetical statement we might mistakenly think that a particular experimental inference lent considerable weight to our theory, whereas such a foil might disclose that our inference was very weak indeed since the observation fit not only our model but an opposing conjecture as well. Similarly, the existence of an offers us the possibility of designing experiments to distinguish between the two views. Thus, there is particular strength in being able to test one model against another.

Perhaps this can be appreciated better by considering the way in which the equilibrium model helps us formulate tests of the cisternal packagingexocytosis model; using it as a comparative case. In this way, it requires clarification of the paradigm. For example, a diffusion-based model predicts that: (1) the movement of molecules is bidirectional across planar membranes, and is therefore experimentally reversible; and (2) the rate of transport can be altered by altering the chemical activity gradient for the transported substrate. These, of course, are merely two aspects of equilibrium behaviour characteristic of diffusive processes. The exocytosis model can be distinguished from a diffusive construct in terms of these two predictions. It views transport as "" or unidirectional; that is, as a process incapable of reversal after membrane fusion. Moreover, it proposes that the rate of transport is not a function of the extant concentration gradients for the transported molecules, but is instead a function of an external force or event which leads to movement (namely, membrane fusion-fission).

In addition to these two rather general predictions, there is a third that can help us distinguish between a diffusion-based and exocytosis explanation for transport, at least for the particular cell that we will be considering. In the acinar cell much of the digestive enzyme is stored in secretion or zymogen granules, and therefore secretion, as envisioned in an equilibrium construct would occur, at least in part, as the result of the movement of protein from zymogen granule to cytosol, and then out of the cell. That is, secretion can involve three compartments, at least when there is a substantial amount of enzyme in a storage pool. The exocytosis model, on the other hand, is clearly different in this regard and proposes that the transfer of matter only involves two compartments, zymogen granule and extracellular space (the duct system in the case of the pancreas).

Thus, as our starting point, we can define the equilibrium hypothesis simply as the application of the diffusion equation to a three-compartment system. More complex constructs may have to be considered, but only as the evidence requires. As already discussed, the cisternal packagingexocytosis model presents us with a much more complex picture. Unlike a simple diffusion hypothesis, it offers an incomplete structural framework that is often rather difficult to 'put to the test. Predictions are not always, or perhaps not even usually, clear or explicit, even if they are implicitly so. This vagueness protects the paradigm from ready falsification. Although most of the predictions that will consider below have been put forth explicitly in one place or another, some are based necessarily implicit paradigmatic belief to be.

To the degree that my view does not accurately reflect that generally held in one specific area or another, it has seemed that way to me. In any event, setting predictions for the paradigm can still be a useful tool to help us organize these ideas and define what the paradigm does indeed predict about events. If predictions about one aspect or another of the system's behaviour cannot be made in the paradigm's terms, then this reflects upon its weakness as a model.

The is more than a complicated grouping of hypotheses, and seems to involve deeply embedded beliefs, not unlike other widely held scientific theories. Perhaps this aspect of the paradigm can be appreciated by considering the role of conjecture in maintaining a paradigmatic perspective in the face of what appears to be negative tests of its validity. Such negative tests are often countered by ad hoc hypotheses; that is, hypotheses that are proposed to account for discrepant observations within the borders of the paradigm. They modify the paradigm, almost invariably by making the model more complex. It is usually and perhaps uniformly true that one can explain negative tests of a paradigm by the addition of such new, ad hoc elements, and in this way protect the larger model from falsification. Of course, at some point such modifications can no longer be integrated and the model remain intact. It is interesting that these conjectures often are not actually tested or thought necessary to test. The mere fact that one can imagine a way to explain a discordant observation in paradigmatic terms may be thought sufficient in itself.

9

Quantitative Estimation of Hydrolases

Colour changes are the most convenient way to follow enzymatic reactions, but most enzymatic reactions do not directly cause a colour change. (Reactions that consume or generate NAD(P)H are a notable exception, making them among the most widely used enzyme assays.) Chromogenic substrates conveniently introduce a colour change into many reactions, but at the cost of using an artificial rather than a true substrate. One of the best solutions is to use the true substrate and then to couple the reaction of the substrate to a colour change reaction. Many such coupled reaction assays have been developed. The focus in this chapter will be on reactions linked to pH changes, which are conveniently detected using pH indicators.

For many applications, substrate selectivity is the characteristic that distinguishes a useful enzyme from a non-useful one. To measure substrate selectivity, one must compare the reactions of two substrates. There are two ways to do this using colour change reactions – separately or in the same reaction mixture.

Measuring the rates of reactions of two substrates separately and then comparing these rates is the simplest way to measure selectivity. However, this approach only yields an estimate of selectivity. Enzyme selectivity stems from differences in the binding of the two substrates to the active site and from differences in the reaction rates of the two substrates. Measuring the reaction rate of each substrate separately may overlook differences in the binding of the two substrates to the active site.

Measuring selectivity by simultaneously measuring the reaction rates of both in the same reaction mixture seems at first to be a very special case, which would rarely apply to real world examples. However, by using a reference compound and two measurements, this approach can be applied to a wide range of compounds and, unlike separate measurements, can yield accurate selectivity values. The idea is to first measure the selectivity of the one true substrate and the reference compound. This reference compound is chosen to be convenient to detect and may be a chromogenic substrate. Reaction of the true substrate may be detected by one colour (e.g. a pH indicator), while reaction of the reference compound is detected by another colour (the chromophore released upon reaction). This competitive experiment gives the true selectivity for this pair of substrates. The second step is to measure the selectivity of the second substrate toward the same reference compound. This yields a second true selectivity. Finally, since the reference compound was the same in both experiments, dividing the two gives the correct selectivity for the two true substrates. We refer to such colourimetric methods where the substrate selectivity is determined not by direct competition of the substrates with each other but in a two-step procedure by comparing reaction of each substrate with a reference compound, as "Quick selectivity methods." This chapter will provide examples of using this Quick selectivity approach to measure enantioselectivity, diastereoselectivity as well as substrate selectivity.

DIRECT ASSAYS USING CHROMOGENIC SUBSTRATES

The simplest and most convenient assays are those involving a chromogenic substrate, that is, one where the reaction yields a product of a different colour from the starting material. The most common chromogenic substrates to assay hydrolase activity are the *p*-nitrophenyl derivatives. In each case the *p*-nitrophenyl derivatives mimic the natural substrate of the hydrolase while adding a *p*-nitrophenyl moiety at the cleavage site. For lipases, phospholipases and esterases, these are esters of *p*-nitrophenol; for proteases, these are amides of *p*-nitroaniline, and for glycosidases, these are glycosides of *p*-nitrophenol.

Other common chromogenic substrates are resorufin or 1-naphthol derivatives. Resorufin analogs have been used previously to spectrophotometrically monitor the activity of enzymes including galactosidases, cellulases, lipases, proteases, esterases, and phosphol-

Lipase or esterase

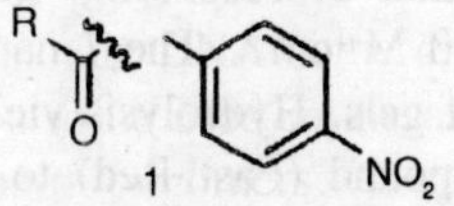

ρ- Nitrophenyl ester

Phospholipase C

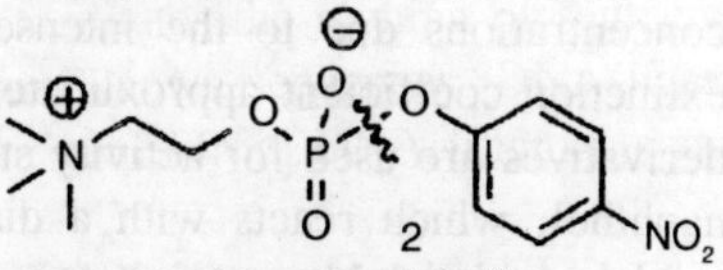

r- Nitrophenylphosphoryl choline

Trypsin

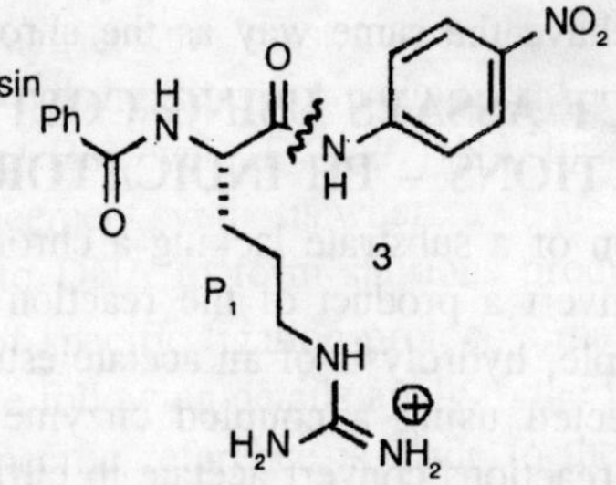

N-a- Benzoyl-L-arginine 4-nitrophenyl anilide

Chymotrypsin, subtilisin

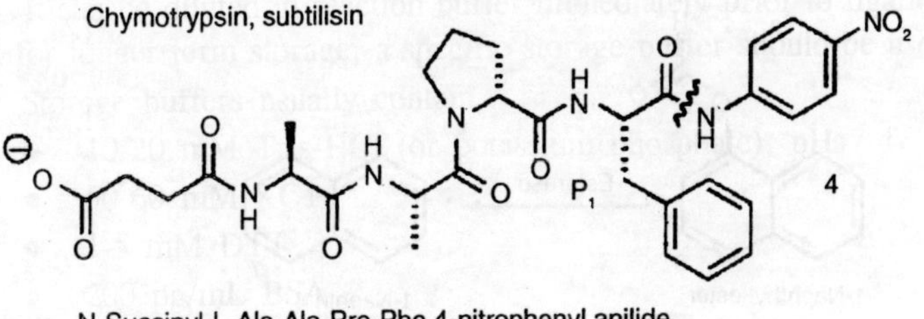

N-Succinyl-L-Ala-Ala-Pro-Phe 4-nitrophenyl anilide

β-Glucosidase

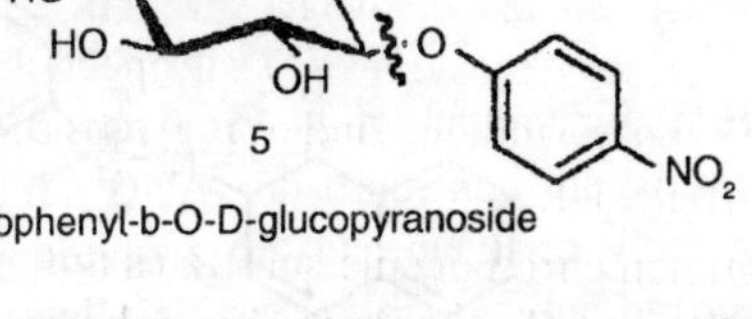

r-Nitrophenyl-b-O-D-glucopyranoside

Figure 9.1 : Typical p-nitrophenyl derivatives for chromogenic assay of hydrolase activity. Each substrate includes a p-nitrophenyl moiety in the normal substrate in such a way that hydrolysis releases p-nitrophenol or p-nitro-aniline, which can be detected by their yellow colour. The wavy line marks the bond cleaved during hydrolysis.

ipases and are particularly useful for detecting activity at low enzyme concentrations due to the intense absorption of resorufin, with an extinction coefficient approximately 70 000 $M^{-1}cm^{-1}$. The 1-naphthol derivatives are used for activity staining of gels. Hydrolysis yields 1-naphthol, which reacts with a diazo compound (Fast Red) to yield red-brown insoluble precipitate.

The disadvantage of chromogenic substrates is that they only mimic the true substrate of interest. One is never sure whether the true substrate will behave the same way as the chromogenic mimic.

INDIRECT ASSAYS USING COUPLED REACTIONS – PH INDICATORS

To detect reaction of a substrate lacking a chromogenic moiety, it is necessary to convert a product of the reaction into a coloured compound. For example, hydrolysis of an acetate ester by a lipase or esterase can be detected using a coupled enzyme assay. Several sequential enzymatic reactions convert acetate to citrate accompanied by the formation of NADH, which absorbs at 340 nm. Another example, suitable for monitoring enzyme-catalysed transesterifications

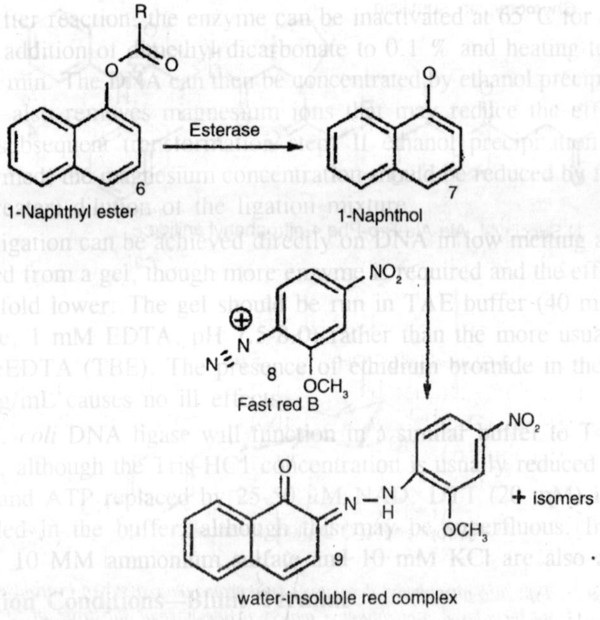

R
O
O
Esterase
O
6
7
1-Naphthyl ester
1-Naphthol
NO_2
N
N
8
OCH_3
Fast red B
NO_2
O
N – N
H
+ isomers
OCH_3
9
water-insoluble red complex

Figure 9.2 : Action of an esterase releases 1-naphthol which reacts with the diazo compound Fast Red B to give a water-insoluble azo dye.

in organic solvent, detects the acetaldehyde released upon transesterification of an alcohol with a vinyl ester to form ester and acetaldehyde. The acetaldehyde reacts with 4-hydrazino-7-nitro-2,1,3-benzoxadiazole to yield a strongly fluorescent compound. Both of the examples above are limited to certain substrates: acetate esters in the first case, transesterifications with vinyl esters in the second case.

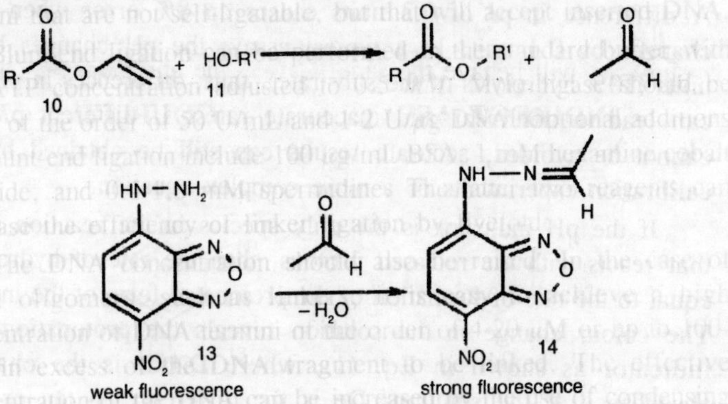

Figure 9.3 : Assaying the acylation of an alcohol with a vinyl ester in organic solvent relies on the detection of acetaldehyde.

A more general assay relies on pH indicators. Hydrolysis of an ester at neutral pH, for example, solketal butyrate (butanoate ester of 2,2-dimethyl-1,3-dioxo-lane-4-methanol) to form solketal below, releases a proton. Detecting this proton release with a pH indicator allows one to monitor a wide range of ester hydrolyses. Researchers have used pH indicators to monitor the progress of enzyme-catalysed reactions that release or consume protons since the 1940s. For example, researchers have monitored reactions catalysed by amino acid decarboxylase, carbonic anhydrase, cholinesterase, hexokinase and ester hydrolysis by proteases. In some cases, researchers used a pH indicator assay qualitatively, but in other cases, the colour change was proportional to the number of protons released. This quantitative use requires a calibration with additional experiment or a careful choice of reaction conditions.

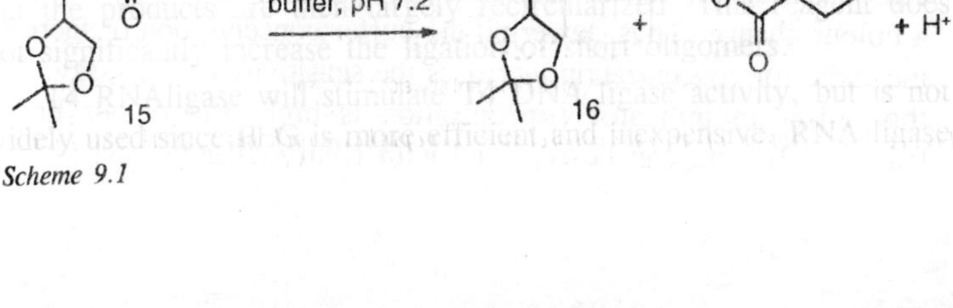

Scheme 9.1

Overview of Quantitative Use of pH Indicator Assay

The most important variables in the pH indicator assay are the choices of buffer and pH indicator. Both the buffer and the indicator must have the same affinity for protons (pK_a values within 0.1 unit of each other) so that the relative amount of buffer protonated and indicator protonated stays constant as the pH shifts during the reaction. A difference in pK_a of 0.3 units causes an 8% error when the pH changes by 0.1 unit. In a typical assay, the pH changes by 0.05 units (10% hydrolysis of the substrate), thus, differences in pK_a values can lead to nonlinear and inaccurate rates. If different pK_a values cannot be avoided, accurate results can still be obtained by using calibration experiments or a more complex equation.

If the pH indicator is the only species in the reaction solution that reacts with the released protons, then the rate of hydrolysis is equal to the rate of formation of the protonated form of the indicator. The colour change in the solution reveals the protonation of the indicator as shown in Eq. (1), where D*A*/d*t* is the change in absorbance of the solution, De is the difference in the extinction coefficient of the protonated and nonprotonated forms of the indicator and *l* is the path-length of the solution.

$$\text{rate} = \Delta[\text{indicator}\cdot\text{H}^+]/\text{time} = \frac{\Delta A/dt}{\Delta_\varepsilon \cdot l} \quad (1)$$

In practice, solutions containing only indicator are difficult to use because the colour changes too readily due to small variation in conditions. For this reason, researchers usually add some buffer to the reaction mixture. Under these conditions, the rate of hydrolysis equals the rate of formation of the sum of the protonated indicator and the protonated buffer.

$$\text{rate} = \Delta[\text{indicator}\cdot\text{H}^+]/\text{time} + \Delta[\text{buffer}\cdot\text{H}^+]/\text{time} \quad (2)$$

When the pK_a values of the indicator and buffer are the same, then released protons partition between the indicator and buffer according to the ratio of their concentrations. The highest sensitivity (largest d*A*/d*t*) occurs with less buffer. The protons released add to either the buffer, giving no colour change, or to the indicator, giving a colour change. Thus, lowering the buffer concentration or increasing the indicator concentration increases the sensitivity of the assay. When the buffer to indicator concentration is high, the (1 + [buffer]/[indicator]) term can be replaced with ([buffer]/[indicator]):

$$\text{rate} = \frac{\Delta A/dt}{\Delta\varepsilon \cdot l} + \frac{[\text{buffere}]}{[\text{indicator}]}\frac{\Delta A/dt}{\Delta\varepsilon \cdot l} = \frac{\Delta A/dt}{\Delta\varepsilon \cdot l}\left(1 + \frac{[\text{buffer}]}{[\text{indicator}]}\right) \quad (3)$$

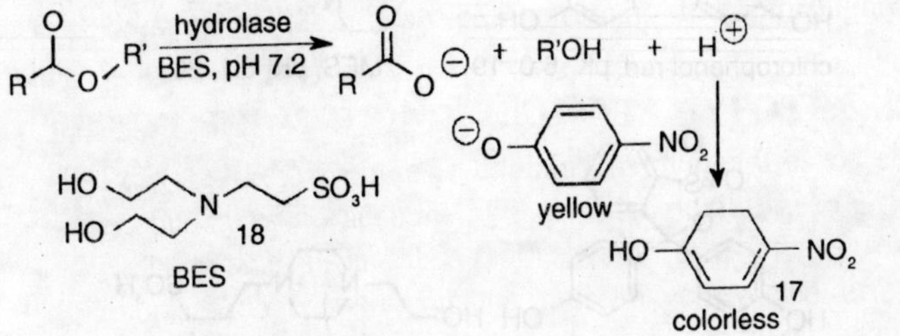

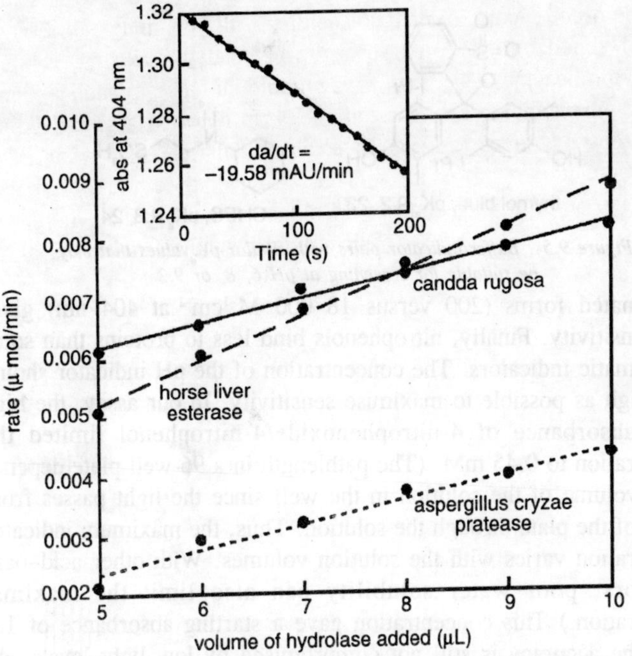

Figure 9.4 : 4-Nitrophenol as a pH indicator to detect hydrolysis of esters. Hydrolysis of an ester releases a proton, which protonates the yellow 4-nitrophenoxide ion.

Since most hydrolases have maximal activity near neutral pH, we developed the assay for pH 7.2. As a pH indicator, we chose 4-nitrophenol 17.

The similarity of its pK_a(7.15) to the pH of the reaction mixture ensures that changes in pH give a large and linear colour change. The large difference in the extinction coefficients of the protonated and

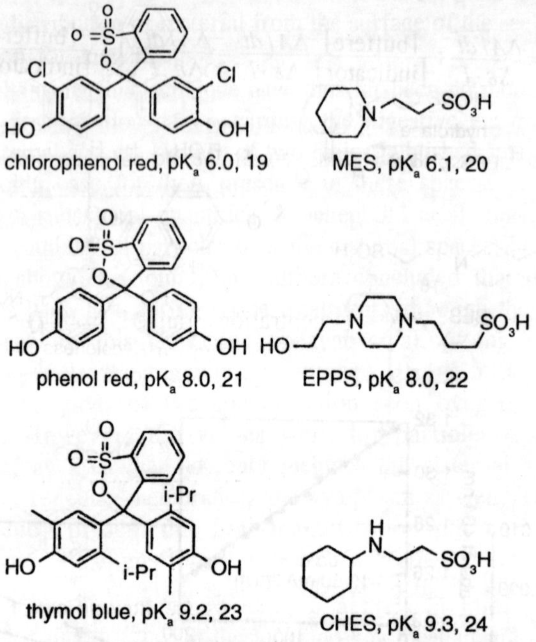

chlorophenol red, pK_a 6.0, 19

MES, pK_a 6.1, 20

phenol red, pK_a 8.0, 21

EPPS, pK_a 8.0, 22

thymol blue, pK_a 9.2, 23

CHES, pK_a 9.3, 24

Figure 9.5 : Buffer/indicator pairs with similar pK_a values that may be suitable for screening at pH 6, 8, or 9.2.

deprotonated forms (200 versus 18 000 $M^{-1}cm^{-1}$ at 404 nm) gives good sensitivity. Finally, nitrophenols bind less to proteins than some polyaromatic indicators. The concentration of the pH indicator should be as high as possible to maximise sensitivity. In our assay, the high initial absorbance of 4-nitrophenoxide/4-nitrophenol limited the concentration to 0.45 mM. (The pathlength in a 96-well plate depends on the volume of the solution in the well since the light passes from the top of the plate through the solution. Thus, the maximum indicator concentration varies with the solution volumes. With other acid–base indicators, poor water solubility can also limit the maximal concentration.) This concentration gave a starting absorbance of 1.2 where the accuracy is still not compromised by low light levels. As a buffer, we chose BES (*N,N*-bis[2-hydroxyethyl]-2-aminoethanesulfonic acid) because its pK_a (7.15) is identical to that of 4-nitrophenol. We chose a buffer concentration of approximately 5 mM as a compromise between low buffer concentrations to maximise sensitivity, and high buffer concentrations to ensure accurate measurements and small pH changes throughout the assay (< 0.05 pH units for 10% hydrolysis at our conditions). The small pH changes are important because kinetic constants can change with changing pH.

Other buffer/indicator pairs may be suitable for screening at other pHs. For example, at pH 6 chlorophenol red (pK_a6.0) and MES (2-[*N*-morpholino]-ethanesulfonic acid, pK_a6.1) may be suitable; at pH 8 phenol red (pK_a 8.0) and EPPS (*N*-[2-hydroxylethyl]piperazine-*N*'-[3-propanesulfonic acid], pK_a 8.0) may be suitable; at pH 9 thymol blue (pK_a9.2) and CHES (2-[*N*-cyclohexylamino]ethanesulfonic acid, pK_a9.3) may be suitable.

Suitable substrate concentrations ranged from 0.5 to 2 mM, typically 1 mM. At substrate concentration below 0.5 mM, the absorbance changes are too small to be detected accurately. For example, hydrolysis of 5% of a 0.25 mM substrate concentration at our standard conditions (pH 7.2, 0.45 mM 4-nitrophenol, 5 mM BES), changes the absorbance by only 0.005 absorbance units. Solubility in water sets the upper limit of substrate concentration because spectrophotometric measurements require clear solutions. Typical organic substrates dissolve poorly in water, so we added organic cosolvent–7 vol% acetonitrile. For very insoluble substrates, we prepared clear emulsions using detergents. The assay tolerates small changes in reaction conditions, such as the addition of 7% acetonitrile. Indeed, the p$K_a$of 4-nitrophenol (**17**) changes only slightly from 7.15 to 7.17 upon addition of 10% ethanol. This result suggests that cosolvent concentrations below 10% do not compromise the accuracy of the assay. Also, small amounts of salts present in the hydrolase solutions (buffer salts in commercial hydrolase preparations, 2 mM $CaCl_2$in the protease solutions) did not affect the accuracy.

Applications

The pH indicator-based assay is a quick way to measure whether an enzyme accepts a particular substrate. This measurement quickly eliminates unsuitable enzymes or unsuitable substrates from more complicated subsequent experiments. It is most useful with unusual substrate/enzyme combinations where only a few are likely to work.

Searching for an Active Hydrolase

One application of this assay is for a rapid survey of commercial enzymes to find those that accept an unnatural or unusual substrate. In one project, we screened a library of 100 commercial hydrolases to find the two that catalysed hydrolysis of hindered spiro compound **25** that was useful as a chiral auxiliary. This screening identified several candidate hydrolases. Subsequent confirmation of this activity and optimisation of the reaction led to a subtilisin-catalysed preparative scale resolution of this auxiliary. In a second project, we screened a library of 72 commercial hydrolases (lipases, esterases, and proteases) towards solketal butyrate, a chiral building block. Most hydrolases reacted with this less hindered substrate so this initial screen eliminated only 20 of the 72 candidate enzymes from the next stage of screening. In a third project, we again assayed a library of 100 hydrolases to find the few that catalysed hydrolysis of the amide link in *N*-acyl

(±)- cis, cis-isomer, rac-25 2.8 g —subtilisin cartsberg, E > 200→ (1R,5R,6R)-25 1.3 g (44% yield) 99% ee + (1S, 5S, 6S)-26 1,1 g (46% yield) 99% ee

sulfinamides to form sulfinamides. Although hydrolysis of amides usually does not work in this assay because it does not release a proton, the *N*-acyl sulfinamide was a special case. The sulfinamide released is not a strong base like an amine and does not take up the proton released from the ionisation of the acid, so the pH indicator assay was suitable for this special amide hydrolysis.

N-acyl sulfinamide, 27 —hydrolase→ 28 + $^{\ominus}$O(C=O)R + $H^{\oplus}$

Other researchers recently used similar pH indicator assays to measure the activity of glycosyl transferase, haloalkane dehalogenase, and nitrilase.

Substrate Mapping of New Hydrolases

Another application of this assay is mapping the substrate range of newly discovered hydrolases, such as those from thermophiles. Advances in microbiology and molecular biology have made hundreds of new enzymes available from unusual microorganisms such as thermophiles. These enzymes may be useful for organic synthesis because they allow the use of higher temperatures, where reactions are faster and substrates are more soluble. In addition, these enzymes may also tolerate other unusual conditions such as high concentrations of organic solvents. To apply these new enzymes in organic synthesis, one needs some idea of their substrate range and selectivity. The pH indicator-based assay is quick way to map the substrate range of new enzymes. Recently, we, and others, used this pH indicator assay to map over 20 different esterases using the substrate library of more than 50 esters.

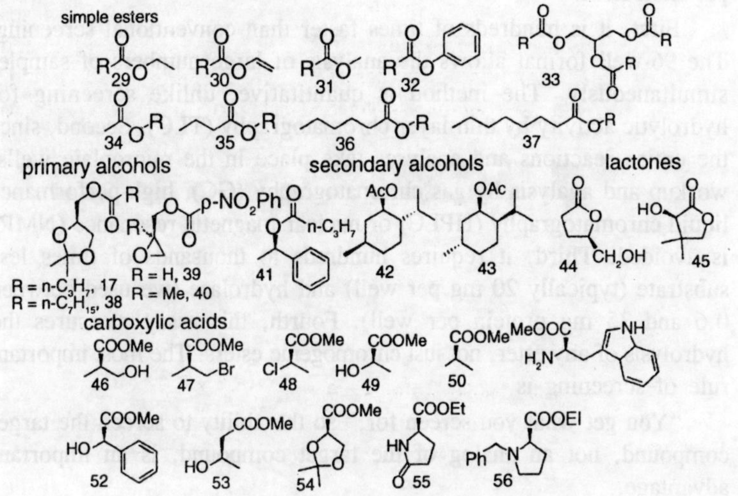

Figure 9.6 : Examples of esters used to survey the substrate selectivity of esterases.

This substrate mapping first grouped each substrate/esterase combination as slow (£ 1 mmol ester hydrolysed mg^{-1} protein min^{-1}), good (1–10) or very good (> 10). The 19 esterases and 31 substrates gave a total of 589 substrate/esterase combinations. Examination of these data yielded several conclusions. The best substrates were activated esters, which were also chemically the most reactive in the library: vinyl esters and phenyl esters. Among the vinyl and ethyl esters with different chain lengths, most of the esterases appear to favor the hexanoate or octanoate, while one esterase (E018b) appeared to strongly favor acetate esters. Esters with a sterically hindered acyl group (vinyl pivaloate **30**, *R* =*t*-butyl or vinyl benzoate **30**, *R* = Ph) and polar esters, e.g. methyl 2-hydroxyacetate **31**, *R* = CH_2OH, were usually poor substrates. Three pairs of esterases (E010/E011, E013/E014, E019/E020) showed similar reaction rates with all substrates and therefore might be very similar or even identical enzymes. Thus, this initial substrate mapping provided a good overview of substrate range and ideas for the further refinement of these general conclusions.

This further refinement usually involves a quantitative measure of substrate selectivity, where one includes reference compounds in the screening.

Comparison with Other Methods

There are several advantages with this screening method using pH indicators.

First, it is hundreds of times faster than conventional screening. The 96-well format allows the analysis of large numbers of samples simultaneously. The method is quantitative, unlike screening for hydrolytic activity by thin-layer chromatography (TLC). Second, since the entire reactions and analyses take place in the microplate wells, workup and analysis by gas chromatography (GC), high-performance liquid chromatography (HPLC) or nuclear magnetic resonance (NMR) is avoided. Third, it requires hundreds to thousands of times less substrate (typically 20 mg per well) and hydrolase (we used between 0.6 and 35 mg protein per well). Fourth, this assay measures the hydrolysis of any ester, not just chromogenic esters. The most important rule of screening is

"You get what you screen for," so the ability to screen the target compound, not an analog of the target compound, is an important advantage.

There are also few disadvantages with our screening method. This assay is approximately seven times less sensitive than one using hydrolysis of 4-nitrophenyl esters. For example, if the rate of hydrolysis towards a nonchromogenic ester and a 4-nitrophenyl ester were identical, then our assay would require seven times more hydrolase to observe the same change in absorbance.

The assay with 4-nitrophenyl esters releases one molecule of 4-nitrophenol (53% of these will be deprotonated at pH 7.2), while our assay protonates one 4-nitrophenoxide for every 12 protons released. Second, this assay requires clear solutions. To obtain clear solutions with water-insoluble substrates, experimentation is sometimes required to find the best cosolvent or emulsion conditions.

Third, the pH indicator method requires a reaction that generates or consumes protons, so it cannot be used for most amide hydrolysis or glycoside hydrolysis.

ESTIMATING AND MEASURING SELECTIVITY

One of the most useful characteristics of enzymes is their selectivity. The true selectivity of an enzyme toward a pair of substrates is the ratio of the specificity constants (k_{cat}/K_m) for each substrate. For example, the enantioselectivity (or enantiomeric ratio, *E*) of an enzyme is the ratio of the specificity constants for the enantiomers.

$$\text{Enantiomeric ratio} - E - \frac{(K_{cat}/K_m)_{\text{fast enantiomer}}}{(K_{cat}/K_m)_{\text{slow enantiomer}}} \quad (4)$$

Other examples of selectivity are substrate selectivity (e.g. selectivity for different acyl chain length among esters), regioselectivity (e.g. selectivity for one ester of several esters in a nucleoside derivative), and diastereoselectivity (e.g. selectivity for a cis isomer over a trans isomer). It is rarely convenient to measure selectivity by measuring the kinetic parameters for each substrate. Such measurement requires multiple kinetic studies, which are slow and tedious. In addition, combining four such measured values, each with an uncertainty of measurement, yields selectivities with large uncertainties.

The best methods to measure selectivities are competitive experiments. A mixture of both substrates competes for the enzyme active site and is converted to product. Measuring the amounts of each product formed reveals the selectivity.

One such method is the enantioselectivity determination method developed by C. J. Sih's group. This method is the "gold standard" to which other methods are compared. To measure *E*, researchers run a test resolution, work up the reaction and measure two of the following: enantiomeric purity of the starting material (ees), enantiomeric purity of the product (eep), or conversion (c). The difficulty with this method is mainly the measurement of enantiomeric purity, which can be time-consuming. Screening hundreds of commercial enzymes or cultures of microorganisms by this method is difficult without laboratory automation.

Estimating Selectivity without a Reference Compound

In some cases, a qualitative estimate of selectivity is sufficient. In these cases, the measured rates of reaction in two experiments are simply compared.

For example, we estimated the enantioselectivity of several lipases toward 4-ni-trophenyl-2-phenylpropanoate by measuring the initial rates of hydrolysis of each enantiomer. The true *E*-values came from a competitive experiment where the enantiomeric purity of the products and starting materials were measured. The estimated *E* came from a simple comparison of the initial rate of reaction for the two enantiomers separately. In this case, a pH indicator was not needed to detect hydrolysis because hydrolysis yields the yellow *p*-nitro-phenoxide directly. The agreement is qualitatively good since the estimated *E* identified enantioselective and nonenantioselective enzymes. However, there is clearly no quantitative agreement. (The Quick *E*-values, which do agree quantitatively with the true *E*-values will be discussed below.) rate of fast enantiomer measured separately estimated *E* rate of slow enantiomer measured separately.

$$\text{estimated } E = \frac{\text{rate of fast enatiomer measured separately}}{\text{rate of slow enantiomer measured separately}} \quad (5)$$

In another example, we estimated the enantioselectivity of the 52 hydrolases toward solketal butyrate by separately measuring the initial rates of hydrolysis of the pure enantiomers. We used pH indicators to measure the rates of hydrolysis and then presented the ratio of these rates as an estimated enantio-selectivity. The true enantio- selectivity and the estimated enantioselectivity agreed to within a factor of 2.3. This qualitative screening identified horse liver esterase as a new hydrolase for the resolution of solketal butyrate with modest enantioselectivity.

In a last example, estimated selectivities differed so much from the true selectivities that they were useless. In this case we estimated the diastereoselectivity of 91 commercial hydrolases towards the *cis*-(2*S*, 4*S*) and *trans*-(2*R*, 4*S*) dioxolanes by measuring the initial rates of hydrolysis of the two pure diaste-reomers separately. The ratio of the two rates is the estimated diastereoselectivity. Three hydrolases – cholesterol esterase, a-chymotrypsin and subtilisin from *Bacillus licheniformis* – showed excellent (> 100) estimated diastereo-selectivity. Unfortunately, the true diastereoselectivity was much lower (4.4–17), so this estimated selectivity screening was not accurate enough to identify selective enzymes.

The main advantages of measuring estimated selectivity are speed and simplicity, while the main disadvantage is the risk that the estimate is so far off from the true value that it is misleading. Typical screening and data workup times for a library of 100 enzymes are several hours. In practice, we often include an estimated selectivity screening before the quantitative screen involving a reference compound (see below) to group the substrates into fast-, medium-, and slow-reacting. This information helps choose a good reference compound.

Quantitative Measure of Selectivity Using a Reference Compound

These estimated selectivities are only estimates because they do involve competition between the substrates. By measuring initial rates of the two substrates separately, we eliminate competitive binding between the two diastereomers.

Both k_{cat} and K_m contribute to the overall selectivity of an enzyme, thus eliminating competitive binding can lead to inaccuracies, as the example in Table 1.3 above showed. To understand the contribution of binding to selectivity, imagine a hypothetical case where both substrates have the same k_{cat} values, but different K_m values. In a competitive experiment, the enzyme will bind and transform the better binding substrate. The reaction is selective. However, if the hydrolysis of the two substrates occurs in separate vessels, the result depends on the substrate concentration. At saturating amounts of substrate, both reactions will proceed at the same rate and estimated selectivity will incorrectly indicate that the reaction is nonselective. At substrate concentrations well below K_m, the estimated selectivity will be close to the true selectivity. One can imagine other combinations of reaction conditions and kinetic parameters that can lead to either overestimation or underestimation of the true selectivity.

Measuring the true selectivity requires a competitive experiment. Adding two substrates to the reaction mixture creates a competitive experiment, but complicates analysis. One must be able to distinguish the reaction of each substrate individually. Methods like HPLC or GC can distinguish the two substrates, but are much slower than colourimetric methods. An alternative is to use one true substrate, whose hydrolysis can be measured by the pH indicator and a second, chromogenic reference compound, whose hydrolysis can be measured by formation of a colour different from the pH indicator. Such an experiment would yield the correct selectivity for true substrate 1 and the reference compound (Eq. 6).

$$\frac{\text{substrate 1}}{\text{reference}}\ \text{selectivity} = \frac{\text{rate of substrate 1 reaction}}{\text{rate of reference reaction}} \cdot \frac{[\text{reference}]}{[\text{substrate 1}]} \qquad (6)$$

In a second experiment, a second substrate competes against the same reference compound to yield the correct selectivity for true substrate 2 and the reference compound (Eq. 7).

$$\frac{\text{substrate 2}}{\text{reference}}\ \text{selectivity} = \frac{\text{rate of substrate 2 reaction}}{\text{rate of reference reaction}} \cdot \frac{[\text{reference}]}{[\text{substrate 2}]} \qquad (7)$$

Finally, dividing these two selectivities yields the desired selectivity of substrate 1 versus substrate 2 (Eq. 8).

$$\frac{\text{substrate 1}}{\text{substrate 2}}\ \text{selectivity} = \frac{\frac{\text{substrate 1}}{\text{reference}}\ \text{selectivity}}{\frac{\text{substrate 2}}{\text{reference}}\ \text{selectivity}} \qquad (8)$$

We define Quick selectivity methods (Quick *E* for enantioselectivity, Quick *D* for diastereoselectivity, Quick *S* for substrate selectivity) as methods that, instead of letting the substrates compete directly against one another, do so indirectly by competing each substrate against a reference compound.

Chromogenic Substrate

In the first Quick *E* experiments, we used chromogenic substrates and a chromogenic reference compound. Hydrolysis of pure enantiomers of 4-nitro-phenyl 2-phenylpropanoate liberates the yellow *p*-nitrophenoxide ion.

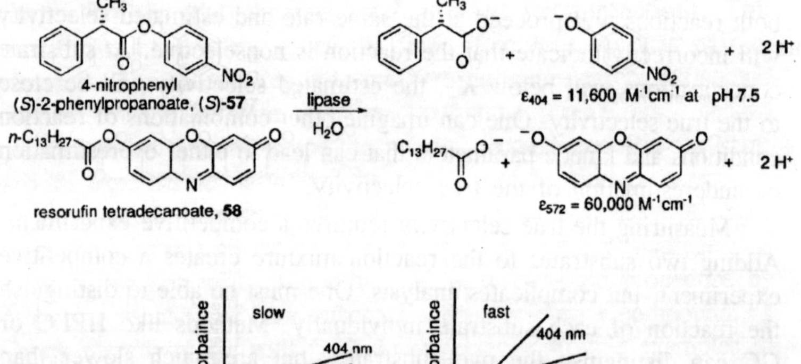

Figure 9.7 : The first step of the Quick E *measure of enantioselectivity of lipases toward 4-nitrophenyl 2-phenylpropanoate.*

The increase in absorbance at 404 nm revealed the initial rates of hydrolysis of each enantiomer, but the ratio of these rates gave only an enantiomeric ratio. The ratio of rates over- or underestimated *E* by as much as 70%.

To reintroduce competition, we added resorufin tetradecanoate as a reference compound. We monitored the initial rates of hydrolysis of 4-nitrophenyl (*S*)-2-phenylpropanoate ((*S*)-57) at 404 nm and the reference compound at 572 nm in the same solution.

Two experiments, one with (*S*)-enantiomer of competing against the reference compound and the second with the (*R*)-enantiomer of competing against the reference compound yielded the data to calculate the Quick *E*-values according to Eqs (6) to (8). These Quick *E*-values agreed with the values measured by the slower methods involving direct competition of the enantiomers (*R*)- or (*S*)-57 and analysis of the enantiomers by HPLC. We measured low (E = 1.4), average (E= 27), and excellent (E = 210) enantioselectivities correctly by this technique. Each hydrolysis experiment requires 30 s; thus, the measurement time for E was only 1 min.

Lipase-catalysed hydrolysis of 4-nitrophenyl (*S*)-2-phenylpropanoate ((*S*)-**57**) and the reference compound, resorufin tetradecanoate (**58**), yields yellow and pink chromophores, respectively. The solution turns deep orange if both substrates are hydrolysed; pink if only the reference compound is hydrolysed.

The second step of the Quick E is the same, except that it uses the (*R*)-enantiomer of the chiral ester. Equations (6) to (8) yield the enantioselectivity.

pH Indicators

To apply the Quick E ideas to nonchromogenic substrates, we used pH indicators to detect the hydrolysis of the substrate. The reference compound remained a chromogenic resorufin ester, acetate. We first demonstrated this to measure diastereoselectivity of the dioxolane stereoisomers.

The Quick D measurements agreed with the true values determined by NMR using a direct competitive hydrolysis of the two diastereomers.

Application

Substrate Mapping of Hydrolases

Mapping the substrate selectivity of hydrolases using Quick S screening was dramatically faster than traditional methods. For example, the mapping of the acyl chain length selectivity of a group of esterases from thermophiles confirmed the high acetyl preference of one esterase (E018b).

We determined the true selectivity of the thermophile esterases toward different acyl chain length using a competitive experiment usually using resorufin acetate as the reference compound. However, accurate measurement of the selectivities requires that the substrate and the reference compound react at comparable rates to measure both reaction rates accurately. When the substrate hydrolysis was much slower than resorufin acetate hydrolysis, we replaced resorufin acetate with the slower reacting resorufin pivaloate or resorufin isobutyrate as reference compounds and extended the measurement time. The trends for the true selectivities were similar to the estimated selectivities. Among the vinyl esters, the hexanoate or octanoate was the best substrate for all except E018b, where the acetate was the best substrate.

We confirmed this high acetyl selectivity of E018b by a direct competition between vinyl buty-rate and acetate monitored by ^{1}H-NMR. The resonances of both substrate vinylesters, and the products, butyric and acetic acid, were monitored over time and revealed a 17-fold acetyl preference. In comparison the Quick *S* measurements showed a 24-fold acetyl preference. Surprisingly, acetyl esterase from orange peel showed only a fourfold preference for acetyl.

Screening of Mutants in Directed Evolution

An important application of Quick *E* screening was search for more enantioselective mutants in a library of randomly mutated esterases. For this application, the ability to measure enantioselectivity accurately was important to find those mutants that only increased enantioselectivity moderately. To increase the enantios-electivity of PFE toward hydrolysis of methyl 3-bromo-2-methylpropanoate (MBMP) (E_{WT}= 12 favoring (*S*)-**64**), we used random mutagenesis by an *Es- cherichia coli* mutator strain. Screening of 288 crude cell lysates with Quick *E* revealed that most of the mutants had enantioselectivities near that of the wild

The selectivity ratio of the *trans*-(2*R*, 4*S*) diastereomer versus resorufin acetate is 1.64 while the selectivity ratio of the *cis*-(2*S*, 4*S*) diastereomer versus resorufin acetate is 0.128. The ratio of the two selectivity ratios equals the diastereoselectivity, 12.8. Conditions in the well during assay: 2.0 mM *cis* or *trans* dioxolane methyl ester, 0.1 mM resorufin acetate, 4.65 mM BES, 0.434 mM 4-nitrophenol, 0.86% Triton X-100, 7% acetonitrile.

Fig. 1.9 Acyl length selectivity of thermophile esterases toward vinyl esters. Darker squares correspond to higher activity. Most esterases favor either vinyl hexanoate or octanoate, but E018b and AcE (acetyl esterase from orange peel) favor vinyl acetate. To show selectivity clearly, the absolute enzyme activities were scaled as indicated. This grayscale array representation was created as described by Reymond and coworkers. AChE = acetylcholine esterase.

type (Quick *E* = 12), but one mutant, MS6-31, showed significantly higher enantioselectivity (Quick *E*= 21). DNA sequencing revealed a $C_{689}T$ transition, which changes Thr230 to isoleucine. This mutant showed *E*= 19 in a scale-up reaction in good agreement with Quick *E*-value. For this and the Thr230Pro mutant in Table 1.4, the differences between Quick *E* and the true *E* are

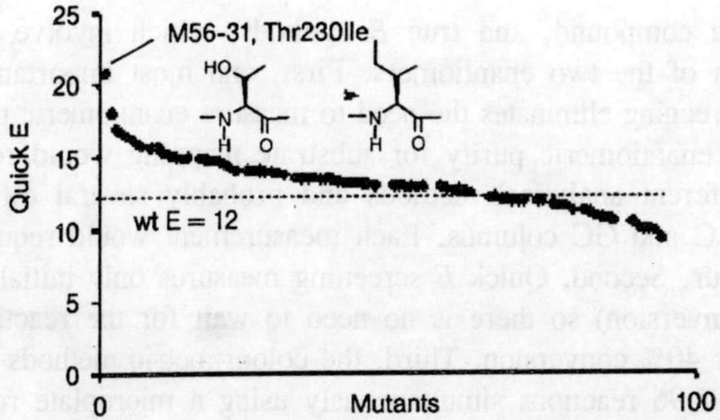

Figure 9.8 : Mutants of Pseudomonas fluorescens esterase generated by random mutagenesis of the entire protein and ordered from highest to lowest enantioselectivity. The flat central part of the curve represents colonies where there was little or no change in enantioselectivity because they either contained no mutations or mutations that did not affect enantioselectivity.

within experimental error even though we measured Quick *E*-values on cell lysates, but measured endpoint method values on purified enzyme.

The next generation of this directed evolution project involved focusing mutations closer to the substrate-binding site. The much more enantioselective mutants at positions 28 and 121 identified in this project also highlighted a weakness of the Quick *E* screening method. The Quick *E*-values proved more difficult to measure because the rate reaction of the slow enantiomer was so slow that it was difficult to measure accurately. First, we increased the concentration of substrate, but solubility limited how much we could add. We could not simply add more enzyme because then the reference compound reacted too quickly. Our solution was a slower reacting reference compound – resofurin iso-butyrate – in place of resorufin acetate. This reference compound **63** reacted about seven times more slowly with *Pseudomonas fluorescens* esterase than the acetate **60**. With new reference compound **63**, we could measure the higher enantioselect-ivities. The errors for measuring the higher enantioselectivity remained higher than that for low enantioselectivity. For this application, this accuracy was sufficient to identify the highly enantioselective mutants. For applications that require measuring high enantioselectivities accurately with Quick *E*, we are developing even slower-reacting reference compounds.

Advantages and Disadvantages

The main advantage of Quick *E* is speed, which comes from four differences between Quick *E*, which involves competition with a reference compound, and true *E* methods, which involve direct competition of the two enantiomers. First, and most important, the Quick *E* screening eliminates the need to measure enantiomeric purity. Measuring enantiomeric purity for substrate mapping would require several different analytical methods and probably several different chiral HPLC and GC columns. Each measurement would require at least an hour. Second, Quick *E* screening measures only initial rates (> 5% conversion) so there is no need to wait for the reaction to reach 30 or 40% conversion. Third, the colourimetric methods allow us to follow 96 reactions simultaneously using a microplate reader. Fourth, the small scale of the reaction (typically 0.1 mmol) allows us to use more enzyme per mole of substrate and thus the reaction is faster.

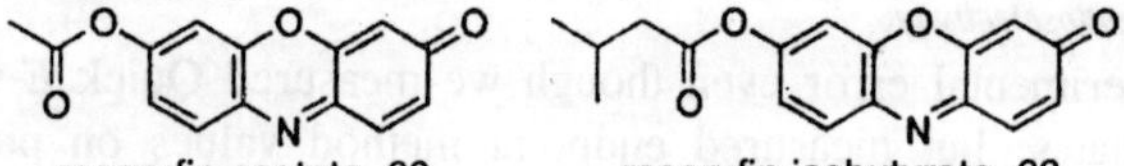

resorufin esetate, 60 resorufin isobutyrate, 63

Figure 9.9 : Resorufin isobutyrate reacts approximately seven times slower than resorufin acetate, which makes it a better reference compound for slow-reacting substrates.

These advantages all contribute to making Quick *E* a much faster method than true *E* methods. Quick *E* is ideal for rapidly measuring the enantioselectivity of large numbers of samples and it is based on the same equations and assumptions as the true *E* methods.

One disadvantage of Quick *E* is that it requires pure enantiomers, albeit in small amounts. In many cases, small amounts are available from preparative HPLC with chiral stationary phases or other methods. Another disadvantage is the need for clear aqueous solutions of substrate. Dissolving hydrophobic substrates in aqueous solution, even with the help of surfactants, is sometimes difficult.

Measuring high enantioselectivity is challenging for both the true *E* methods and the Quick *E* methods. For the true *E* methods, measuring high enantio-selectivity requires the slow-reacting enantiomer to be detected accurately. For the Quick *E* methods, measuring high enantioselectivity requires the rates of reaction of the slow-reacting enantiomer and a reference compound to be measured simultaneously. This measurement will normally require a slower reacting reference compound. Another limit to measuring high enantioselectivity with Quick *E* is the enantiomeric purity of the starting slow enantiomer. If the slow-reacting enantiomer is contaminated with the fast-reacting one, then the measured Quick *E*-value will be lower than the true *E*.

10

Chemistry of Biological Catalyst

A covalent chemical bond involves the sharing of a pair of electrons between the atoms (X : Y). If the bond breaks by *homolytic fission*, each atom separates as a highly *reactive free radical*, possessing an unpaired electron (X· + Y·). Such reactions are very uncommon in aqueous solutions, where most bonds are broken by *heterolytic fission*, leaving one of the atoms with both electrons. If both electrons are retained by a carbon atom, a *carbanion* is produced:

$$R_3C:X \rightleftharpoons \underset{\text{carbanion}}{R_3C:^-} + X^+$$

If neither electron is retained by the carbon atom, a *carbonium ion* is produced:

$$R_3C:X \rightleftharpoons \underset{\text{carbanium ion}}{R_3C:^+} + X^-$$

Carbanions and carbonium ions may be stabilized by *delocalisation* of the charge over other atoms in the molecule. For example:

$$CH_2 = CH - \dot{C}H_2 \leftrightarrow \dot{C}H_2 - CH{=}CH_2$$

The curved arrow indicates the flow of a pair of electrons which would take place if one of these *canonical structures* of the carbonium ion was converted to the other. The actual structure is somewhere in between the two canonical structures written down: it should not be imagined that the two forms are being rapidly interconverted. *Conjugated* double bond systems (i.e. where double and single bonds occur alternately) are of particular importance for stabilisation, for a charge can be delocalised over the entire system.

The formation of carbanions and carbonium ions is likely to occur as part of a complete reaction sequence. Organic compounds can participate in four main types of reaction: displacement (or substitution) reactions, addition reactions, elimination reactions and rearrangements.

Substitution reactions may be *nucleophilic*; when the group attacking the carbon atom is an electron donor called a nucleophile since it is attracted to nuclei; alternatively they may be *electrophilic*, when the attacking group is an electron acceptor, or electrophile. Electrophilic substitution reactions often involve the displacement of hydrogen. For example:

$$C_6H_5\text{–}H + NO_2^+ \longrightarrow C_6H_5\text{–}NO_2 + H^+$$

benzene nitrobenzene

In *nucleophilic substitution*, an atom other than hydrogen is usually displaced. Such reactions may have a unimolecular or a bimolecular mechanism.

In unimolecular nucleophilic substitution (S_N1) reactions, the rate-limiting step is the ionisation of a single molecule to form a carbonium ion which then reacts with a nucleophile, as in the following example:

$$R'R''R'''C\text{–}Y \xrightarrow{-Y^-} [R'R''R'''C^+] \xrightarrow{+X^-} R'R''R'''C\text{–}X \text{ or } X\text{–}CR'R''R'''$$

The carbonium ion is planar and, if the resulting product has four different groups attached to that particular carbon atom (as above), a racemic mixture of optical isomers wiil be obtained. The overall rate of the reaction is determined by the rate of ionisation, and thus on the basicity of the leaving group (Y^-): weak bases are good leaving groups and depart readily; strong bases are poor leaving groups. The strength of the attacking nucleophile does not usually affect the rate of the reaction.

In *bimolecular nucleophilic substitution* (**S_N2**) reactions, the attacking nucleophile adds to the carbon atom at a point diametrically opposite the leaving group, which it displaces in one rapid step.

It can be seen that the reaction involves an inversion of configuration at the carbon atom. The rate of the reaction depends, among other factors, on the relative strengths of X^- and Y as nucleophiles.

It is unnecessary here to go into details about other types of

$$X^- + R'R''R'''C{-}Y \longrightarrow \left[X\cdots C(R')(R'')(R''')\cdots Y\right]^{-} \longrightarrow X{-}C(R'')(R')(R''') + Y^-$$

transition-state

reaction. *Addition reactions* involve the addition of an electrophile or a nucleophile to form a relatively stable compound without any group being displaced, while ***elimination reactions*** involve' the removal of a group, usually in a strongly basic solution, without it being replaced by another group. *Rearrangements* are reactions where bonds are formed and broken within a single molecule.

Information about the mechanism of a particular reaction can be obtained from the *stereochemistry* of the products, as discussed above for the S_N1 and S_N2 reactions. Another useful technique is the investigation of the effect of *isotopic substitution* on the reaction rate. For example, it is usually found that a C-D bond is broken more slowly than a C-H bond, so the use of deuterium-labelled reactants can give information as to which C-H bonds are normally broken as part of the rate-limiting step of the overall process.

MECHANISMS OF CATALYSIS

Acid-base Catalysis

Acids can catalyse reactions by temporarily donating a proton; bases can do the same by temporarily accepting a proton. In section 6.2.3 we saw how a base could catalyse the hydrolysis of an ester by stabilizing the transition-state. Bases may also increase reaction rates by increasing the nucleophilic character of the attacking groups. Thus, hydroxide ions displace halides from alkyl halides more rapidly than do neutral water molecules. Similarly, acids facilitate the removal of leaving groups, where these are strong bases.

Similarly, in the case of ester hydrolysis, protonation of the substrate leads to a transition-state which avoids the unstable two-charge arrangement occurring under neutral conditions, thus increasing the rate of reaction:

$$\overset{+}{O}H{=}C(R')(OR'') + H_2O \rightleftharpoons H{-}O{-}C(R')(OR''){-}\overset{+}{O}(H){-}H \rightleftharpoons \text{products}$$

In general, acid or base catalysis may be shown to be operating by determining the rate constants of a reaction at different concentrations of acid or base.

Electrostatic Catalysis

A transition-state may be stabilized by electrostatic interaction between its charged groups and charged groups on a catalyst. Thus, the positive charge on a carbonium ion can be stabilized by interaction with a negatively charged carboxylate ion; similarly, the negative charge on an oxyanion can be stabilized by a positively charged metal ion. For example, the hydrolysis of glycine esters:

$$H_2N.CH_2C(=O).OCH_3 + H_2O \rightleftharpoons H_2N.CH_2C(=O).OH + CH_3OH$$

may be catalysed by cupric ions, the mechanism probably involving the following steps:

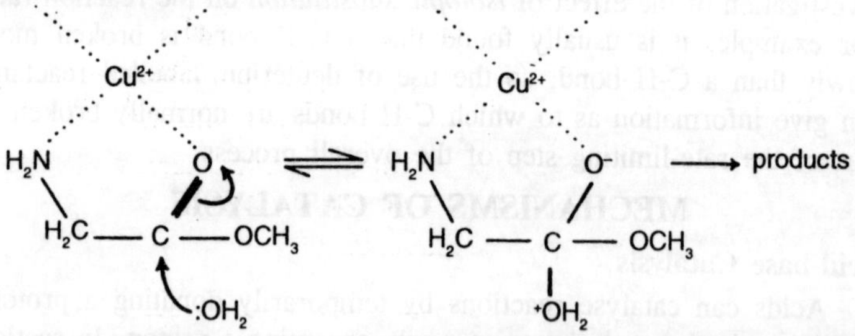

$$Cu^{2+}\cdots H_2N{-}H_2C{-}C(=O\cdots Cu^{2+})({-}OCH_3) + :OH_2 \rightleftharpoons Cu^{2+}\cdots H_2N{-}H_2C{-}C(O^-\cdots Cu^{2+})(OCH_3)({-}{}^{+}OH_2) \longrightarrow \text{products}$$

Covalent Catalysis

In contrast to acid-base and electrostatic catalysis, where the transition-state is merely modified, covalent catalysis introduces a different reaction mechanism and is sometimes termed *alternative-*

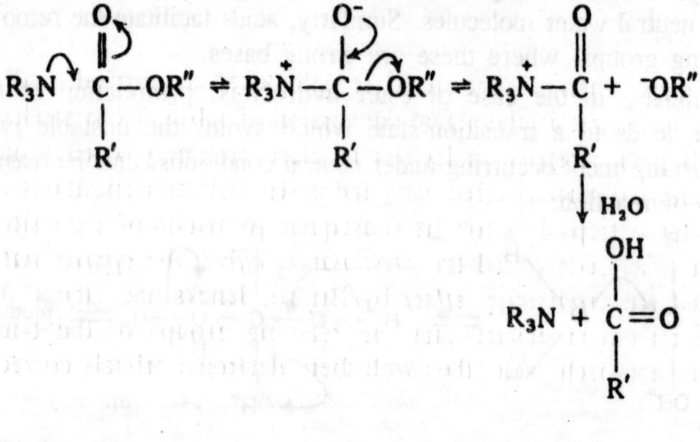

$$R_3N + R'C(=O){-}OR'' \rightleftharpoons R_3\overset{+}{N}{-}C(O^-)(R'){-}OR'' \rightleftharpoons R_3\overset{+}{N}{-}C(=O)R' + {}^{-}OR''$$

$$R_3\overset{+}{N}{-}C(=O)R' \xrightarrow{H_2O} R_3N + R'C(=O)OH$$

pathway catalysis. In the case of nucleophilic catalysis, the catalyst is more nucleophilic than the normal attacking groups and so rapidly forms an intermediate which itself rapidly breaks down to give the products. For example, Bender and co-workers showed that a variety of tertiary amines catalyse the hydrolysis of esters.

In contrast, electrophilic catalysts act by withdrawing electrons from the reaction centre of an intermediate and may be termed *electron sinks*. This type of catalysis is found in the reactions of the coenzymes *thiamine pyrophosphate* and *pyridoxal phosphate*. Sometimes, metal ions may also be regarded as acting in this way, where Cu^2 facilitates the withdrawal of electrons from the reacting carbon atom.

Enzyme Catalysis

All of the above mechanisms of catalysis are seen in enzyme-catalysed reactions. However, enzymes, because of their great size and range of properties, are able to impose their presence on a reaction to a far greater extent than most catalysts.

All catalysts act by reducing the energy of activation, or more correctly the free energy of activation *($\Delta G^{\#}$)*, of the reaction being catalysed. Free energy is made up of an entropic and an enthalpic component, since $\Delta G = \Delta H - T\Delta S$. Hence $\Delta G^{\neq} = \Delta H^{\neq} - T\Delta S^{\neq}$, where $\Delta H^{\neq}$ is the enthalpy of activation and $\Delta S^{\neq}$ is the entropy of activation, and so anything which decreases *AH#* or makes $AS^{\#}$ more positive will lower the value of $\Delta G^{\neq}$ and help the reaction proceed.

There is a great loss of entropy when reactants leave their random existence in free solution to become bound in the transition-state, so $-T\Delta S^{\neq}$ will generally have a large positive value, possibly making up about half of the total free energy of activation where there is more than one reactant. In enzyme-catalysed reactions there is inevitably a similar loss of entropy at some stage, but it occurs largely in the binding steps when enzyme-substrate complexes are formed, and not to the same degree in the actual conversion of substrates to products: the binding of substrate molecules in close proximity to each other on the enzyme surface effectively increases their concentrations and reduces the entropy loss for the subsequent formation of a transition-state; this has been called the *proximation effect*, the *approximation effect* and the *propinquity effect* by Bruice, Jencks and others. The enzyme may also ensure that the reacting groups of the bound substrates approach each other with their electronic orbitals correctly orientated, thus ensuring that the reaction takes place under optimal conditions; Koshland and others have termed this property of enzymes *orbital steering*.

The contribution of the catalytic sites on the enzyme to the reaction will generally be to the enthalpic factor by stabilizing the transition-state; it appears that the stability of the enzyme-bound transition-state may be the most important single factor in determining whether the reaction proceeds. It is possible that several different catalytic processes may be involved in the same enzyme-catalysed reaction. Another way in which the enthalpy factor may be reduced occurs if the positions of the binding sites on the enzyme do not correspond exactly to those on a substrate: as the substrate binds, some of its bonds may be distorted and weakened, resembling those found in the transition-state; this is the Haldane and Pauling concept of *strain*. Alternatively, the enzyme-substrate complex may be under stress from internal interactions which are absent from the enzyme-bound transition-state complex: again, this would facilitate the forward reaction. A distortion of the protein to fit the substrate (the Koshland induced-fit hypothesis,) would not have any effect on enthalpy, but would not preclude stress factors from playing a part.

Some reactions proceed better in organic rather than aqueous solutions. Most enzymes contain regions of non-polar character and these may be of importance in the catalysis of such reactions.

So, enzymes create a suitable environment for the reaction to proceed; they reduce the total free energy of activation required and also spread out the free energy requirement over several stages. Nevertheless, some activation energy is still required for each stage of the reaction sequence. This cannot come from the substrate because the translational energy that these possess in free solution is lost the moment they become bound to the enzyme. It seems likely that the required energy is obtained from collisions between solute molecules and the enzyme-substrate complex, but the mechanism for this is far from clear.

MECHANISMS OF REACTIONS CATALYSED BY ENZYMES WITHOUT COFACTORS

Enzymes which operate without cofactors tend to be relatively small and have relatively straightforward reaction mechanisms. For these reasons, such enzymes were amongst the first to be investigated in detail. Some examples are given below.

Chymotrypsin

Chymotrypsin is formed by the cleavage of several peptide bonds in the inactive monomeric protein chymotrypsinogen, which is synthesised and secreted by mam-malian pancreas; the active enzyme thus produced consists of three non-identical polypeptide chains.

Chymotrypsin catalyses the cleavage of peptide bonds at the carboxyl side of aromatic amino acid (phenylalanine, tyrosine or tryptophan) residues; it will also hydrolyse a variety of amides and esters, and these artificial substrates have been used to investigate the enzyme in great detail. We discussed kinetic evidence showing that the chymotrypsin-catalysed hydrolysis of an ester proceeded via the formation of an acyl-enzyme. This is also true for the hydrolysis of amides.

We discussed evidence that the substrate can bind to serine-195; and histidine-57 was also implicated in the reaction mechanism. X-ray diffraction studies revealed a hydrophobic binding pocket at the active site for aromatic side chains, and showed that aspartate-102, buried in the interior of the molecule, could be closely linked to the action of histidine-57 and serine-195, possibly setting up what has been termed a charge relay system, with aspartate-102 removing a proton from histidine-57, making it easier for the latter to remove a proton from serine-195 during the course of the reaction. Detailed ^{1}H, ^{13}C and ^{15}N nmr studies showed that histidine-57 has a relatively normal pK_a value near 7, at least in the free enzyme, casting doubt on the charge relay theory. However, investigations using site-directed mutagenesis have shown that the three amino acids, if not setting up a charge relay system as such, nevertheless act in a concerted fashion, perhaps explaining why several enzymes have a 'catalytic triad' of this type at the active site.

An outline reaction mechanism for the hydrolysis of an ester or amide by chymotrypsin (or other serine protease) is as follows. Histidine-57 acts as a base catalyst to enable the oxygen of serine-195 make a nucleophilic attack on the carboxyl group of the enzyme-bound substrate. An unstable tetrahedryl intermediate is formed whose negatively charged oxygen atom may be stabilized by hydrogen bonding with the backbone -NH of glycine-193. Similarly, the positive charge on the imidazole ring of histidine-57 is stabilized by electrostatic interaction with aspartate-102. The imidazole group of histidine-57 then acts as an acid catalyst to facilitate the liberation of the first product (YH), leaving behind the acyl enzyme. (Note the covalent bond linking the acyl group to serine-195.)

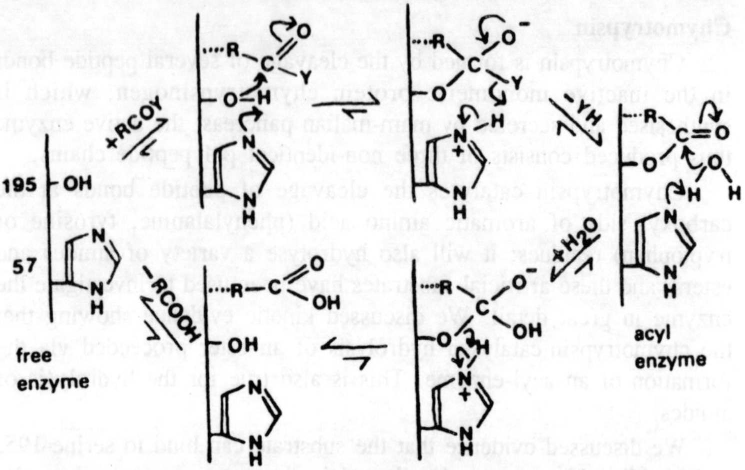

The second stage of the reaction, like the first, is initiated by a nucleophilic attack, this time by water; the liberation of the product (RCOOH) is again assisted by acid catalysis.

The same mechanism is thought to occur for peptide substrates R′.NH.C(R″).CO.NH.R″: if the substrate is again written in the form R.CO.Y, to be compatible with the illustration above, R- represents R'.NH.C(R″)-(where R″ is the side chain which fits into the binding pocket, the -NH forming a hydrogen bond with the backbone carboxyl group of serine-214) and -Y represents -NH.R‴.

Ribonuclease

Bovine pancreatic ribonuclease A has been studied in great detail by enzymologists and was the first enzyme to have its complete amino acid sequence determined. It catalyses the cleavage of the phosphodiester backbone of ribonucleic acids by a reaction involving transfer of a phosphate group from the 5'-position of one nucleotide to the 3'-position of the next nucleotide in the chain.

We discussed evidence that histidine-12 and his-tidine-119 both contribute to the catalytic process, one possibly acting as a proton donor and the other as a proton acceptor; similar investigations have suggested that lysine-41 is also involved. The three-dimensional structure of the enzyme, as revealed by the X-ray diffraction studies of Kartha and colleagues, shows these three amino acids are sited close together near a substrate-binding cleft in the molecule.

In Rabin and co-workers suggested the following mechanism for the reaction.

His–119 His–12 RNA (substrate) ⇌ His–119 His–12 transition-state

⇌ R′–OH (first product)

His–119 His–12 (second product) ⇌ His–119 His–12.

Histidine-12 removes a proton from the substrate, producing a nucleophilic oxygen atom which attacks the P. Histidine-119 donates a proton to the complex, and the first product, R′OH, is liberated. The rest of the reaction is the reverse of the first part, with water replacing the R′OH.

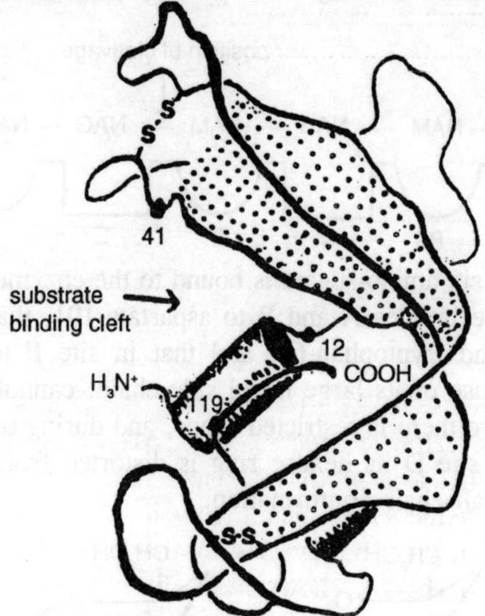

Figure 10.1 : A simplified representation of the three-dimensional structure of ribonuclease A, as revealed by the X-ray diffraction studies of Kartha and colleagues.

This mechanism is consistent with the evidence now available. It is likely that lysine-41 helps stabilize the complex by an electrostatic interaction with the negative charge of the phosphate.

Lysozyme

Lysozyme catalyses the hydrolysis of the polymer consisting of alternating N-acetylglucosamine (NAG) and N-acetylmuramic acid (NAM) units which gives shape and rigidity to bacterial cell walls.

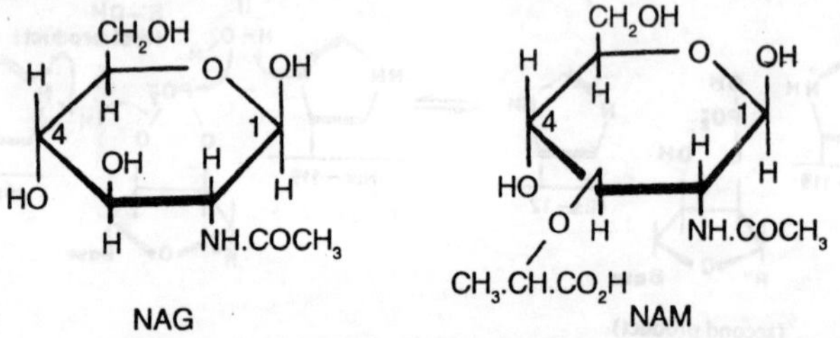

NAG NAM

The lysozyme which has been most studied is that from hen egg-white; this was the first enzyme to have its complete three-dimensional structure determined by X-ray crystallography (Blake, Phillips and colleagues,). These and subsequent inves-tigations revealed that the enzyme has a large cleft which can accommodate six of the substrate's amino-sugar units. This may be represented diagram-matically as follows:

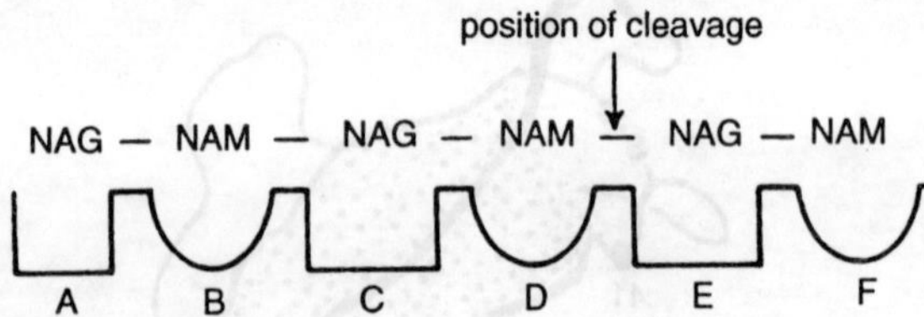

Each of the six amino-sugars is bound to the enzyme by hydrogen bonds, e.g. those in sites A and B to aspartate-101, that in site B to tryptophan-62 and tryptophan-63, and that in site F to asparagine-37. NAM, because of its large lactyl side chain, cannot bind to sites A, C or E, where there is restricted space, and during or immediately after binding to site D its hexose ring is distorted from the chair to the less-stable half-chair conformation.

chair conformation half-chair conformation

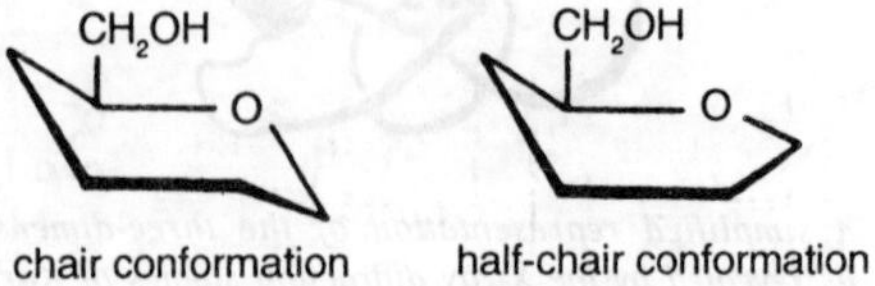

The hydrolysis takes place between the units in sites D and E, and therefore must always involve the β-1,4 glycosidic link between C1 (the reducing end) of NAM and C4 of NAG. The amino acids near this bond are glutamate-35 and aspartate-52; evidence that these are involved in the reaction.

Vernon suggested the following mechanism for the reaction (many bonds and atoms not involved in the reaction being omitted from the diagram for the sake of clarity):

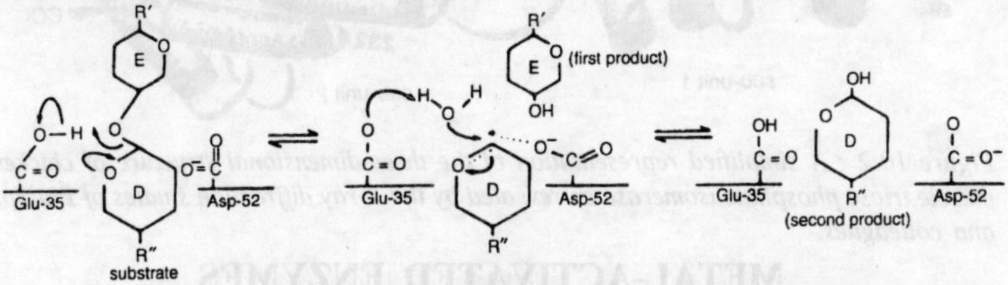

The substrate binds to the enzyme and ring D is distorted to the half-chair conformation, which facilitates the formation of a carbonium ion also having the half-chair conformation. Glutamate-35, which is in a non-polar environment and thus protonated at pH 5, can act as a general acid catalyst and donates its proton to the oxygen of the glycosidic bond, causing the bond to break. Aspartate-52 is in a more polar environment than glutamate-35 and is negatively charged at pH 5; therefore, it can stabilize the carbonium ion by electrostatic interaction. The first product leaves, and the reaction is completed by a nucleophilic attack on the carbonium ion by water.

Triose phosphate isomerase

Muscle triose phosphate isomerase is a dimeric enzyme which catalyses the inter-conversion of glyceraldehyde 3-P(G3P) and dihydroxyacetone-P (DHAP). X-ray crystallography has revealed that each sub-unit has an inner cylinder of eight strands of parallel β-pleated sheet linked by predominantly helical regions which form an outer cylinder. Each of the two identical binding sites contains lysine- 13, histidine-95, glutamate-165 and glycine-232 from one sub-unit and phenylalanine-74 from the other. The reaction mechanism apparently involves glutamate-165 acting as a base catalyst and histidine-95 as an acid one:

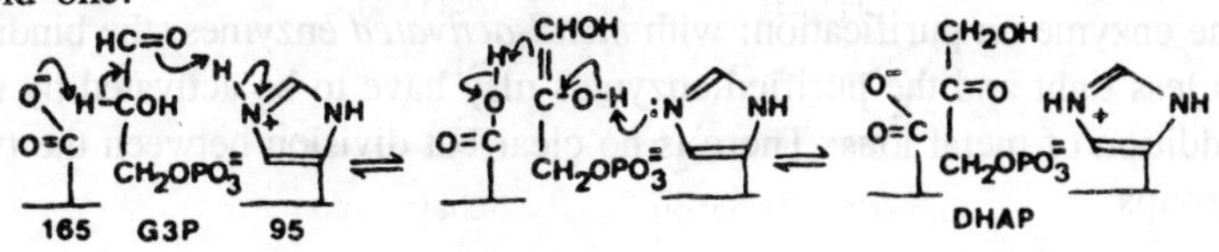

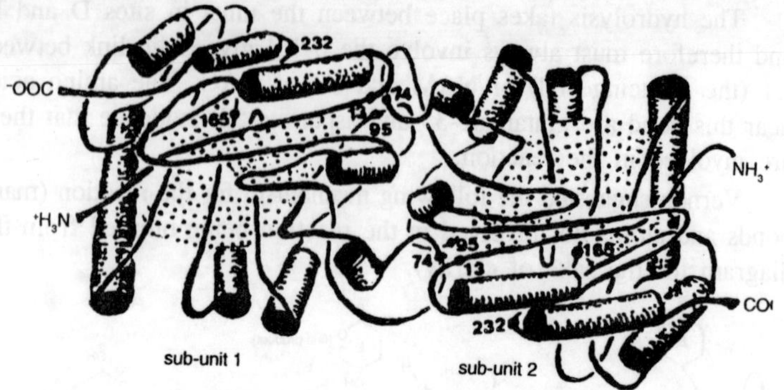

Figure 10.2 : A simplified representation of the three-dimensional structure of chicken muscle triose phosphate isomerase as revealed by the X-ray diffraction studies of Phillips and colleagues.

METAL-ACTIVATED ENZYMES AND METALLOENZYMES

More than a quarter of all known enzymes require the presence of metal atoms for full catalytic activity. Metal atoms usually exist as cations and often have more than one oxidation state, as with ferrous (Fe^2) and ferric (Fe^3) iron. We have noted that this positive charge can stabilize transition-states by electrostatic interactions, giving one mechanism for catalysis by metals. However, irrespective of the oxidation state and charge carried, a metal ion can bind a particular number of groups *(ligands)* by accepting free electron pairs to form co-ordinate bonds in specific orientations.

Therefore, metal ions can be involved in enzyme catalysis in a variety of ways: they may accept or donate electrons to activate electrophiles or nucleophiles, even in neutral solution; they themselves may act as electrophiles; they may mask nucleophiles to prevent unwanted side reactions; they may bring together enzyme and substrate by means of co-ordinate bonds, possibly causing strain to the substrate in the process; they may hold reacting groups in the required three-dimensional orientation; they may simply stabilize a catalytically active conformation of the enzyme.

With *metalloenzymes*, the metal is tightly bound and retained by the enzyme on purification: with *metal-activated* enzymes, the binding is less tight and the purified enzymes may have to be activated by the addition of metal ions. There is no clear-cut division between the two groups.

Mildvan pointed out that ternary complexes formed between an enzyme (E), metal ion (M) and substrate (S) may be enzyme bridge complexes (M-E-S), substrate bridge complexes (E-S-M) or metal bridge complexes (E-M-S). Metalloenzymes cannot form substrate bridge complexes since the purified enzyme exists as E-M.

The involvement of metal ions in enzymes may be investigated by nmr, esr and proton relaxation rate (PRR) enhancement techniques. Examples are given below, some of these, as in chapter 10, being of historical importance, providing a basis for later investigations.

Activation by Alkali Metal Cations (Na^+ and K^+)

Alkali metal cations bind only weakly to form complexes with enzymes, but K^+, the most abundant intracellular cation, is known to activate a great many enzymes, particularly those catalysing phosphoryl transfer or elimination reactions. It appears that the role of K^+ is largely to bind to negatively charged groups on an inactive form of the enzyme and thus cause a change in conformation to a more active form.

However, in some cases it may also aid substrate binding. For example, muscle pyruvate kinase, a tetrameric enzyme which catalyses the reaction:

$$\underset{\substack{|\\ OPO_3^{2-}}}{CH_2 = C}.CO_2^- + H^+ + ADP \rightleftharpoons \underset{\substack{||\\ O}}{CH_3C}.CO_2^- + ATP$$

phonsphoenolpyruvate (PEP) pyruvate

has a requirement for alkali metal cations and for Mn^{2+} (or Mg^{2+}), all of which bind in the region of the active site, Various studies have indicated that the carboxyl group of PEP binds to the enzyme-bound K^+ whereupon a conformational change takes place which facilitates the progress of the reaction via an E-Mn^{2+} -PEP complex.

Activation by Alkaline Earth Metal Cations (Ca^{2+} and Mg2$^+$)

Oxygen atoms are often involved in the bonds of both alkali metal and alkaline earth metal cations, bonds of the latter being relatively stronger. The divalent cations, Ca^{2+} and Mg^{2+}, can form six co-ordinate bonds to produce octahedral complexes.

MW is accumulated by cells in exchange for transport of Ca^{2+} in the opposite direction. As might be expected, therefore, enzymes requiring Ca^{2+} for activation are mainly *extracellular* ones, e.g. the salivary and pancreatic a-amylases: the Ca^{2+} appears to play a role in maintaining the structure required for catalytic activity. A variety of *intracellular* enzymes require Mg^{2+} for activity and, in most cases, this requirement can be replaced *in vitro by* one for Mn^{2+}. Unlike Mg^{2+}, Mn^{2+} is paramagnetic, which enables the system to be more easily investigated. (Note that Mn^{2+} may, in its own right, be a component of a metalloenzyme, e.g. arginase, where it stabilizes a reactive hydroxide ion, ensurini that an activated nucleophile is available for catalysis. In the rat liver enzyme, Mn^{2+} is bound to histidine-101, aspartate-124, and aspartate-232 in each of the three sub-units, and has a catalytic interaction with aspartate- 128.)

It has been shown that all possible types of ternary bridge complexes involving divalent cations can exist. Most *kinases* form E-S-M complexes, where S is the reacting nucleotide. Let us consider, as an example, the reaction catalysed by muscle *creatine kinase*:

$$\text{creatine} + \text{MgATP} \rightleftharpoons \text{MgADP} + \text{phosphocreatine} + H^+$$

The true substrate is MgATP and the reaction proceeds via the formation of the complex creatine-E-ATP-Mg. Cohn and colleagues (1971) showed that the divalent cation binds to the a- and n-phosphates of the nucleotide, but not to the terminal (y) phosphate, which is transferred to creatine. The cation helps in the orientation of the complex and may also assist in the breaking of the pyrophosphate bond by withdrawing electrons from the P-phosphate. The overall reaction has a random-order rapid-equilibrium kinetic mechanism, and dead-end complexes may be formed; identification of some of these, e.g. E-ADP-Mn^{2+}, helped in the elucidation of the mechanism of the reaction.

In contrast, the reaction catalysed by *pyruvate kinase* involves a cyclic metal bridge complex:

```
        Mg
      / |
E ——— ATP
      \
        pyruvate
```

Metal bridge complexes are found with many enzymes which use pyruvate or phosphoenolpyruvate as substrate.

Enolase is a dimeric enzyme, two Mg^{2+} ions being required to stabilize the active dimer; a further two Mg^{2+} ions are required if each of the two active sites binds a substrate. Nowak, Mildvan and colleagues demonstrated that the enzyme-bound cation probably binds to water, forming a co-ordinated hydroxyl group which can attack the phosphoenolpyruvate:

$$H_2C{=}C{-}CO_2^- \quad OPO_3^{2-} \quad H{-}O{-}H \quad E{-}Mg$$

A few enzymes, e.g. *E. coli glutamate-ammonia ligase*, have a mechanism involving an enzyme bridge complex. Here the divalent cation presumably has a purely structural role.

Activation by Transition Metal Cations (Cu, Zn, Mo, Fe and Co Cation)

Transition metal ions bind to enzymes much more strongly 'than the metal ions discussed above, and usually form metalloenzymes; this makes their involvement in enzyme-catalysis relatively easy to investigate. They are found in only trace amounts in living organisms, for larger amounts can be toxic.

The trace metals Mo and Fe are found in *nitric-oxide reductase*, the nitrogen-activating complex of nitrogen-fixing bacteria, and Fe is a component of haemoglobin, the oxygen-carrying haemoprotein of the erythrocytes of vertebrates. Another trace metal, Co, is found in vitamin B_{12}. We will now consider, in a little more detail, an example of a Cu- and a Zn-metalloenzyme.

Superoxide dismutase is a copper-metalloenzyme which catalyses the removal of the highly reactive Oz^- produced, for example, by oxidation of xanthine by molecular oxygen in the presence of *xanthine oxidase*. *Fridovich and colleagues demonstrated that the dismutase reaction proceeds as follows*:

$$2O_2^- + 2H^+ \rightleftharpoons H_2O_2 + O_2$$

Bovine erythrocyte superoxide dismutase is a dimeric protein containing two Cu^{2+} ions and two Zn^{2+} ions. The Zn^{2+} ions appear to have a structural rather than a catalytic role, while the Cu^{2+} ions are involved in the reaction sequence:

$$E - Cu^{2+} + O_2^- \rightarrow E - Cu^+ + O_2$$

$$E - Cu^{2+} + O_2^- \xrightarrow{+2H^+} E - Cu^{2+} + H_2O_2$$

In contrast to the above, Zn^{2+} has a catalytic role in the reaction catalysed by *carboxypeptidase* A, where the C-terminal amino acid of a polypeptide is removed, provided it has a non-polar side chain. Carboxypeptidase A from bovine pancreas is a monomeric enzyme which contains one atom of zinc. X-ray crystallography studies (by Lipscomb and colleagues,) and the determination of the complete amino acid sequence (by Bradshaw and co-workers,) revealed that the active site contains the co-ordinated Zn^{2+} ion bound to histidine-69, glutamate-72, histidine-196 and H_2O, as well as a groove for the polypeptide substrate and a hydrophobic pocket for binding the side chain of the C-terminal amino acid the terminal carboxyl group of the substrate forms an electrostatic interaction with arginine-145.

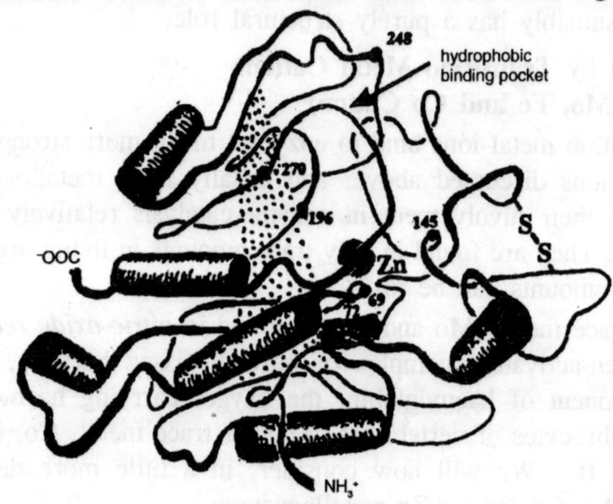

Figure 10.3 : A simplified representation of the three-dimensional structure of carboxypeptidase A. as revealed by the X-ray diffraction studies of Lipscomb and colleagues.

During the reaction, the carbonyl oxygen of the peptide bond being hydrolysed replaces the water molecule bound to Zn^{2+}; the metal ion probably facilitates cleavage of the peptide bond by withdrawing electrons from this carbonyl group. Vallee proposed a general mechanism for the reaction which involved acid and base groups on the enzyme; X-ray diffraction studies have shown that tyrosine-248 is located in such a position in the enzyme-substrate complex that it could donate a proton to the nitrogen of the peptide bond being hydrolysed and there is also evidence that the carboxyl group of glutamate-270 acts as a general base catalyst to make the attacking water molecule nucleophilic. However, site-directed mutagenesis experiments by Gardell and colleagues, in which tyrosine-248 was replaced by phenylalanine without affecting catalytic activity, have suggested that this tyrosine residue may simply have a binding function. Thus the reaction appears to involve:

.... Zn^{2+} ....

side chain binding pocket

Glu - 270 ⊢ CO_2^- H–O–H   O   R

$R - \overset{\|}{C} - N - \overset{|}{C} - CO_2^-$ .... $\overset{+}{NH_2}$ ⊣ Arg-145

H   H

HO ⊣ Tyr-248

where R″ is the side chain of the C-terminal amino acid, and R' represents the rest of the peptide molecule.

X-ray diffraction studies have also shown that substrate-binding results in arginine-145, glutamate-270 and especially tyrosine-248 moving to new positions close to the substrate, forcing water molecules out of the active site and thus creating a hydrophobic environment; the substrate is so tightly enclosed that it would not have been able to get into this position but for the mechanism of the conformational change.

THE INVOLMENT OF COENZYMES IN ENZYME-CATALYSED REACTIONS

Coenzymes are organic compounds required by many enzymes for catalytic activity; they are often vitamins, or derivatives of vitamins. Sometimes they can act as catalysts in the absence of enzymes, but not so effectively as in conjuction with an enzyme.

As with metal-enzyme linkages, there is a range of bond strengths for coenzyme-enzyme links, the point of distinction between tightly bound cofactor (prosthetic group) and loosely bound cofactor being arbitrary. Coenzymes which are prosthetic groups form an integral part of the active site of an enzyme and undergo no net change as a result of acting as a catalyst; for this reason, some would exclude these from classification as coenzymes, simply calling them organic cofactors. Conversely, loosely bound coenzymes can be regarded as *co-substrates* since they often bind to the enzyme-protein together with the other substrates at the start of a reaction and are released in an altered form at the end of it. They are generally regarded as coenzymes since they usually bind to the enzyme before the other substrates are bound, since they participate in many reactions, and since they may be reconverted to their original form by many enzymes present within cells.

Some important coenzymes are discussed below.

Nicotinamide nucleotides (NAD^+ and $NADP^+$)

These are derived from the vitamin *niacin*, which is nicotinamide or nicotinic acid. The structure of nicotinamide adenine dinucleotide (NAD^+) in its oxidised and reduced forms is given below:

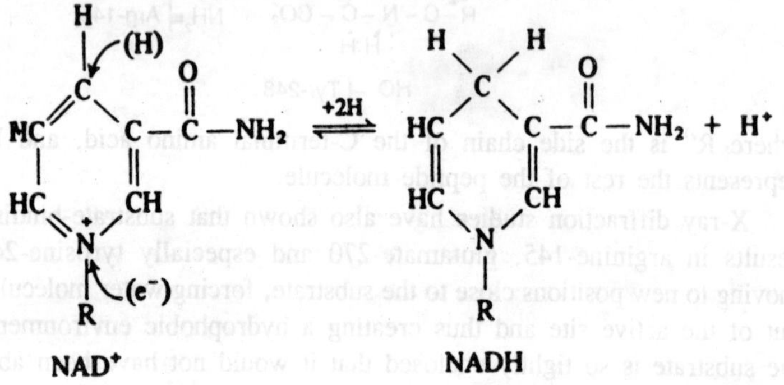

$$NAD^+ \underset{}{\overset{+2H}{\rightleftharpoons}} NADH + H^+$$

NAD⁺ **NADH**

where —R represents D-ribose-phosphate-phosphate-D-ribose-adenine.

It can be seen that the reduction of NAD^+ to NADH requires two reducing equivalents per molecule; one electron (e) and one hydrogen atom ($H = H^+ + e^-$), which together may be regarded as a hydride ion (H^-), add to the pyridine ring of nicotinamide. The pyridine ring is conjugated, so the positive charge may be delocalised, making several points vulnerable to nucleophilic attack. However, it seems that the reaction mechanism usually involves:

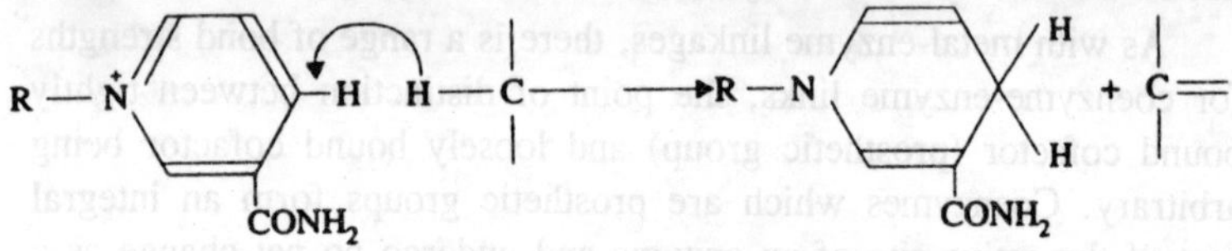

$$R-\overset{+}{N}(\text{ring})(CONH_2) \cdots H \; H-\overset{|}{\underset{|}{C}}- \longrightarrow R-N(\text{ring})(H)(H)(CONH_2) + C=$$

and the transfer of the hydride ion shows stereochemical specificity: with some enzymes, the transferred hydride ion becomes the H_R atom of nicotinamide (that in the diagram coming out of the plane of the paper towards the reader), whereas with others it becomes the H_S atom (that going away from the reader).

Nicotinamide adenine dinucleotide phosphate ($NADP^+$) is identical to NAD^+, except that the 2'-position of the D-ribose unit attached to adenine is phosphorylated. This does not affect the oxidation/reduction properties, but results in NAD^+ and $NADP^+$ acting as coenzymes for different enzymes: enzymes utilizing NAD^+ usually have a catabolic function, the NADH produced being an energy source for the cell; anabolic enzymes, in contrast, frequently involve NADPH as coenzyme.

The names and abbreviations given above are those currently recommended by the International Union of Biochemistry and Molecular Biology. NAD^+ has also been known as diphosphopyridine nucleotide (DPN^+) or Coenzyme I; $NADP^+$ as triphosphopyridine nucleotide (TPN^+) or Coenzyme II.

Needless to say, NAD^+ and $NADP^+$ act as coenzymes for oxidation/reduction reactions; they are loosely bound, and leave the enzyme in a changed form at the end of the reaction.

The kinetic mechanism of horse liver *alcohol dehydrogenase*, an enzyme which involves NAD^+ as coenzyme in the catalytic oxidation of primary or secondary alcohols. It was noted that NAD^+ is the first substrate to be bound and NADH the last product to leave, the dissociation of NADH from the enzyme being the rate-limiting step of the overall reaction; this is one reason why NAD^+ is regarded as a coenzyme rather than simply as a substrate. Horse liver alcohol dehydrogenase is a dimer, each sub-unit containing one binding site for NAD^+ and two sites for Zn^{2+}; only one of these zinc ions is directly involved in catalysis. From results of X-ray diffraction studies, Branden and colleagues deduced that the ternary complex was of the form:

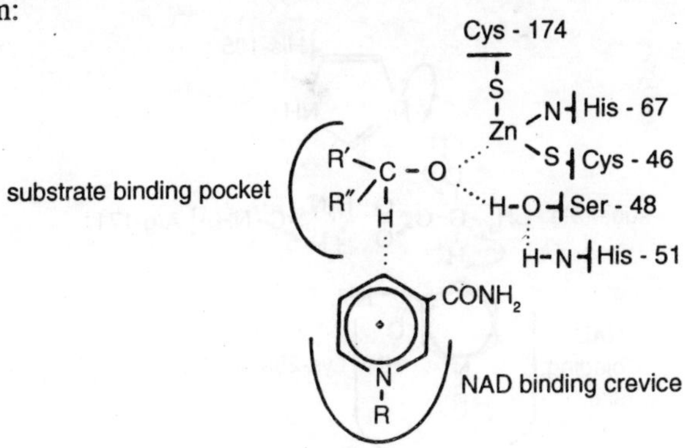

Cys - 174
S
Zn
N His - 67
S Cys - 46
R′ C - O
R″ H
H-O Ser - 48
H-N His - 51
substrate binding pocket
$CONH_2$
N
R
NAD binding crevice

and the reaction mechanism involves:

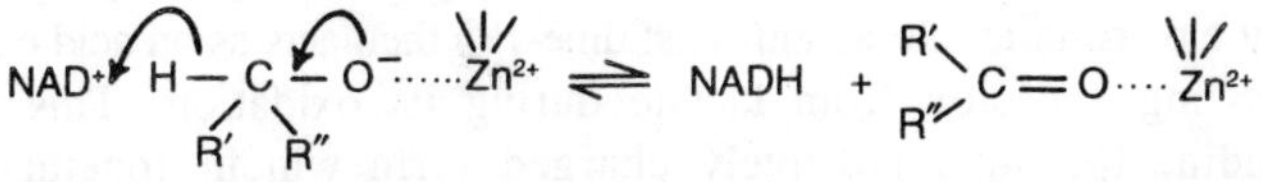

$$NAD^+ \; H-C(R')(R'')-O^- \cdots Zn^{2+} \rightleftharpoons NADH + R'R''C{=}O \cdots Zn^{2+}$$

The initial proton loss from the substrate in the hydrophobic environment of the active site to the surface of the enzyme may be by means of a shuttle system involving histidine-51, the 2'-hydroxyl group of the coenzyme and serine-48. The zinc ion then acts as an electrostatic catalyst to stabilize the negatively charged transition state, whilst a hydride ion is transferred to NAD^+, becoming the H_R atom of NADH. For the reverse reaction, the zinc ion acts as an electrophilic catalyst to enhance the polarisation of the carbonyl group of the substrate and facilitate hydride transfer from the reduced coenzyme.

Dogfish muscle *lactate dehydrogenase*, which catalyses the reaction:

$$\underset{\text{(S)-lactate}}{CH_3.\underset{\underset{OH}{|}}{CH}.CO_2^-} + NAD^+ \rightleftharpoons \underset{\text{pyruvate}}{CH_3\underset{\underset{O}{||}}{C}.CO_2^-} + NADH + H^+$$

has also been investigated in detail. It is a tetrameric enzyme, each sub-unit having a binding site for NAD^+; no metal ions are bound. Like alcohol dehydrogenase, the reaction has a compulsory-order mechanism, the coenzyme binding first and bringing about a conformational change in the enzyme which enables the substrate to bind. Substrate-binding, in turn, causes a peptide loop of the enzyme to close over the active site. From the studies of Adams and others in the early 1970s, it was concluded that the reaction mechanism involves:

His-195
substrate
Arg-171.
NAD binding site
Lys-250

Arginine-171 binds the carboxyl group of the substrate by a two-point electrostatic interaction. Histidine-195 then acts as an acid catalyst, removing a proton from lactate during its oxidation. This leaves liistidine-195 iq a positively charged form which, together with arginine-109, can stabilize the transition-state by electrostatic catalysis.

The positive charge on histidine-195 is itt turn stabilized by electrostatic interaction with aspartate-168. The reaction is completed by the transfer of a hydride ion to NAD^+, in the same orientation as with alcohol dehydrogenase.

Flavin Nucleotides (FMN and FAD)

Flavin nucleotides are derived from riboflavin, vitamin B_2; like the nicotinamide nucleotides, they function in oxidation/reduction reactions, the reducing equivalents being carried by the fused three-ringed system of flavin as shown below:

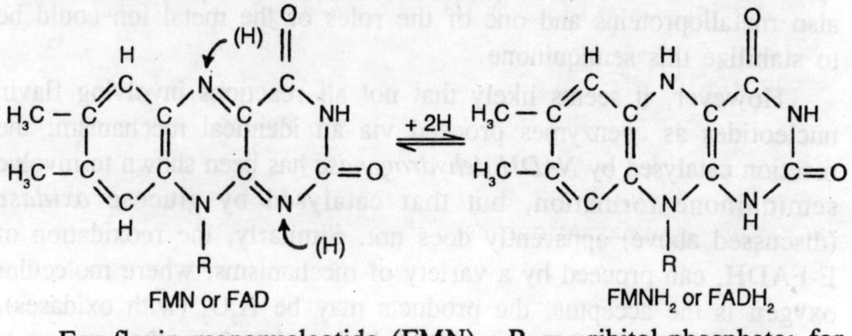

FMN or FAD $\underset{}{\overset{+2H}{\rightleftharpoons}}$ $FMNH_2$ or $FADH_2$

For flavin mononucleotide (FMN), -R = -ribitol-phosphate; for flavin adenine dinucleotide (FAD), -R -ribitol-phosphate-phosphate-D-ribose-adenine.

In contrast to NAD^+ and $NADP^+$, FMN and FAD are prosthetic groups and cannot generally be separated from the protein without denaturing it, the protein-flavin nucleotide complex being termed a *flavoprotein*. Hence, some enzymologists do not accept the classification of FMN and FAD as coenzymes.

Because the flavin nucleotide does not have an independent existence, reactions catalysed by flavoproteins usually involve the transfer of reducing equivalents from a donor via the flavin to some specific external acceptor. For example, *glucose oxidase*, which catalyses the reaction

$$\text{D-glucose} + O_2 \rightleftharpoons \text{D-glucose-}\delta\text{-lactone} + H_2O_2$$

utilizes FAD as prosthetic group and O_2 as hydrogen acceptor. Bright and Gibson showed that this is a two-stage reaction:

$$\text{E–FAD} + \text{D-glucose} \rightleftharpoons \text{E–FADH}_2 + \text{D-glucose-}\delta\text{-lactone}$$

$$\text{E–FADH}_2 + O_2 \rightleftharpoons \text{E–FAD} + H_2O$$

With some flavoproteins, the reduction of the flavin has been shown to be a two-step process, involving an unstable free radical *semiquinone* as intermediate:

FMN or FAD $\underset{}{\overset{+H}{\rightleftharpoons}}$ flavosemiquinone $\underset{}{\overset{+H}{\rightleftharpoons}}$ $FMNH_2$ or $FADH_2$

flavosemiquinone
(FMNH or FADH)

The unpaired electron, which can be delocalised about the ring system, has been revealed by esr studies. Many flavoproteins are also metalloproteins and one of the roles of the metal ion could be to stabilize this semiquinone.

However, it seems likely that not all reactions involving flavin nucleotides as coenzymes proceed via an identical mechanism: the reaction catalysed by *NADH dehydrogenase* has been shown to involve semiquinone formation, but that catalysed by glucose *oxidase* (discussed above) apparently does not. Similarly, the reoxidation of E-$FADH_2$ can proceed by a variety of mechanisms: where molecular oxygen is the acceptor, the products may be H_2O_2 (with oxidases), H_2O and hydroxylated products (with hydroxylases), or the superoxide anion (OZ.) and flavin semiquinone.

Flavin-dependent disulphide oxidoreductase enzymes, of which *glutathione reductase* is an example, use FAD to shuttle reducing equivalents from NADPH (or sometimes NADH) to a cysteine residue, which generally involves the cleavage of a disulphide bridge. Usually, two electrons are transferred simultaneously to FAD, so it never passes through the semiquinone form. Finally, the substrate is reduced by the cysteine.

X-ray diffraction analysis of human erythrocyte *glutathione reductase*, a large dimer consisting of identical 478-residue sub-units, was carried out by Schulz, Schirmer and colleagues during the 1980s. These studies revealed that each sub-unit consists of NADPH, interface and FAD domains, the last-mentioned incorporating an N-terminal domain. The two sub-units link along the interface domains, with a catalytic site, consisting of residues from both sub-units, at each end. In each sub-unit, NADPH and FAD bind to their respective domains, which results in the catalytically active components of each coenzyme being positioned in the region of the catalytic site. Each catalytic site consists of active residues cysteine-58, cysteine-63 and tyrosine-197 from the sub-unit providing the NADPH and FAD, together with glutamate-472 and histidine-467 from the other sub-unit.

In the absence of NADPH, the side chain of tyrosine-197 prevents solvent obtaining access to the NADPH-binding site and to the flavin ring of FAD. When NADPH binds, it does so in such a position that its active pyridine ring lies parallel, and in close proximity, to the flavin ring of FAD. Equally close to the flavin ring, but on the opposite side to NADPH, lies the sulphur atom of cysteine-63.

A hydride ion is transferred from NADPH to FAD (the H, atom of NADPH being the one involved), resulting in the formation of the FADH - anion. The electrons are then immediately passed on from $FADH^-$ to cysteine-63, causing the cleavage of its disulphide link with cysteine-58. Cysteine-63 has a transient covalent linkage with the flavin ring, but this changes to a non-covalent linkage as the amino acid accepts a hydrogen atom to finish in the -SH form, whilst cysteine-58 picks up a proton from solution to finish in the same form. Overall, the formation of the reduced cysteine residues may be written:

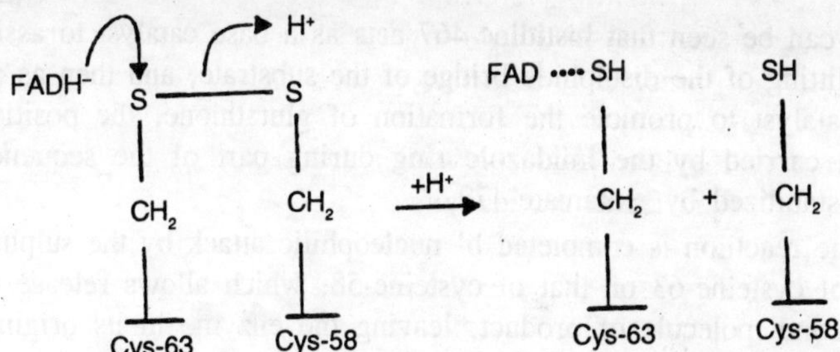

$NADP^+$ may then be released from its binding site to leave the enzyme in its reduced form. Again, the side chain of tyrosine-197 blocks access to the vacant coenzyme-binding site, thus preventing re-oxidation by the solvent before the enzyme has had chance to act upon the substrate, glutathione disulphide, and convert it to glutathione. The product, glutathione, is y-glutamylcysteinylglycine, whilst the substrate consists of two glutathione molecules whose cysteine residues are linked by a disulphide bridge.

When glutathione disulphide binds to the reduced form of the enzyme, cysteine-58, made more nucleophilic by the concerted action of histidine-467 and glutamate-472 (operating rather like histidine-57 and aspartate-102 in chymotrypsin) initiates the attack on the substrate which results in the breaking of the disulphide bond. One half of the substrate becomes attached to cysteine-58, whilst the other half picks up a proton from histidine-467 and is released as the first of the two product molecules:

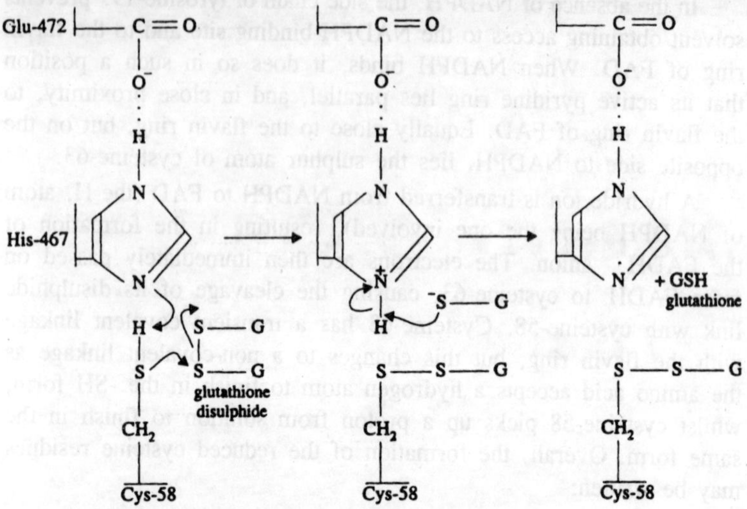

Glu-472 — C=O … His-467 … S—G glutathione disulphide … CH$_2$ … Cys-58 → … $^-$S—G … Cys-58 → … + GSH glutathione … S—S—G … Cys-58

It can be seen that histidine-467 acts as a base catalyst to assist the splitting of the disulphide bridge of the substrate, and then as an acid catalyst to promote the formation of glutathione, the positive charge carried by the imidazole ring during part of the sequence being stabilized by glutamate-472.

The reactiion is completed b' nucleophilic attack by the sulphur atom of cysteine-63 on that of cysteine-58, which allows release of the second molecule of product, leaving the enzyme in its original oxidised form.

Adenosine Phosphates (ATP, ADP and AMP)

The nucleoside phosphates ATP, ADP and AMP are involved in phosphate transfer reactions.

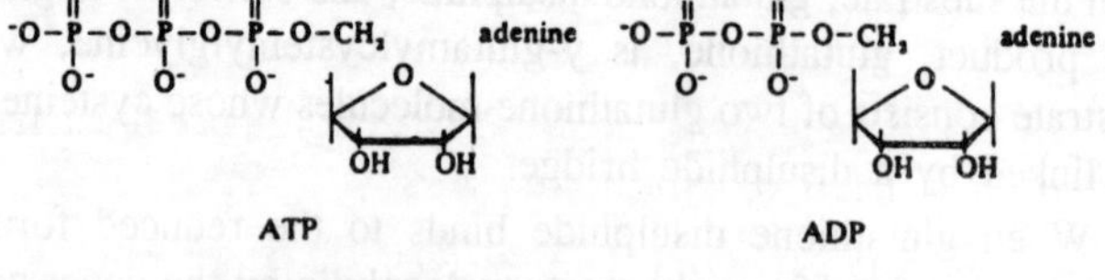

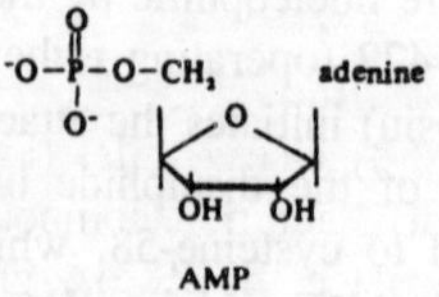

ATP

ADP

AMP

ATP and ADP may be interconverted by the reaction

$$ATP + H_2O \rightleftharpoons ADP + P_i$$

This tends to go strongly in the forward direction as written ($\Delta G'$ = – 31.0 kJ mol^{-1}) because the four negative charges which are in close proximity on ATP make it an unstable molecule (although the product, ADP, itself has three negative charges close together), and because the reverse reaction requires negatively charged ADP to react with negatively charged P_i. The reaction can be coupled to others, so that phosphate may be transferred between ATP and other organic compounds without ever being present as free P_i.

The importance of ATP in energy metabolism is that, by comparison to other organic phosphates, it is only moderately unstable. Hence it may be synthesized by the transfer of phosphate to ADP from a more unstable organic phosphate (e.g. phosphoenolpyruvate) by substrate level phosphorylation, or by oxidative phos-phorylation; however, ATP is sufficiently unstable to be able to force the transfer of the phosphate to a whole variety of other compounds, thus driving such processes as biosynthesis, active transport and muscular contraction.

In the cell, adenosine phosphates are stabilized by binding to Mg^{2+} ions, and their metabolism is strictly mediated by enzymes; in some instances two phosphate groups, rather than one, may be removed from ATP to liberate inorganic pyrophosphate:

$$ATP + H_2O \rightleftharpoons AMP + (PP)_i$$

Adenosine phosphates, like the nicotinamide nucleotides, are loosely bound by enzymes and may be regarded both as coenzymes and as co-substrates/co-products of the reactions in which they participate.

Coenzyme A (CoA.SH)

Coenzyme A has the sructure:

$HS.CH_2CH_2NH$-pantothenic acid-$OPO_3^-.PO_3^-$ -ribose-3-P-adenine

With carboxylic acids it can form thioesters:

$$RCO_2H + HS.CoA \rightleftharpoons \underset{\text{acyl-CoA}}{\dot{R}.COSCoA} + H_2O$$

These thioesters are of great importance in biochemical metabolism since they can be attacked by electrophiles (including other acyl-CoA molecules and CO_2) to form addition compounds, and by nucleophiles (including water) to displace the -SCoA group:

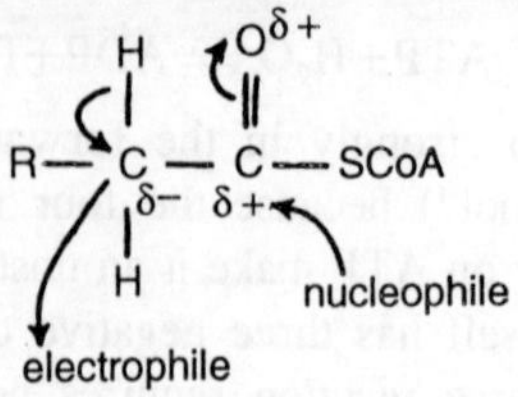

Thiamine Pyrophosphate (TPP)

Thiamine pyrophosphate is derived from vitamin B_1, thiamine, and has the structure:

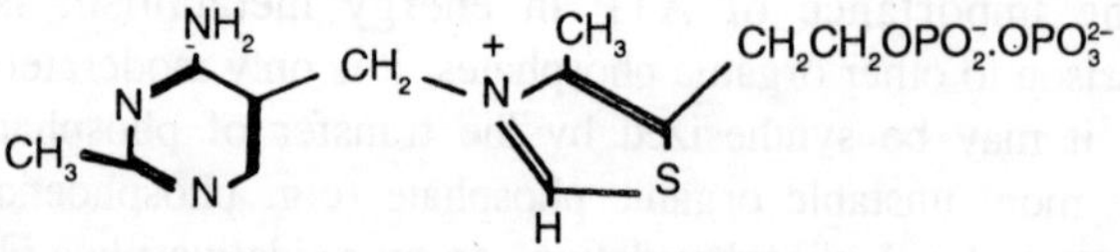

The **thiazole ring** can lose a proton to produce a negatively charged carbon atom:

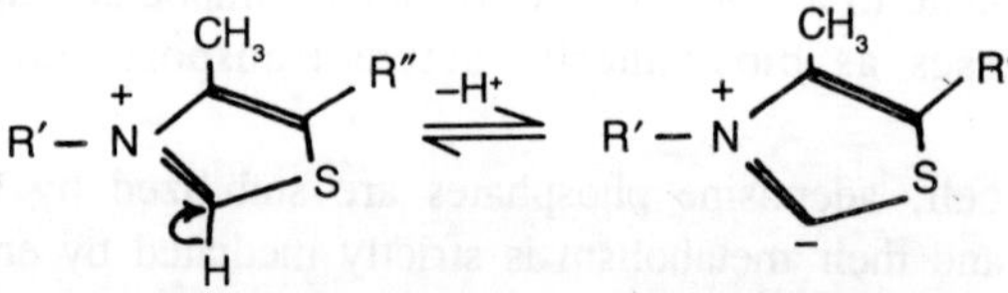

This is a potent nucleophile and can participate in covalent catalysis, particularly with a-keto (oxo) acid decarboxylase, a-keto acid oxidase, transketolase and phosphoketolase enzymes.

For example, *pyruvate decarboxylase*, found in yeast and some other micro-organisms, utilizes TPP to catalyse the production of acetaldehyde from pyruvate. Breslow proposed the following reaction mechanism:

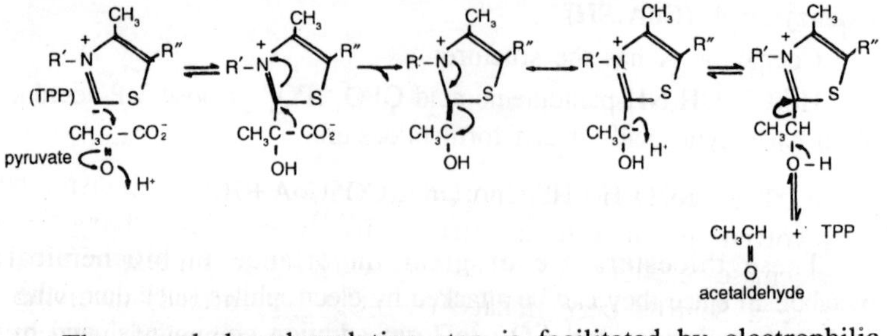

The actual decarboxylation step is facilitated by electrophilic catalysis as the thiazole ring withdraws electrons. The reaction will proceed in the absence of enzyme, but the acetaldehyde formed tends to react with the complex to produce acetoin as the final product. It is likely that the enzyme stabilizes the TPP-acetaldehyde complex and prevents this condensation from occurring.

The multienzyme complex known as *pyruvate dehydrogenase* also catalyses the decarboxylation of pyruvate, but it utilizes a second coenzyme, *lipoic acid*, to introduce an oxidation step and a third coenzyme, *coenzyme A (CoA.SH)*, to react with the acetyl-lipoamide complex, giving acetyl-CoA as the final product. The complex is formed as above, and then the reaction is thought to proceed as follows:

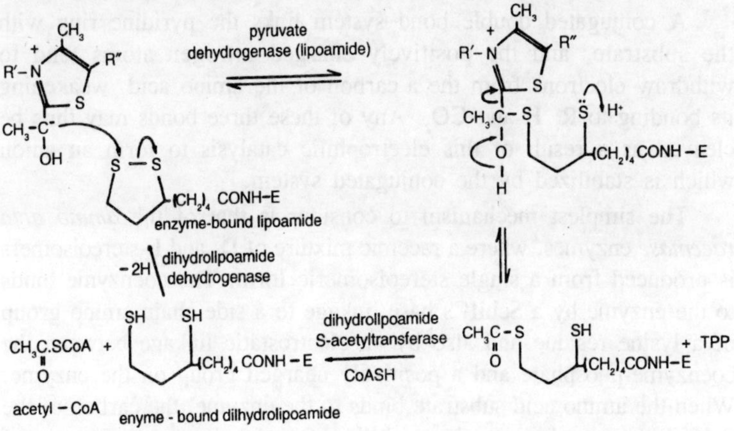

pyruvate dehydrogenase (lipoamide)

$(CH_2)_4$ CONH–E
enzyme-bound lipoamide

–2H dihydrolipoamide dehydrogenase

dihydrolipoamide S-acetyltransferase

$CH_3C.SCoA$ + TPP

CoASH

acetyl – CoA

enyme - bound dilhydrolipoamide

Pyridoxal Phosphate

Pyridoxal 5'-phosphate is derived from pyridoxal, pyridoxine or pyridoxamine, vitamin B_6; it has the structure:

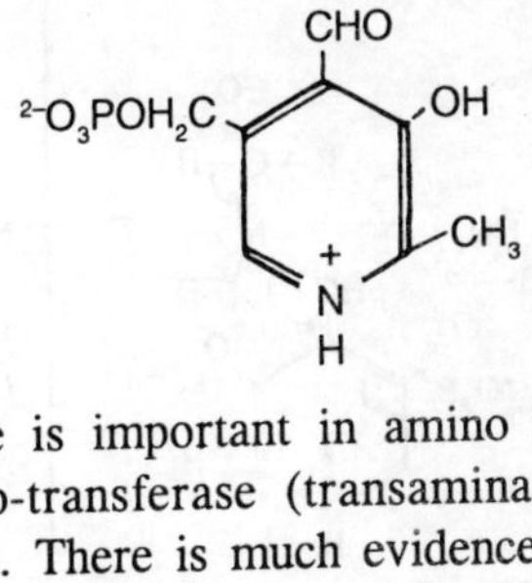

CHO, OH, $^{2-}O_3POH_2C$, CH_3, $\overset{+}{N}$–H

The coenzyme is important in amino acid metabolism, being involved in amino-transferase (transaminase), decarboxylase and racemase reactions. There is much evidence to suggest that in each case a Schiff's *base* linkage (-N=CH-) is formed, involving the aldehyde group of the coenzyme. The phenolic groupd of the coenzyme is also important, as it may help to stabilize the Schiff's base intermediate:

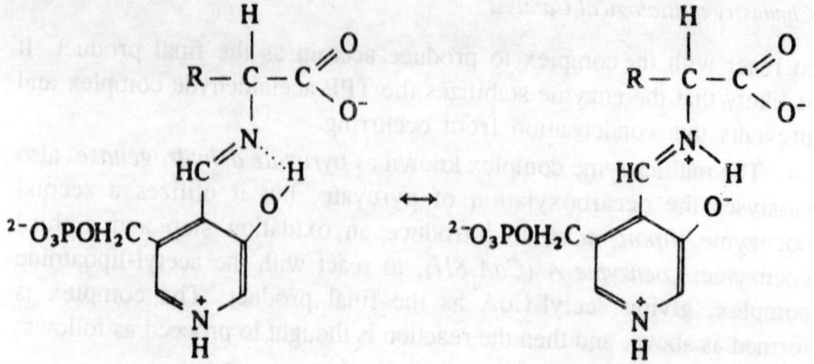

A conjugated double bond system links the pyridine ring with the substrate, and the positively charged nitrogen atoms tend to withdraw electrons from the a-carbon of the amino acid, weakening its bonding to R, H and CO_2. Any of these three bonds may thus be cleaved as a result of this electrophilic catalysis to form an anion which is stabilized by the conjugated system.

The simplest mechanism to consider is that of the *amino acid racemase* enzymes, where a racemic mixture of D- and L-stereoisomers is produced from a single stereoisomeric form. The coenzyme binds to the enzyme by a Schiff's base linkage to a side chain amino group of a lysine residue and also by an electrostatic linkage between the coenzyme-phosphate and a positively charged group on the enzyme. When the amino acid substrate binds to the enzyme, the carbon of the Schiff's base undergoes electrophilic attack by the a-amino group of the substrate and a new Schiff's base linkage is formed, this time between coenzyme and substrate. The weak C-H bond in the complex is then broken to release a proton:

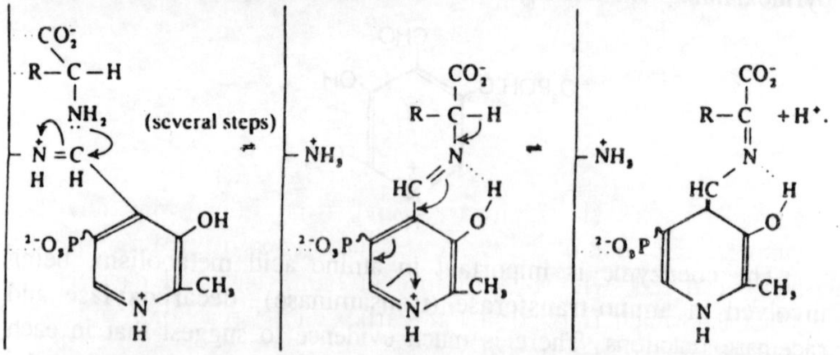

The whole process is reversible, so a proton could add on to the complex and re-form the free amino acid. However, the initial loss of the proton also results in the loss of asymmetry of the α-carbon atom, so the amino acid produced is not necessarily the same stereoisomer as that present at the start.

Aminotransferase (transaminase) reactions proceed as above, with loss of a proton from the Schiff's base. However, the resulting complex is then attacked in a different place by a proton and undergoes hydrolysis to form pyridoxamine phosphate, with the liberation of an oxo acid:

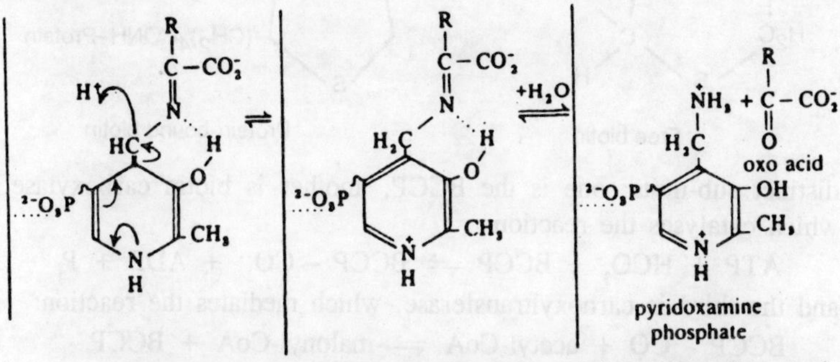

oxo acid

pyridoxamine phosphate

The reverse sequence may then take place, initiated by the attack on pyridoxamine phosphate by a different oxo acid, and so producing a different amino acid from that initially present. A mechanism of this type was established for *aspartate transaminase* by Jenkins and co-workers The two-stage reaction sequence is:

L-glutamate + E-pyridoxal-P $\rightleftharpoons$ 2-oxoglutarate + E-pyridoxamine-P

oxaloacetate + E-pyridoxamine-P $\rightleftharpoons$ L-aspartate + E-pyridoxal-P

Kinetic studies have indicated a ping-pong bi-bi mechanism, which is consistent with the above chemical mechanism.

Decarboxylase reactions involving amino acid substrates commence in much the same way, but the Schiff"s base complex loses the α-carboxyl group rather than the hydrogen atom; the resulting complex is then hydrolysed to reform the coenzyme and give an amine (RCH_2 NH_2) as product.

Biotin

Biotin, like lipoic acid, is always found firmly bound to a side chain amino group of one of the lysine residues of a protein. Protein-bound biotin can link with CO_2 to form a biotin carboxyl carrier protein (BCCP), which is important in carboxylation reactions, e.g. that catalysed by acetyl-CoA carboxylase.

Acetyl-CoA carboxylase from *E. coli* has been shown, by Lane and co-workers and Vagelos and colleagues, to dissociate into three

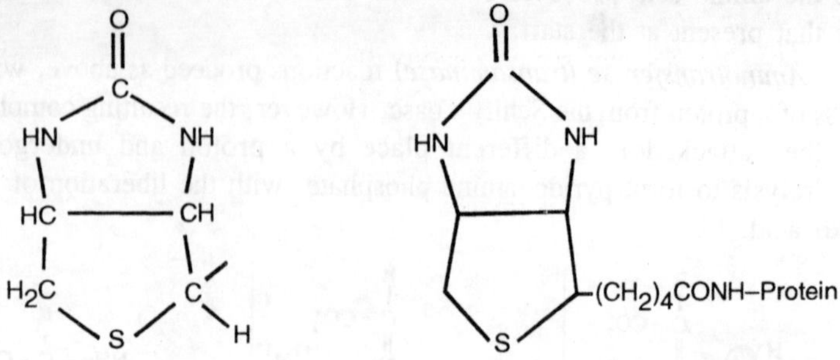

Free biotin

Protein-bound biotin

distinct sub-units: one is the BCCP, another is biotin carboxylase, which catalyses the reaction:

$$ATP + HCO_3 + BCCP \rightleftharpoons BCCP - CO_2^- + ADP + P_i$$

and the third is carboxyltransferase, which mediates the reaction:

$$BCCP - CO + \text{acetyl-CoA} \rightleftharpoons \text{malonyl-CoA} + BCCP$$

The BCCP appears to act as a flexible arm, transporting the CO_2^- from the active site of biotin carboxylase to that of carboxyltransferase, where it is presented to the acetyl-CoA. The chemical mechanism may involve:

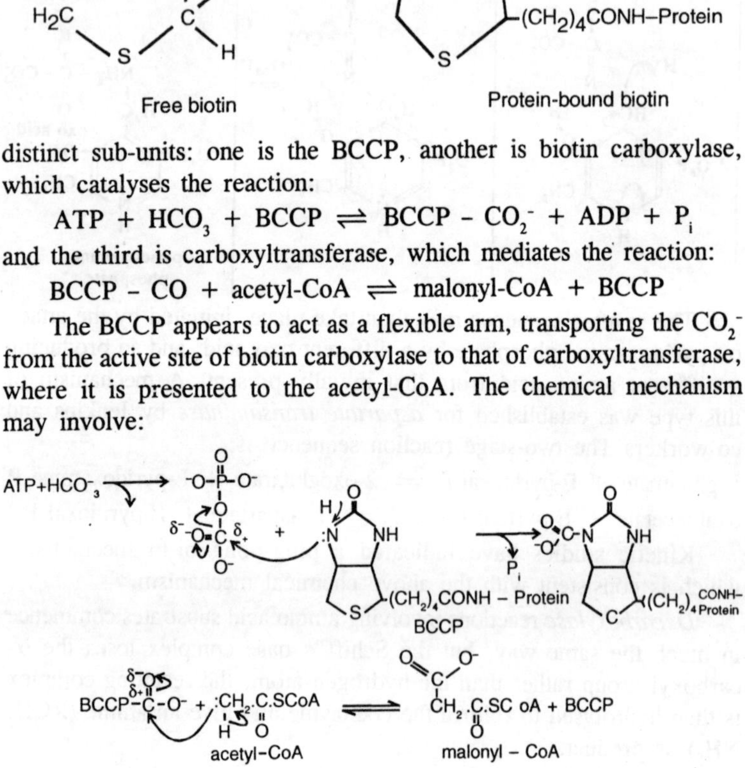

$ATP+HCO_3^-$ → $-O-P-O^-$ ... $(CH_2)_4CONH$ – Protein BCCP ... P_i ... $(CH_2)_4$ CONH–Protein

$BCCP-C-O^-$ + $:CH_2.C.SCoA$ ⇌ $CH_2.C.SCoA$ + BCCP

acetyl-CoA malonyl – CoA

Tetrahydrofolate

Tetrahydrofolate (FH_4) is derived from the vitamin folic acid and has the structure:

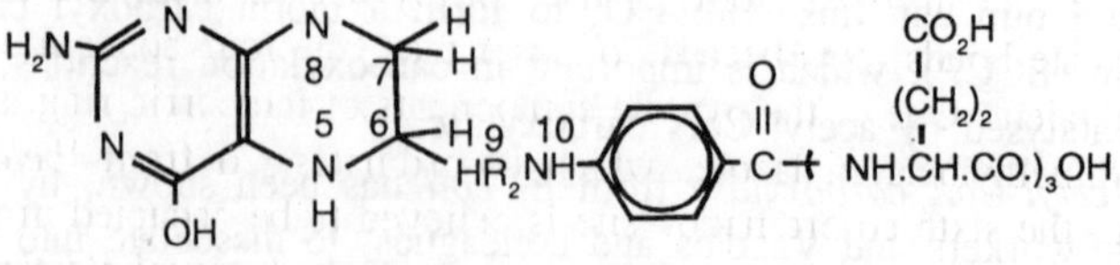

H_2N ... N ... 8 7 ... 5 6 ... H 9 10 HR_2 NH ... C ... $NH.CH.CO.)_3OH$ ... CO_2H ... $(CH_2)_2$ ... OH

Atoms N5 and N10 can carry *one-carbon units* for transfer to a suitable acceptor. *Formate-tetrahydrofolate ligase* (formerly formyltetrahydrofolate synthetase) catalyses the addition of formate to FH_4 by the reaction:

$$ATP + formate + FH_4 \underset{}{\overset{Mg^{2+}}{\rightleftharpoons}} 10\text{-formyl-}FH_4 + ADP + P_i$$

The mechanism for this is likely to resemble that for the carboxylation of biotin, except that it involves addition of a formyl group and not a carboxyl group. 10-formyl-FH_4 can then react further to produce a variety of one-carbon groups carried by the coenzyme:

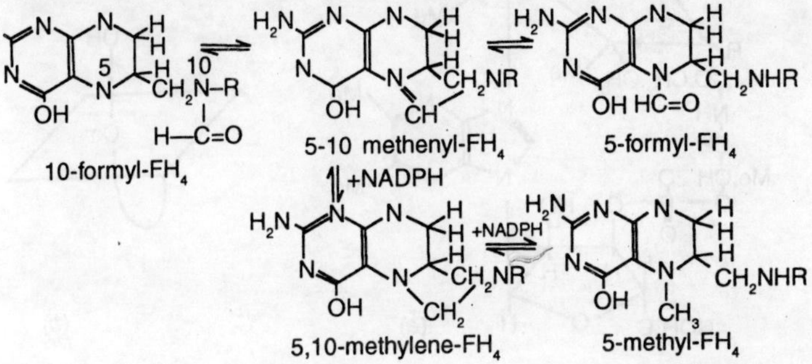

10-formyl-FH_4 ⇌ 5-10 methenyl-FH_4 ⇌ 5-formyl-FH_4

5-10 methenyl-FH_4 ⇌ (+NADPH) 5,10-methylene-FH_4 ⇌ (+NADPH) 5-methyl-FH_4

These, and other, one-carbon units may then be transferred to an acceptor. Consider, for example, the reaction catalysed by *glycine hydroxymethyltransferase*:

$$\underset{\text{glycine}}{H_2N.CH_2CO_2H} + 5,10\text{-methylene-}FH_4 \overset{+H_2O}{\rightleftharpoons} \underset{\text{L-serine}}{H_2N\overset{CH_2OH}{\overset{|}{C}}H.CO_2H} + FH_4$$

Pyridoxal phosphate is also involved and, as is usual with reactions involving this coenzyme, the substrate (glycine) binds by a Schiff's base linkage and then goes on to lose a proton from the α-carbon atom: the methylene-FH_4 complex then presents the one-carbon unit in a suitable orientation and the reaction is completed.

Coenzyme B_{12}

Hodgkin showed that *cobalamin* (*vitamin* B_{12}) has a central core of a cobalt ion surrounded by a *corrin ring*: the metal ion is linked to the four nitrogen atoms of the ring by one covalent and three co-ordinate bonds, but, because of resonance, the four bonds are almost equivalent. One of the pyrrole components of the corrin ring is joined to dimethylbenzimidazole, which co-ordinates Co from '*below*' the ring; the sixth co-ordination site is believed to be occupied in vivo by OH^-. In this naturally occurring compound, *hydroxocobalamin* (B_{12a}),

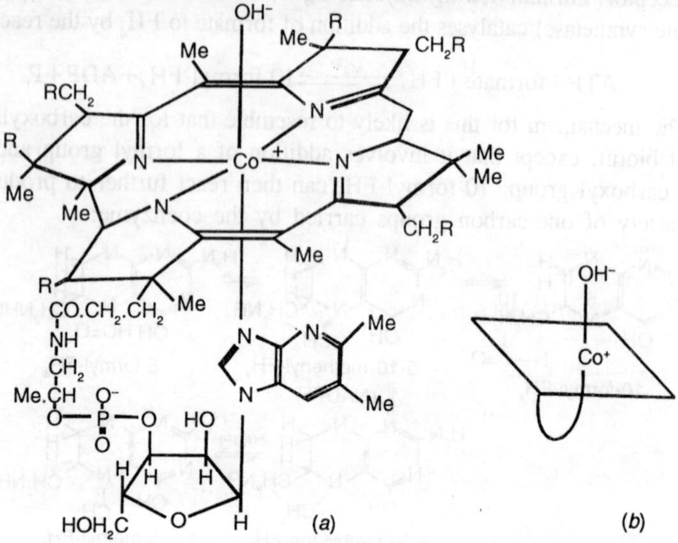

Figure 10.4 : Vitamin B_{12} (hydroxocobalamin): (a) complete structure ($-R = -CH_2CONH_2$, $-Me = -CH_3$); (b) simplified representation.

the metal ion has an oxidation state of +3, but possesses only a single positive charge because of its interactions with various groups; this is indicated as $Co(III)^+$.

The coenzyme forms are obtained after activation of the vitamin by NADH-linked reducing systems:

$$\underset{B_{12a}}{Co(III)^+} \xrightarrow[.OH]{e} \underset{B_{12r}}{Co(II)} \xrightarrow{e} \underset{B_{12s}}{\ddot{C}o(I)}$$

The Co(I) in B_{12s} is strongly nucleophilic and can be alkylated, a bond of largely covalent character being formed. Reaction with ATP, mediated by an adenosyl-transferase, results in the formation of *5-deoxyadenosylcoenzyme* B_{12}, also called *5′-deoxyadenosylcobalamin*. Alternatively, methylation may take place, with the formation of *methylcobalamin*.

5′-Deoxyadenosylcobalamin acts as a coenzyme for a variety of transfer reactions, usually those of an intramolecular kind, in animals and bacteria.

There is evidence, as with all reactions of this type, that the hydrogen atom is transferred via the coenzyme.

Methylcobalamin acts as a coenzyme in several methyl transfer reactions, e.g. in the conversion of homocysteine to methionine in some bacteria and mammals:

$$\text{N5-methyl-FH}_4 + \text{HS.CH}_2\text{CH}_2.\underset{\text{homocysteine}}{\text{CH}(\overset{+}{\text{N}}\text{H}_3)\text{CO}_2} \rightarrow \text{FH}_4 + \underset{\text{methionine}}{\text{H}_3\text{C.S.CH}_2\text{CH}_2\text{CH}(\overset{+}{\text{N}}\text{H}_3)\text{CO}_2}$$

The methyl group is transferred via the cobalt ion of B_{12}: in other bacteria, as well as in fungi and higher plants, this reaction does not involve B_{12}.

The detailed mechanisms of action of 5'-deoxyadenosyl- and methyl-cobalamin are not known for certain; they may well involve further changes in the oxidation state of cobalt.

SPECIFICITY OF ENZYME ACTION

Types of Specificity

A characteristic feature of enzymes is that they are specific in action. Some enzymes exhibit *group specificity*, i.e. they may act on several different, though closely related, substrates to catalyse a reaction involving a particular chemical group. An example of this kind of enzyme is alcohol dehydrogenase, which will catalyse the oxidation of a variety of alcohols. Another is hexokinase, which will assist the transfer of phosphate from ATP to several different hexose sugars. Other enzymes will act only on one particular substrate, when they are said to exhibit *absolute specificity*. For example, glucokinase catalyses the transfer of phosphate from ATP to glucose and to no other sugar (although the specificity of the so-called 'glucokinase' found in liver is less clear-cut, in contrast to the true glucokinases of bacteria and invertebrates).

Uncatalysed reactions often give rise to a wide range of products, but enzyme-catalysed reactions are *product-specific* as well as being *substrate-specific*. Also, in addition to showing chemical specificity, enzymes exhibit *stereochemical specificity*: if a substrate can exist in two stereochemical forms, chemically identical but with a different arrangement of atoms in three-dimensional space, then only one of the isomers will undergo reaction as a result of catalysis by a particular enzyme. For example, L-amino acid oxidase mediates the oxidation of L-amino acids to oxo acids; a separate enzyme, D-amino acid oxidase, is required.for the corresponding oxidation of D-amino acids.

Even greater specificity is shown by the fungal enzyme glucose oxidase, which catalyses the reaction:

$$\beta\text{-D-glucose} + O_2 \rightleftharpoons \text{D-gluconolactone} + H_2O_2$$

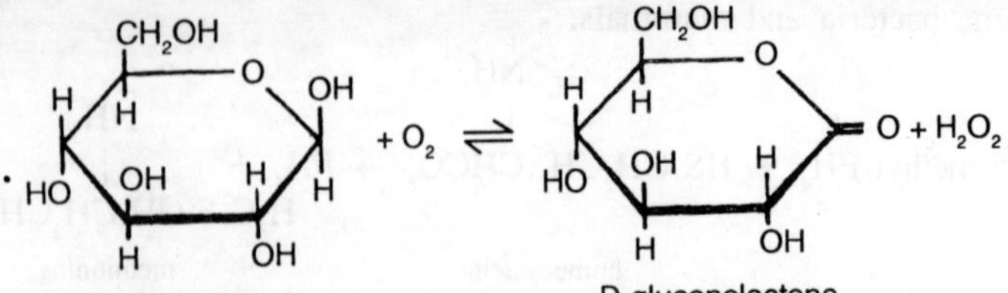

β-D-glucose D-gluconolactone

No other naturally occurring sugar, including α-D-glucose and β-D-galactose, can be acted upon to any appreciable extent.

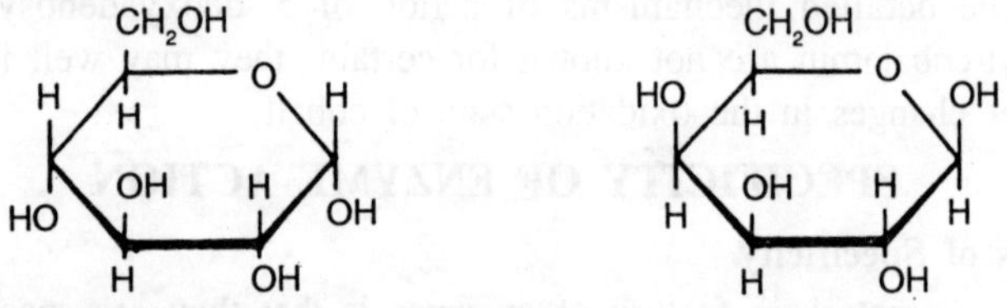

α-D-glucose β-D-galactose

The only enzymes which act on both stereoisomeric forms of a substrate are those whose function is to interconvert L- and D-isomers. An example is alanine racemase, which catalyses the reaction:

$$\text{L-alanine} \rightleftharpoons \text{D-alanine}$$

Enzyme-catalysed reactions may yield stereospecific products even when the substrate possesses no asymmetric carbon atom. For example, the action of glycerol kinase on glycerol always results in the production of L-glycerol-3-phosphate (sn-glycerol-3-phosphate):

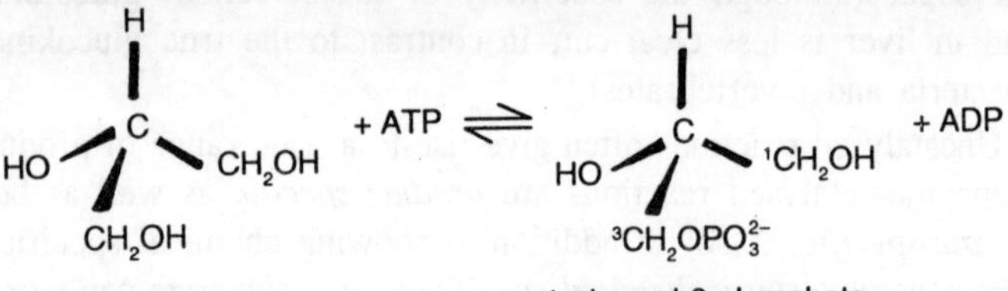

$$\text{glycerol} + ATP \rightleftharpoons \text{1-glycerol-3-phosphate} + ADP$$

$HOCH(CH_2OH)CH_2OH$ → $HO\text{–}CH(^{1}CH_2OH)(^{3}CH_2OPO_3^{2-})$

gloyucerol 1-glycerol-3-phosphate

No L-glycerol-l-phosphate is formed, even though the two -CH_2OH groups of glycerol are chemically identical.

THE ACTIVE SITE

In order to explain the stereochemical specificity of enzymes, Ogston pointed out that there must be at least three different points of *interaction* between enzyme and substrate.

These interactions can have either a binding or a catalytic function: *binding sites* link to specific groups in the substrate, ensuring that the enzyme and substrate molecules are held in a fixed orientation with respect to each other, with the reacting group or groups in the vicinity of *catalytic sites*. For example, sites A″ and A‴ might represent binding sites for R″ and R‴ respectively, and A′ a catalytic site for a reaction involving R′. Thus, even if R′ and R″ are chemically identical (as with glycerol in the glycerol kinase reaction), the asymmetry of the enzyme-substrate complex means that only R′ can react, providing binding site A′ is specific for R‴: R″ can never undergo reaction under these conditions, since it is not brought into the vicinity of site A′ even when R′ binds to site A″.

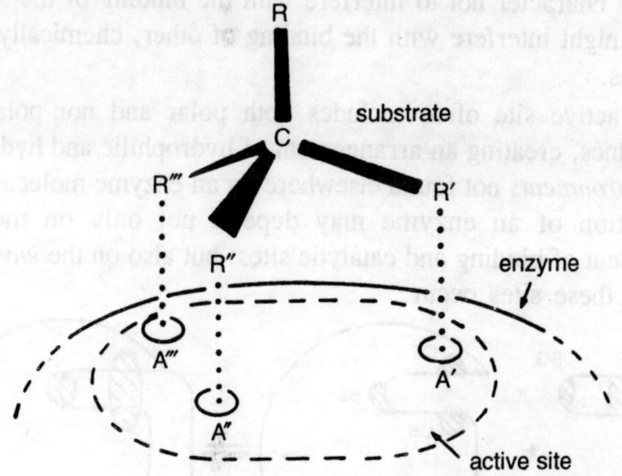

Figure 10.5 : Diagrammatic representation of three-point interaction between enzyme and substrate. A′. A″. and A′ are sites on the enzyme which interact with groups R′. R″ and R′. respectively. of the substrate. Each point of interaction may have a binding or a catalytic function.

Generally similar considerations apply to enzymes catalysing reactions involving more than one substrate. In this case, the reacting groups of each substrate are brought together in the vicinity of one or more catalytic sites.

The region which contains the binding and catalytic sites is termed the *active site*, or *active centre*, of the enzyme. This comprises only a small proportion of the total volume of the enzyme and is usually at or near the surface, since it must be accessible to substrate molecules. In some cases, X-ray diffraction studies have revealed a clearly defined pocket or cleft in the enzyme molecule into which the whole or part of each substrate can fit.

Although the active site is given a planar, it should be realised that it has, in fact, a three-dimensional structure since it consists of portions of a polypeptide chain. The amino acid residues involved may be widely separated in the primary structure, being brought together in space because of the twists and turns within the molecule. The binding and catalytic sites must be either amino acid residues or cofactors, the latter being themselves bound to amino acid side chains. Substrate-binding may involve a variety of linkages, but the bonds formed are usually relatively weak (i.e. non-covalent).

Those amino acid residues in the active site which do not have a binding or catalytic function may nevertheless contribute to the specificity of the enzyme. Their side chains must be of suitable size, shape and character not to interfere with the binding of the substrate, but they might interfere with the binding of other, chemically similar, substances.

The active site often includes both polar and non-polar amino acid residues, creating an arrangement of hydrophilic and hydrophobic *microenvironments* not found elsewhere on an enzyme molecule. Thus, the function of an enzyme may depend not only on the spatial arrangement of binding and catalytic sites, but also on the *environment* in which these sites occur.

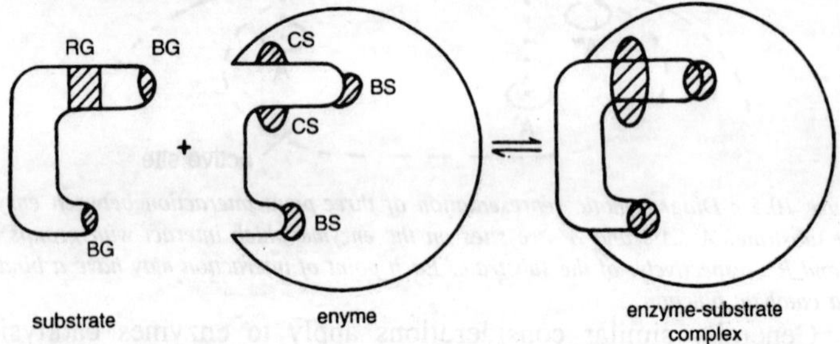

Figure 10.6 : Diagrammatic representation of the interaction between an enzyme and its substrate, according to the lock-and-key model. In the example illustrated the single substrate is bound at two points, bringing the reacting group in the vicinity of two different catalytic sites. (BS = a binding site on the enzyme, CS = a catalytic site, BG = a binding group on the substrate and RG = a reacting group, i.e. a group undergoing enzyme-catalysed reaction.)

Thus it can be seen that the three-point interaction theory provides only a limited explanation of enzyme specificity, a more complete view coming from consideration of a whole range of interactions in three-dimensional space.

THE FISCHER "LOCK-AND-KEY" HYPOTHESIS

As early as 1890, Fischer suggested that enzyme specificity

implied the presence of complementary structural features between enzyme and substrate: a substrate might fit into its complementary site on the enzyme as a key fits into a lock. This is entirely consistent with the more detailed aspects of active site structure. According to the *lock-and-key* model, all structures remain fixed throughout the binding process.

THE KOSHLAND 'INDUCED-FIT' HYPOTHESIS

The lock-and-key hypothesis explains many features of enzyme specificity, but takes no account of theknown flexibility of proteins. X-ray diffraction analysis and data from several forms of spectroscopy, including nuclear magenetic resonance (nmr), have revealed differences in structure between free and substrate-bound enzymes. Thus the binding of a substrate to an enzyme may bring about a *conformational change*, i.e. a change in three-dimensional structure but not in primary structure. This is not necessarily surprising, for the bonds formed between a substrate and its binding sites may have replaced previously existing linkages between each binding site and neighbouring groups on the enzyme. Also, the presence of a substrate at the active site may. exclude water molecules and thus make the region more non-polar. Both of these factors could be responsible for some degree of change in tertiary structure taking place.

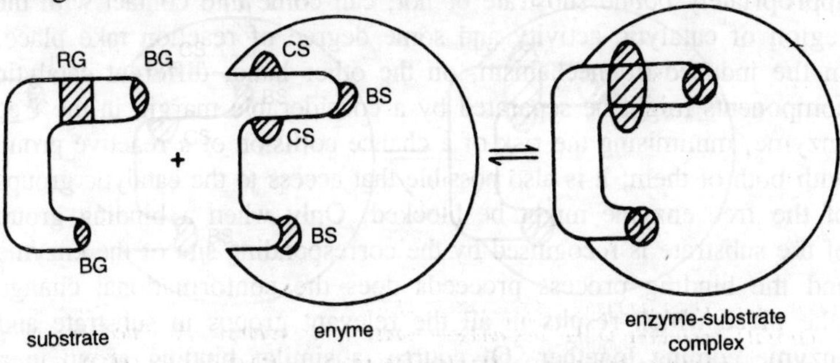

Figure 10.7 : Diagrammatic representation of the interaction. between an enzyme and its substrate, according to the induced-fit model.

Koshland, in his *induced-fit* hypothesis of 1958, suggested that the structure of a substrate may be complementary to that of the active site in the enzyme-substrate complex, but not in the free enzyme: a conformational change takes place in the enzyme during the binding of substrate which results in the required matching of structures. The induced-fit hypothesis essentially requires the active site to be floppy and the substrate to be rigid, allowing the enzyme to wrap itself around the substrate, in this way bringing together the corresponding catalytic sites and reacting groups. In some respects, the relationship between a substrate and an active site is similar to that between a hand and a woollen glove: in each interaction the structure of one component (substrate or hand) remains fixed and the shape of the second component (active site or glove) changes to become complementary to that of the first.

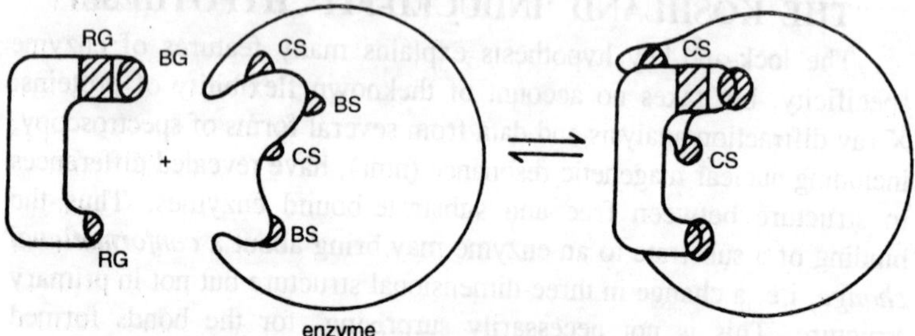

Figure 10.8 : Diagrammatic representation of the non-productive interaction between an enzyme and a compound resembling its substrate, according to the induced-fit model.

Such a mechanism could help to achieve a high degree of specificity for the enzyme. In the lock-and-key mechanism, the active site is always structurally intact, with the catalytic sites aligned and freely accessible. Thus a suitable reacting group, whether part of an appropriately bound substrate or not, can come into contact with the region of catalytic activity and some degree of reaction take place. In the induced-fit mechanism, on the other hand, different catalytic components might be separated by a considerable margin in the free enzyme, minimising the risk of a chance collision of a reactive group with both of them; it is also possible that access to the catalytic groups of the free enzyme might be blocked. Only when a binding group of the substrate is recognised by the corresponding site of the enzyme and the binding process proceeds does the conformational change take place which results in all the relevant groups in substrate and enzyme coming together. Of course, a similar binding group in a substance other than the substrate might trigger off a conformational change but, in general, this would not result in catalytic groups being brought together in the vicinity of an appropriate reacting group, so no reaction would take place. This would be termed *non-productive binding*.

An example of a reaction which appears to proceed via an induced-fit mechanism is that catalysed by *yeast hexokinase*:

$$\text{D-hexose} + \text{ATP} \rightleftharpoons \text{D-hexose-6-P} + \text{ADP}$$

In the absence of hexose, bound ATP is hydrolysed extremely slowly, even though, in chemical terms, this hydrolysis could be brought about by the action of a water molecule in the solvent just as well as by an -OH group in the hexose; this, together with X-ray diffraction evidence, suggests that the binding of the hexose causes a conformational change in the enzyme which activates the ATP.

Conformational changes of the type discussed in this section have been shown to play a part in the mechanism of action of several other enzymes, e.g. *carboxypeptidase* A. They have also been useful in explaining the behaviour of *allosteric enzymes*.

HYPOTHESIS INVOLVING STRAIN OR TRANSITION-STATE STABILISATION

Although the lock-and-key and induced-fit models can explain enzyme specificity, neither suggests any direct mechanism by which the catalysed reaction may be driven forward. Substrate-binding often involves the expenditure of a considerable amount of energy and, although it serves a very useful purpose in bringing reacting and catalytic groups together, further energy must be supplied before the reaction can proceed. Haldane, in 1930, pointed out that if the binding energy was used to distort the substrate in such a way as to facilitate the subsequent reaction, then less energy would be required for the reaction to take place; this concept was developed further by Pauling, in 1948.

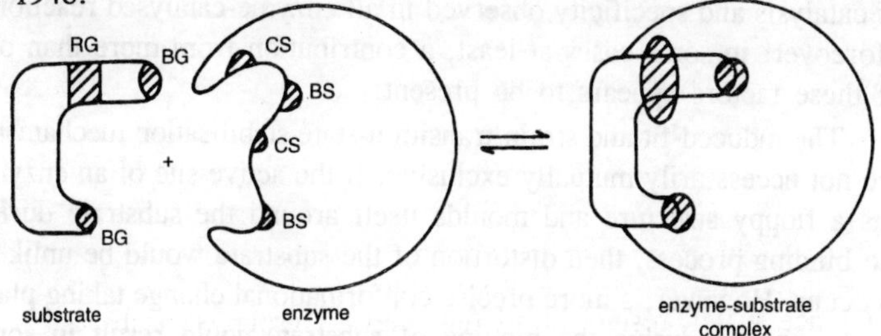

Figure 10.9 : Diagrammatic representation of the interaction between an enzyme and its substrate, incorporating a 'strain' effect.

Let us assume, for example, that the structure of the active site is almost complementary to that of a substrate, but not exactly so. If the structure of the active site is rigid, the substrate must be distorted slightly in order to bind to the enzyme.

This distortion might result in the stretching, and thus weakening, of a bond which is subsequently to be cleaved, thus assisting the forward reaction.

In fact, little clear-cut evidence has been obtained for the occurrence of distorted binding. An alternative, and possibly more likely, mechanism for driving the reaction forward is ***transition-state stabilization***. This assumes that the substrate is bound in an undistorted form, but the enzyme-substrate complex possesses various unfavour-able interactions. These tend to distort the substrate in such a way as to favour the following reaction sequence: enzyme-substrate complex - + transition-state products. As the reaction proceeds, the unfavourable interactions diminish, and are absent from the transition-state.

Thus, the overall effects of ***strain*** and ***transition-state stabilisation*** are very similar, but the sequence of events is slightly different in two cases. An example of an enzyme-catalysed reaction proceeding via a transition-state stabilisation mechanism is the hydrolysis of peptides by *chymotrypsin*. *Lysozyme* is often cited as an example of an enzyme which operates by a strain mechanism, but even in this case the true mechanism may be transition-state stabilisation.

FURTHER COMMENTS ON SPECIFICITY

Specificity of enzyme action is determined by two separate factors: the relative ability of a potential substrate to bind to the enzyme and, once bound, its relative ability to undergo a reaction to form products. Only the overall rate of product formation indicates whether the enzyme can utilize a particular potential substrate.

No single one of the hypotheses is able to account for the features of catalysis and specificity observed in all enzyme-catalysed reactions. Moreover, in some cases at least, a contribution from more than one of these factors appears to be present.

The induced-fit and strain/transition-state stabilisation mechanisms are not necess-arily mutually exclusive. If the active site of an enzyme has a floppy structure and moulds itself around the substrate during the binding process, then distortion of the substrate would be unlikely to occur. However, a more precise conformational change taking place in the protein during the binding of substrate could result in some degree of strain being present in the latter. In either case the enzyme-substrate complex formed could possess internal stress and thus assist the subsequent reaction by transition-state stabilisation. The mechanism for the hydrolysis of peptides by *papain* appears to involve a conformational change in the enzyme in addition to either a strain or transition-state stabilisation factor.

In general, irrespective of the mechanism of an enzyme-catalysed reaction, the major factor governing specificity is the stability of the enzyme-bound transition-state which exists during the conversion of enzyme-bound substrate to products. A potential substrate which can form a relatively stable transition-state when bound to the enzyme will be converted to products at an appreciable rate. This was first pointed out by Pauling in 1948 and has since been confirmed by experiments with ***transition-state analogues***. These are stable compounds which resemble the transition-state com-pounds thought to be formed as part of a reaction sequence. Such analogues have been shown to bind very tightly to the active sites of the appropriate enzymes, more tightly in fact than the corresponding substrate or products.

The investigation of transition-state structure is difficult because it occurs only transiently. under normal conditions. However, if a reaction is carried out at low temperatures (e.g. at -21 °C in aqueous dimethyl sulphoxide), the lifetimes of intermediates are extended and their structures may be studied by techniques such as nmr. This is termed *cryoenzymology*.

11

Enzymes of Nucleic Acid Synthesis

Ligases are a class of enzymes that catalyze the joining of nucleic acid molecules by the formation of phosphodiester bonds between their termini. The nucleic acid substrate to be linked may be DNA or RNA depending on the type of ligase involved.

The enzymes are widespread and have been identified in a range of organisms, including bacteria, phage-infected bacteria, yeasts, amphibians, and mammals, including *Homo sapiens*. Many ligase-deficient and ligase-overproducing mutants have been isolated, and levels of enzyme in the *E. coli* bacterium can vary 1000 fold (from 0.01 to 10 × normal) without deleterious effect.

DNA LIGASES

The role of DNA ligases in vivo is believed to include the joining of short DNA fragments formed during DNA replication, and so enabling DNA synthesis to progress in an overall 3'-5' direction on the antiparallel strand of the double helix, while continual 5'-3' synthesis proceeds on the other strand.

Other possible functions of DNA ligases include roles during genetic recombination and in the repair of UV-damaged DNA, though ligase bacterial mutants can in general perform these functions sufficiently well. In yeast, however, there is evidence to support the idea of a role for DNA ligases in UV damage repair.

In vitro a variety of activities have been detected. These include the joining of hydrogen-bonded cohesive DNA termini (with either

5' or 3' projecting single-stranded regions), the joining of blunt-ended double-stranded DNA molecules, the resealing of single-stranded "nicks" in DNA duplexes, and the interconversion of ATP and AMP by the exchange of pyrophosphate groups.

The two most intensively studied and widely used DNA ligases are that from the *E. coli* bacterium and that occurring in *E. coli* that has been infected with bacteriophage T4. This section of the chapter will concentrate largely on these two enzymes.

E. coli DNA ligase has been purified from several strains that overproduce this enzyme. The T4 enzyme is prepared from *E. coli* that has been infected with either wild-type or replication-deficient T4 phage. Both the *E. coli* and T4 DNA ligase genes have been cloned into recombinant ? vectors.

The Enzyme

E. coli DNA ligase is a monomeric enzyme of asymmetric shape ($S_{20^{,}},,,$ = 3.9S) with a mol mass of 77,000 Da as determined by sedimentation equilibrium, or 74,000 ± 3000 Da by polyacrylamide gel electrophoresis (PAGE). T4 DNA ligase is also an elongated monomer ($S_{20,W}$ = 3.5) of mol mass 68,000 ± 6800 Da determined by gel filtration (18) or 63,000 ± 3200 Da by PAGE (17). T4 DNA ligase exhibits five species on isoelectric focusing. The adenylated form of the enzyme displays a major band with *a pI* of 6.0, and the nonadenylated form has a value of 6.2 for this band.

Enzyme Reaction

Nucleic Acid Substrate

DNA ligase requires as a substrate two DNA termini, the 5' terminus, carrying a phosphate group, and the 3' terminus, a hydroxyl group. These termini must reside on a double-stranded molecule (DNA: DNA or DNA:RNA). Both strands of the duplex may terminate, in the form of a staggered end or a blunt end, and the ligase then requires a second similar double-stranded terminus and proceeds tojoin the two in an intermolecular reaction. Alternatively, the two termini may be provided by a nick in just one strand of a duplex, which the enzyme will then seal.

The enzyme will not ligate a 3' phosphate group to a 5' hydroxyl, a 3' hydroxyl to a 5' hydroxyl, a 3' dideoxy nucleoside to a 5' phosphate, or a 3' hydroxyl to a 5' triphosphate terminus. If only one of the 5' termini of a double-stranded break is phosphorylated, DNA ligase can only rejoin that one strand and the resultant product is a nicked molecule, since the second strand cannot be ligated. However, this doublestranded DNA molecule is intact and can then be used in transformation reactions enabling the ligation to be completed in vivo.

Only T4 DNA ligase will ligate blunt-ended DNA molecules, the reaction proceeding by way of nicked intermediates. Unlike joining cohesive ends, the rate is not linearly dependent on enzyme concentration. The enzyme can also ligate at a mispaired 3' base and will ligate DNA/RNA hybrids, but with much reduced activity.

The reverse reaction is also catalyzed, the enzyme behaving as an AMP-dependent endonuclease, yielding nicked DNA. The two activities of DNA ligase result in the slow (but eventually complete) relaxation of supercoiled DNA.

The only action on RNA performed by *E. coli* DNA ligase is the ligation of the 3' hydroxyl of an RNA strand to a phosphorylated 5' DNA terminus. T4 DNA ligase has some activity in joining RNA molecules annealed to DNA and even RNA:RNA.

E. coli DNA ligase has a K_m for the 5' phosphate group of 2.5-5.6 $\times$ 10^{-8} M. The value for T4 DNA ligase is 6 $\times$ 10^{-7} M for cohesive ends, 5 $\times$ 10^{-5} M for blunt ends, and 1.5 $\times$ 10^{-9} M for repairing nicks.

Cofactors

DNA ligase requires a nucleotide cofactor for reaction, forming a covalent AMP-enzyme intermediate. *E. coli* DNA ligase utilizes NAD (as do ligases from *B. subtilis, T thermophilus,* and *S. typhimurium)*. The enzyme has a high specificity for NAD with a K_m of 3 $\times$ 10^{-8} to 7 $\times$ 10^{-6} M.

T4 DNA ligase (and T7 and mammalian ligases) requires ATP. It will also utilize dATP (at 0.5% of the rate), which acts as a competitive inhibitor with ATP. The value of K_m *ATP* for the joining reaction is 1.4-10.0 $\times$ 10^{-5} M and the $K_{1\ dATP}$ is 3.5 $\times$ 10^{-5} M. For the pyrophosphate exchange reaction used in the enzyme assay, $K_{m\ ATP}$ is 2 $\times$ 10^{-6} M and $K_{1\ dATP}$ is 1 $\times$ 10^{-5} M. T7 DNA ligase can in fact utilize either ATP or dATP.

Temperature

Rates of reaction are very temperature-sensitive, the optimum depending on the lengths of the oligonucleotides being joined, and being greater than the T_m of the substrate. *E. coli* DNA ligase has an optimum temperature of 10-15°C for ligating cohesive ends.

Another bacterial ligase from the thermophilic bacterium *Thermus thermophilus* is becoming widely used as further applications of the polymerase chain reaction are developed. The enzyme is similar to *E. coli* DNA ligase in size, pH optimum, cofactor and cation requirement, and response to activators. However, its temperature optimum is much higher, being 24-37°C for ligating cohesive termini, and 65-72°C for closing nicks (within a range of 15-85°C), which may result in more widespread applications of the technique of ligation.

T4 DNA ligase has an optimum temperature for ligating cohesive ends of 4°C and for sealing nicks of 37°C. The optimal temperature for blunt-end ligation is 25°C for 16 mers or longer, smaller molecules requiring lower temperatures consistent with their decreased melting temperatures.

pH

Optimal pH for ligation varies to some extent with the reaction catalyzed and the buffer employed. *E. coli* DNA ligase has a pH optimum for joining DNA strands of 7.5-8.0 in Tris-HC1 buffers and of 8.0 in sodium phosphate. For the phosphate exchange reaction, the optimal pH is 6.5 in potassium phosphate, the rate of reaction falling to 50% at pH 5.6 and 7.5. This reaction rate falls to 20% in TrisHCl buffer at pH 8.0.

T4 DNA ligase has a similar pH optimum range for joining DNA, the reaction rates falling to 40% at pH 6.9 and 65% at pH 8.3. The pH optimum for the exchange reaction is also similar to that for *E. coli* DNA ligase. For the sealing of nicks, the optimal pH range is 7.2-7.8 in Tris-HCI, with the rates dropping to 46% at pH 6.9 and to 65% at pH 8.5.

Cations

DNA ligases require a divalent cation, with Mg^{2+} being the most commonly utilized. The *E. coli* enzyme requires Mg^{2+} at an optimum concentration of 1-3 mM. Low concentrations of Mn^{2+} can substitute, and in the range 0.2-1.0 mM cause a slight increase in activity, though higher concentrations are inhibitory. Co^{2+} and Nil are inactive, Zn^{2+} has slight activity, and Cal has been reported to yield 60% activity in some cases and none in others.

T4 DNA ligase has a magnesium optimum of 10 mM, whereas Mn^{2+} is only 25% as efficient. However, in joining DNA:RNA hybrids, Mn^{2+} is twice as effective as Mg^{2+}.

Activators and Inhibitors

Ammonium ions in low concentrations stimulate the *E. coli*

enzyme, in joining DNA termini, and V_{max} can be increased by up to 20-fold. K^+ and Rb^+ show similar stimulation, Cs^+ and Li^+ exert a small effect, and Na^+ is ineffective.

In contrast, the T4 DNA ligase is unaffected by low concentrations of ammonium ions, whereas higher levels (0.2M) of NH_4^+, Na^+, K^+, Cs^+, and Li^+ inhibit almost completely. Blunt-end ligation is inhibited by 25 mM phosphate and 50 mM Na^+.

Polyamines, such as spermine and spermidine, also inhibit, but this can be overcome by increasing the DNA concentration. Spermine is the most effective, inhibiting the joining reaction by 90%. Some workers, however, recommend the use of spermidine.

Sulfhydryl Reagents

E. coli DNA ligase does not need the presence of sulfhydryl reagents. T4, T7, and mammalian DNA ligases do require either β-mercaptoethanol or dithiothreitol (DTT). T4 DNA ligase performs much more efficiently if DTT is the reagent of choice.

Enzyme Assay

There are several types of assays carried out to define the activity of DNA ligase. Functional assays include the ligating of radiolabeled DNA to unlabeled DNA in solution or bound to a matrix, and the conversion of poly A:T to a form resistant to exonuclease III. Direct measurements can be made of the conversion of radiolabeled phosphate monoesters to diesters resistant to alkaline phosphatase activity. Also, the restoration of biological activity to DNA previously nicked by pancreatic DNAse or restriction endonucleases can be used as an assay.

Three types of units are defined commonly. First, a Weiss unit catalyzes the exchange of 1 nmol of ^{32}P from inorganic pyrophosphate to ATP in 20 min at 37°C, 0.015 Weiss units ligate 50 % of the ***Hind*III** fragments from 5 μg λ DNA in 30 min at 16°C. Second, the ModrichLehman unit, based on the exonuclease resistance assay, is equivalent to 5 Weiss units. Finally, cohesive end units are functionally defined by various commercial suppliers, and are generally much smaller than Weiss units and difficult to relate quantitatively.

EXPERIMENTAL PROCEDURES

Uses of DNA Ligases

The most widespread use of DNA ligase is in the construction of recombinant DNA molecules. This may be necessary for the cloning of cDNA or genomic fragments for construction of libraries, or for mapping, sequencing, or use as probes. The use of ligase in the inverse polymerase chain reaction facilitates the cloning of segments of genomic DNA some distance from known sequences. DNA ligase may also be used in the assembly of genes from DNA fragments and synthetic oligomers.

Other uses of DNA ligase include the detection of nicked DNA by the release of AMP, nearest neighbor analysis following kinasing, use in mutagenesis, and the making of affinity columns by attaching DNA to solid matrices. In most cases, the enzyme of choice is T4 DNA ligase. *E. coli* ligase is not widely used, because it is inefficient with blunt ends. However, *E. coli* DNA ligase is used for cDNA cloning by replacement synthesis where a virtue is made of its inability to ligate RNA to DNA to form spurious products.

Examples of specific ligation protocols may be found in Gaastra and Hansen. The following details apply to the use of T4 DNA ligase, except where specific reference is made to the *E. coli* enzyme.

Storage and Stability

T4 DNA ligase is usually supplied at a concentration of 1-5 U/ μL. It can be diluted in reaction buffer immediately prior to ligation, but for longer term storage, a specific storage buffer should be used.

Storage buffers usually contain:

- 10-20 mM Tris-HCl (or potassium phosphate), pH 7.4-7.6
- 50-60 mM KC1
- 1-5 mM DTT
- 200 pg/mL BSA
- 50% Glycerol

High concentrations of enzyme are very stable at -20°C. If diluted to 500 U/mL, the enzyme is 90% stable at this temperature for 6 mo. Lower concentrations are increasingly less stable. For example, a 10 U/mL dilution will lose 40% of its activity in 3 mo. Storage at temperatures lower than -20°C may also be detrimental.

Reaction Conditions-Cohesive Termini

Reaction conditions can vary with the purity of each batch of enzyme, the purity of the DNA substrate, and the presence of buffer components from any previous or subsequent in vitro reaction step. DNA ligase acts in virtually all restriction enzyme buffers and also in the same buffer used to kinase linkers prior to their ligation.

A commonly used buffer for ligation reactions with T4 DNA ligase is given below with a range of acceptable parameters:

20 mM Tris-HC1 pH 7.6	(20-66 mM, pH 7.5-7.8)
10 mM $MgC1_2$	(5-50 *mm)*
10 mM DTT	(5-20 mM)
1 mM ATP	(66 μM- 1 mM)
0.1 Weiss units enzyme	(0.01-1.0 U/μg DNA)

The inclusion of BSA (Fraction V) at 50 μg/mL is optional. DNA concentration:

Reaction vol 20 μL	(10-100 μL)
Temperature 16°C	(4-25°C)
Time 4 h	(20 min-24 h)

If DNA substrates with cohesive ends are being ligated, the DNA should be heated to 70°C for 10 min to melt any hydrogen bonding between the termini of similar molecules before the various DNA substrates are mixed. After heating, the mixture should be allowed to cool slowly to allow intermolecular reannealing, prior to the addition of the reaction buffer and finally the DNA ligase.

Reaction times and temperatures have an inverse relationship. For example, similar results might be achieved by incubation at 25°C for 1 h, at 15°C for 4-6 h, and at 4°C for 16 h.

After reaction, the enzyme can be inactivated at 65°C for 10 min or by addition of dimethyl dicarbonate to 0.1 % and heating to 37°C for 10 min. The DNA can then be concentrated by ethanol precipitation, which also removes magnesium ions that may reduce the efficiency of a subsequent transformation step. If ethanol precipitation is not performed, the magnesium concentration should be reduced by fivefold (or greater) dilution of the ligation mixture.

Ligation can be achieved directly on DNA in low melting agarose excised from a gel, though more enzyme is required and the efficiency is 10-fold lower. The gel should be run in TAE buffer (40 mMTris-acetate, 1 mM EDTA, pH 7.5-8.0) rather than the more usual Tris-borateEDTA (TBE). The presence of ethidium bromide in the gel at 0.5 μg/mL causes no ill effects.

E. coli DNA ligase will function in a similar buffer to T4 DNA ligase, although the Tris-HC1 concentration is usually reduced (20-30 mm) and ATP replaced by 25-50 μM NAD. DTT (20 mM) is often included in the buffer, although this may be superfluous. In some cases, 10 MM ammonium sulfate and 10 mM KCl are also added.

Reaction Conditions—Blunt Termini

The ligation of blunt- or flush-ended DNAs is only possible using

T4 DNA ligase and not *E. coli* DNA ligase. The reaction is slower than the ligation of cohesive ends, but is extremely useful in that it can be used to join virtually any two pieces of DNA, be they naturally occurring or synthetic.

In practice, this ability can be made universal by the infilling or "polishing" of cohesive or incompatible termini. This technique can be adapted to infill partially "ragged" termini leaving short cohesive termini that are not self-ligatable, but that will accept inserted DNA.

Blunt-end ligation can be performed in the standard buffer with the ATP concentration adjusted to 0.5 *mM*. More ligase should be used, of the order of 50 U/mL and 1-2 U/μg DNA. Optional additions for blunt-end ligation include 100 μg/mL BSA, 1 mM hexamine cobalt chloride, and 0.1-1.0 mM spermidine. The latter two reagents can increase the efficiency of linker ligation by fivefold.

The DNA concentration should also be raised. In the case of small oligomers, such as linkers, it is easy to achieve a high concentration of DNA termini of the order of 4-20 μM or up to 100-fold in excess of the DNA fragment to be linked. The effective concentration of the DNA can be increased by the use of condensing agents or volume excluders, such as polyethylene glycol (PEG), Ficoll, and hexamine cobalt chloride, which accelerate the rate of ligation by up to 1000-fold, and permit ligation at lower absolute DNA and enzyme concentrations. Condensing agents alter the distribution of the products and suppress intramolecular reaction (recircularisation), most products being linear multimers.

PEG8000 gives maximum stimulation of ligation when present at a concentration of 15%, this effect being achieved at a concentration of 0.5 mM ATP and 5 mM $MgC1_2$. Even slight increases in the ATP or decreases in the magnesium ion concentrations greatly diminish the effect of PEG. PEG enhances the ligation of blunt-ended oligomers as short as 8 by and also simulates ligation of cohesive ends by 10-100-fold.

Hexamine cobalt chloride stimulation depends on the concentration used, 1.0-1.5 pM having maximal effect. Blunt-end ligation is stimulated 50-fold, but cohesive-end ligation only fivefold. Ligation is possible in the presence of monovalent cations (e.g., 30 mM KCl), but the products are then largely recircularized. This reagent does not significantly increase the ligation of short oligomers.

T4 RNAligase will stimulate T4 DNA ligase activity, but is not widely used since PEG is more efficient and inexpensive. RNA ligase cannot perform blunt-end DNA ligations itself, nor will it activate the *E. coli* enzyme.

Ligation Products

The products of ligation depend largely on the nature of the DNA substrates employed. Removal of the 5'-phosphate groups from one of the substrate molecules (e.g., a vector) by alkaline phosphatase treatment will prevent self-ligation and favor the formation of recombinant products. The use of DNA molecules carrying different cohesive termini at each end (asymmetric cloning) also prevents self-ligation and offers control over the relative orientation of the DNA fragments in the resultant products.

The products obtained also depend on the absolute concentrations and ratios of the DNA substrates present in the mixture. Low DNA concentrations favor intramolecular reaction (recircularisation), and higher concentrations favor intermolecular reaction (oligomerisation and formation of recombinant molecules). Condensing agents exert a similar effect. For example, 10% polyethylene glycol enhances intermolecular ligation. Longer DNA molecules are more prone to recircularisation than shorter ones.

Optimal ratios usually have to be determined empirically, though they are fairly broad, and a 30% variation in starting concentrations makes little difference. Some theoretical guidelines have been suggested by Revie et al. for vectors of 2.5-7.5 kb and inserts of 0.210 kb, which can be briefly summarized as follows;

1. Always use forced directional (asymmetric) cloning and phosphatased vectors where possible.

2. Keep the vector DNA concentration below I gg/mL, unless "scavenging" for minute quantities of insert DNA

3. Under normal conditions, use 3:1 molar ratio of insert:vector, but not more than 5 pg/mL insert DNA.

4. Increase the insert concentration as the vector size decreases.

In practice, a suitable preliminary experiment would use a concentration of 20-60 μg/mL vector (2-3 kb in size) with an equimolar or slightly greater concentration of insert DNA.

The above rules apply only to plasmid ligations. For the formation of λ or cosmid concatamers, insert and vector should be ligated at a 1:1 molar ratio in as high a concentration as possible.

In all cases, the ability to select for the recombinant DNA molecule of choice by, for example, transformation and expression of an antibiotic resistance or by enzymic formation of a coloured product from a chromogenic substrate would offer much more flexibility in the choice of DNA concentrations and reaction conditions of the ligation.

Reaction Protocol

A specific example of a ligation reaction protocol for DNA fragments with cohesive termini is as follows:

Materials Required

1. 10 mM Tris-HCl/1 mM EDTA, pH 7.6.
2. Phenol.
3. Chloroform.
4. Ethanol.
5. 10X Ligation buffet.
   - 200 mM Tris-HC1, pH 7.6
   - 50 mM $MgC1_2$
   - 50 mM DTT
   - 500 μg/mL BSA
6. 10 mM ATP

Protocol

1. Prepare the fragments to be legated by digestion of the source DNA species (e.g., vector and foreign genomic DNA) with restriction endonucleases Purify by phenol/chloroform extraction and ethanol precipitation. Resuspend each DNA in *10 mM* Tris/1 mM EDTA, pH 7.6.

2. Mix aliquots containing 0.1 μg of each DNA species.. Add 15 μL water. Warm to 70°C for 10 min, to melt any annealing of cohesive ends between molecules of the same species. Allow to cool slowly to room temperature, and then place on ice.

3. Add 2 μL of l0X ligation buffer, plus 1 μL 10 mM ATP.

4. Add 0.1 Weiss unit T4 DNA ligase.

5. Incubate at 16°C for 4 h.

6. Terminate the reaction by heating to 70°C for 10 min. Aliquots of the ligation mixture can then be used for bacterial transformation either directly or after ethanol precipitation.

T4 RNA LIGASE

Like T4 DNA ligase, this enzyme is found in *E. coli* after infection with T -even phage. It has also been cloned and is now available from several overproducing strains of *E. coli*. The enzyme forms a phosphodiester bond between a phosphorylated 5' terminus and a 3' hydroxyl similarly to the DNA ligases, except that the substrates in this case are in general RNA moieties.

The in vivo role of the enzyme is unclear, although it is probably involved in host RNA modification and transcript splicing. The protein does have a morphological role in the noncovalent attachment of tail fibers to the base plate of the phage.

The Enzyme

The mol mass of the enzyme has been determined as 48,000 Da by sedimentation equilibrium, 47,000 Da by gel filtration, and 41,00045,000 Da by SDS-PAGE. It is monomeric in solution and has an isoelectric point of 6.1.

Enzymic Reaction

Nucleic Acid Substrate

The normal substrate of RNA ligase is single-stranded RNA, although it will act on a variety of single- or double-stranded RNA or DNA molecules. RNA ligase will act on very small pieces of ribonucleic acid, with 40 mers being the probable upper size limit. The minimum size of the 5' moiety is a ribonucleoside *3'5'-bis* phosphate (i.e., P-base-P), a 5'-monophosphate being ineffective. The nature of the base has some bearing on the reaction rate, pyrimidines reacting 2-10 times faster than purines, and modified purines slower yet. With these reservations, a wide variety of bases will be accepted, including deoxynucleosides, O-methylated-, halogenated-, dihydro-, thio-, and deaza-derivatives, and deaminated purines. The minimum size of the 3' moiety is a trinucleoside bis-phosphate, base-P-base-P-base-OH.

Cofactors

The enzyme requires ATP (although dATP will substitute) and forms an enzyme-AMP complex during reaction. Low levels of ATP are preferred, often provided by an ATP-generating system.

Cations

Magnesium ions are required for reaction, although Mn^{2+} will substitute and often improve the rate of reaction; 1M $(NH_4)_2SO_4$ completely inhibits activity.

Enzyme Assay

The enzyme is usually assayed by measuring the conversion of

5'^{32}P-labeled poly-A to a circular form insensitive to phosphomonoesterase at 37°C. One unit converts 1 nmol of 5'-^{32}P-labeled poly-A at a concentration of 1 μM or 10 μM termini to a phosphataseresistant form in 30 min at 37°C.

Another definition of the unit of activity is that it ligates 1 pmol of ^{32}P-labeled mononucleoside biphosphate to a larger TCA-precipitable RNA species in 30 min at 4°C. This unit is approximately equivalent to one-sixth of that defined by the phosphatase resistance assay.

A third unit can be defined on the basis of the pyrophosphate exchange reaction. One unit of enzyme catalyzes the exchange of 1 nmol of ^{32}Plabeled inorganic pyrophosphate into ATP in 30 min.

EXPERIMENTAL PROCEDURES

Uses of RNA Ligase

The first obvious we of RNA ligase is in synthesizing oligomers. Modification of preexisting RNA molecules is perhaps a more interesting application. This may involve merely the end-labeling of an RNA moiety, or alternatively, bases can be removed by periodate oxidation and n-elimination, and then modified bases can be added by RNA ligase to examine the restored function. Internal modification can also be achieved after nicking or treatment with RNAse H. The enzyme can also be used to join DNAs, although this requires a large concentration of enzyme and a longer time.

Storage and Stability

T4 RNA ligase is available at specific activities of 1000-4000 U/mg protein corresponding to 1-4 U/μL. The enzyme should be diluted to the required concentration in storage buffer, which is similar to that used for T4 DNA ligase. In this case, the Tris-HC1 concentration may be raised to 50 mM, and KC1 may be omitted. The enzyme should be held at -20°C.

Reaction Conditions

As with all manipulations involving RNA, great care should be taken to eliminate the presence of ribonucleases. In practice, it is usually sufficient to ensure that all solutions and vessels have been autoclaved prior to use.

T4 RNA ligase reaction buffers commonly consists of:

- 50 mM HEPES/NaOH pH 7.5 (or 50 mM Tris-HO, pH 8.0)
- 10 mM $MgC1_2$ (10-18 MM)
- 20 mM DTT (3-33 mM)

- 1 mM ATP (0.1-1 0 mM)
- 10% (v/v) DMSO
- 100 μg/mL *BSA* (10-100 μg/mL)

Common alternative concentrations are shown in parentheses.

The substrate termini should be present in a concentration of 1-100 μM (corresponding to 0.5-20.ig RNA), in a reaction vol of 20-50/μL. The amount of RNA ligase required will vary with each application within the range 0.1- 16.0 U.

The temperature of reaction must be optimized for the specific application, conditions ranging from 37°C for 30 min to 4°C overnight.

After this time, the reaction should be terminated by boiling for 2 min.

DNA METHYLTRANSFERASES

DNA methyltransferases (Mtases) catalyze the transfer of the S-methyl group of S-adenosylmethionine (SAM) to deoxycytosine (dC) or deoxyadenine (dA) bases within defined DNA sequences. Individual enzymes are specific for one or the other base, and modify at the 6-NH_2 of dA or at the N^4 or 5-C position of dC depending on the particular enzyme (2). The reaction is predominantly irreversible. Enzymes, such as O^6-methylguanine DNA Mtase, that participate in DNA repair processes will not be discussed here.

Both mammalian and bacterial DNA Mtases have a sequence-specificity component that is an integral part of the reaction. However, sequence recognition is much more relaxed in the case of mammalian Mtases. Moreover, all known mammalian DNA Mtases are dC-specific, whereas both dA and dC specificities are found in bacteria. A minimum of four and a maximum of eight specific bases are found in the DNA recognition sites of bacterial enzymes, whereas mammalian Mtases require only a CpG doublet. (There is some evidence that CpN sequences are also modified in other higher organisms and that plant DNA Mtases can also methylate CpNpG motifs.) The biological consequences of DNA methylation are diverse, but are usually the result of a modulation of protein:nucleic acid interactions induced by methylation. Thus, for example, in prokaryotes, site-specific DNAmethylation is sufficient to render a sequence refractory to endonucleolytic cleavage by a restriction enzyme of the same specificity. This is the molecular basis of restriction and modification. In mammalian cells, cytosine methylation has been implicated in the regulation of gene transcription and in the inactivation of the X chromosome, although the molecular details of these processes remain unresolved. The reaction catalyzed by all DNA Mtases is shown below. SAM invariably acts as the methyl donor, and the natural acceptor is chromosomal DNA, but in vitro, plasmid DNA, phage DNA, or an appropriate oligonucleotide duplex can also act as methyl acceptors.

$$\text{DNA} + \text{SAM} \rightarrow \text{McDNA} + \text{SAH}$$

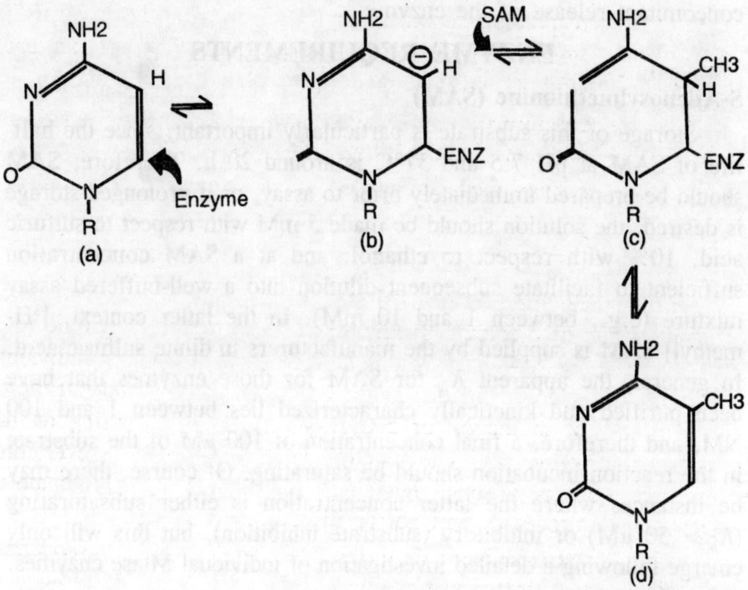

NH2 N H O N R (a) Enzyme NH2 N H ENZ N R (b) SAM NH2 CH3 H ENZ O N R (c) NH2 N CH3 O N R (d)

Figure 12.1 : Proposed mechanism of action of 5' cytosine-specific DNA methyltransferases based on the studies of thymidilate synthase and Hha I DNA methyltransferase.

The products of the reaction are methylated DNA and S-adenosylhomocysteine (SAH). The site of modification is enzyme-specific. Little is known about the mechanism of dA methylation, whereas studies on the mechanistically related enzyme thymidilate synthase have led to the proposed scheme for dC methylation shown in Figure elsewhere in this chapter. Interestingly, a catalytic cysteine flanked on the N-terminal side by a proline residue is found in all of the pyrimidine-specific Mtases, which modify at a carbon atom so far characterized. Indeed, amino acid sequences of the dC DNA Mtases from mouse, human, and a variety of bacteria, although otherwise quite dissimilar, share a pro-cys dipeptide along with a putative specificity determining sequence toward the Cterminus of the polypeptide.

It has been suggested that the first stage of C-5, dC methylation involves the nucleophilic attack by the cysteine-sulfhydryl group at the C-6 position of the heterocycle resulting in the transient formation of a covalent enzyme:nucleic acid adduct. Subsequent abstraction of a proton at the C-5 position following the addition of the activated methyl group of SAM leads to the formation of 5-methyldC with the concomitant release of the enzyme.

ENZYME REQUIREMENTS

S-Adenosylmethionine (SAM)

Storage of this substrate is particularly important, since the half-life of SAM at pH 7.5 and 37°C is around 20 h. Therefore, SAM should be prepared immediately prior to assay, or if prolonged storage is desired, the solution should be made 5 mM with respect to sulfuric acid, 10% with respect to ethanol, and at a SAM concentration sufficient to facilitate subsequent dilution into a well-buffered assay mixture (e.g., between 1 and 10 mM). In the latter context, [^{3}H-methyl] SAM is supplied by the manufacturers in dilute sulfuric acid. In general, the apparent K_m for SAM for those enzymes that have been purified and kinetically characterized lies between 1 and 100 NM, and therefore, a final concentration of 100 μM of the substrate in the reaction incubation should be saturating. Of course, there may be instances where the latter concentration is either subsaturating ($K_m > 50$ μM) or inhibitory (substrate inhibition), but this will only emerge following a detailed investigation of individual Mtase enzymes.

DNA

In order to assay the activity of a DNA Mtase during a purification protocol, [^{3}H-methyl] SAM is often employed as the methyl donor. In this way, a sensitive estimation of methyltransfer can be obtained by liquid scintillation counting. Alternatively, methylation of a DNA substrate can be monitored by subsequent restriction enzyme analysis. In each case, the DNA substrate should be carefully chosen in order to maximize the incorporation of methyl groups during the incubation. For example, the rate of methylation of calf thymus DNA by a vertebrate DNA Mtase will be slower than that of a bacterial DNA preparation, since the frequency of methylated CpG dinucleotides will be much greater in the former substrate. Furthermore, it has been shown that in the case of mammalian and some (but not all) prokaryotic DNA Mtases, hemimethylated DNA is the preferred substrate. Therefore, a judicious choice of DNA substrate can improve the sensitivity of the assay considerably. In the case of dA methylation, a chromosomal DNA preparation from a *dam*⁻ strain of *E. coli* will be free of adenine methylation and is therefore a convenient substrate for monitoring activity during purification.

More defined substrates may be required in those cases where the sequence specificity of the enzyme is known. For example, in order to determine the specific activity of the enzyme *EcoRI* Mtase, a plasmid or oligonucleotide duplex containing one or more GAATTC recognition sequences would be a suitable substrate. Incubation of the latter enzyme preparation with a plasmid containing no *EcoRI* sites would serve as a control against nonspecific methylation. This is especially important during the early stages of a protein purification.

Cofactor Requirements

Restriction and modification enzymes are classically divided into three groups that reflect their structural and functional complexity. The Type I modification enzymes are the most complex and require Mg^{2+} ions (-5 mM), and ATP (100 MM) for optimal activity. In general, Type II and III enzymes catalyze methyltransfer in the absence of ATP and Mg^{2+}, although there are some exceptions, such as *TaqI* Mtase, which do require divalent cations. Mammalian DNA Mtases require the presence of neither of the above cofactors, although it has been reported that Co^{2+} (100 MM) ions stimulate some mammalian enzymes.

Ionic Strength

All enzymes have an intrinsic optimal ionic strength for activity, and this can only be determined on an empirical basis. However, in general, the optimum salt (usually NaCl or KCl) concentration is that of the corresponding restriction endonuclease in the case of prokaryote DNA Mtases. Mammalian dC Mtases have been shown to be extremely sensitive to KCI: a concentration of 25 mM giving optimal activity with the majority of mammalian enzymes

Buffer Components

Again for any given enzyme, the buffer that balances optimum rates with complementary stabilisation of the enzyme can only be determined by a program of empirical experiments. Tris buffers have been traditionally employed for restriction and modification enzymes, although there is no *a priori* reason why this should be the case. Indeed, phosphate buffers are generally easier to titrate to a given pH and are excellent buffers at pH 7.4, the pH optimum of many if not all of the DNA Mtases so far isolated. For convenience, however, the three types of restriction enzyme buffers described by Maniatis et al. (9) are suitable for most purposes. The buffering agent is usually employed at a concentration of between 50 and 100 mM and is often supplemented with dithiothreitol (1 mM) or β-mercaptoethanol (5 mM) and bovine serum albumin (100 μg/mL). The latter reagents are not always necessary and are occasionally inhibitory. If anew Mtase is being isolated, it is best to determine the enzyme activity in the presence and absence of these reagents.

The reaction is usually carried out at 37°C for between 1 and 2 h and quenched by the addition of an equal vol of phenol. The reaction vol should be around 20 μL, although preparative incubations of 200 μL may be required if the DNA is to be analyzed further.

Experimental Procedures

Materials

1. 50 mM Tris-HCI, pH 8.0, 100 mM NaCl, 1 mM EDTA.
2. Bio-Rad A 0.5M gel.
3. 0.3 M Potassium acetate (or ammonium acetate).
4. Ethanol.
5. [^{3}H-methyl]-S-adenosylmethionine (SAM).
6. 100 mM NaCl.

Assay

The quantitative estimation of DNA Mtase activity can be acheived in several ways. However, most methods are based on one of two principles. The first and perhaps the simplest involves the enzyme catalyzed transfer of tritiated methyl groups to DNA from [3H-methyl] SAM followed by separation of the reaction components either by selective precipitation or by chromatography. The tritiated DNA produced in the incubation is subsequently estimated by liquid scintillation counting.

In the tritium transfer method, the DNA Mtase enzyme is incubated with the appropriate substrates, and the reaction mixture quenched with phenol. The tritiated DNA can then be isolated by chromatography through a Pasteur pipet-sized column of Bio-RadAO. 5M gel-filtration medium (or similar). Typically, the reaction mixture is loaded onto the column and eluted with a buffer containing 50 mMTris-HC1, pH 8, 100 mM NaCl, and 1 mM EDTA. The DNA elutes well in advance of the unreacted SAM. Alternatively, the DNA can be precipitated with 3 vol of cold ethanol in the presence of 0.3M potassium (or ammonium) acetate. If precipitation is used, care should be taken not to "carry over" any unreacted SAM. Moreover, stringent controls in the absence of individual reaction components should be included in order to verify the assay results. This is particularly important in assays of crude extracts, which probably contain protein and RNA Mtases.

A different strategy for estimating DNA Mtase activity involves the use of DNA fragments containing specific enzyme recognition sites. This approach is particularly convenient in assaying the DNA Mtases of bacteria. In one method, a DNA fragment containing the methylation site of, for example, EcoRI Mtase is incubated with the enzyme in the presence of unlabeled SAM, and the modified DNA so produced is extracted with phenol and precipitated with ethanol. The DNA is subsequently incubated with EcoRI restriction enzyme. The extent of methylation can be determined by the extent of resistance to endonuclease digestion. This approach can be made highly sensitive by incorporating radioactively labeled DNA into the reaction mixture. The DNA fragments can either be resolved by gel electrophoresis or by HPLC.

APPLICATIONS OF DNA METHYLTRANSFERASES

DNA Mtases have been incorporated into a number of molecular biology protocols and generally bring a greater degree of flexibility to molecular cloning strategies. Some of the uses of DNA MTases are outlined in the following sections.

Construction of Gene Libraries

A partial restriction digest of genomic DNA from any organism can form the starting material for the construction of a genomic library. In order to take advantage of the versatile bacteriophage λ vectors, which contain single unique restriction sites (e.g., λgt11, λgt10, and so on), DNA must be either digested with *EcoR 1* (compatible with the unique cloning site of λgt10 and λgt11) or with any enzyme followed by the addition of *EcoRI* linkers to the ends of the DNA fragments. The "activation" of the linkers requires an *incubation* of the fragments with an excess of EcoRI. As a consequence, DNA that has been digested with, say, *SauIIIA* and therefore probably contains one or more internal *EcoRI* sites will be further reduced in size. In order to prevent this secondary fragmentation of the DNA, the genomic digest is incubated with *EcoRl* Mtase in the presence of SAM prior to the addition of the linkers. In this way, only the tandem linker sites are cleaved, and the DNA fragment remains intact. This procedure can of course be applied to any restriction enzyme, providing a compatible Mtase is available. Moreover, the principle of protection of specific restriction sites can be extended to other related DNA manipulations as required. This strategy is particularly suitable for the construction of cDNA libraries.

Modifying Restriction Sites

Methylation of a DNA sequence that is the target for a restriction enzyme with partially degenerate specificity by a DNA Mtase of overlapping specificity can reduce the number of such sites that are cleaved by the cognate restriction enzyme. For example, as shown in Figure elsewhere in this chapter, the enzyme *HincII* recognizes and cleaves the sequence GTPyPuAC. The two sequences GTCAAC and GTCGAC, which are found in the plasmid pBR322, are both sites for *HincII*. However, the sequence *GTCGAC,* but not the sequence GTCAAC, is a target for the TCGAspecific *TaqI* Mtase. Therefore, methylation of pBR322 by *TaqI* Mtases produces a plasmid with effectively a single *HincII* site. This approach has been promoted by New England Biolabs, who offer an extensive range of DNA Mtases for such purposes.

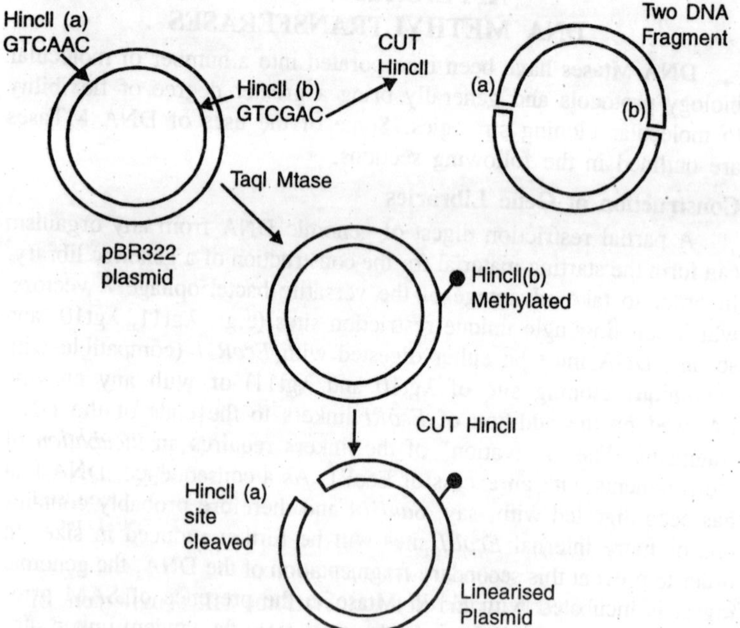

Figure 11.2 : Selective inactivation of partially degenerate restriction sites.

In a second approach, the use of a DNA Mtase to protect a restriction site involves the methylation of bases that are part of overlapping Mtase sites. Thus, in the example shown in Figure elsewhere in this chapter, *BamHI* and *Mspl* sites overlap by two bases in the sequence GGATCCGG. Methylation of the 5' *dC* of the *Mspl* site renders the *BamHI* site refractory to *BamHI* endonuclease. Those *BamHl* sites that are not followed by a GpG dinucleotide are not methylated by *Mspl* Mtase and are therefore not protected.

In Vitro Methylation of Eukaryotic Genes

Methylation of dC bases in mammalian DNA has been implicated in the control of gene transcription. The availability of methylated and unmethylated DNA for the experimental investigation of this phenomenon is highly desirable. In order to reproduce eukaryotic methylation patterns in vitro, the DNA must either be synthesized chemically from a 5-methyldCTP precursor or prepared enzymically using a preparation of a CpG-specific Mtase. Alternative but somewhat less satisfactory Mtases include *Hpall* (CCGG-specific) and *HhaI*

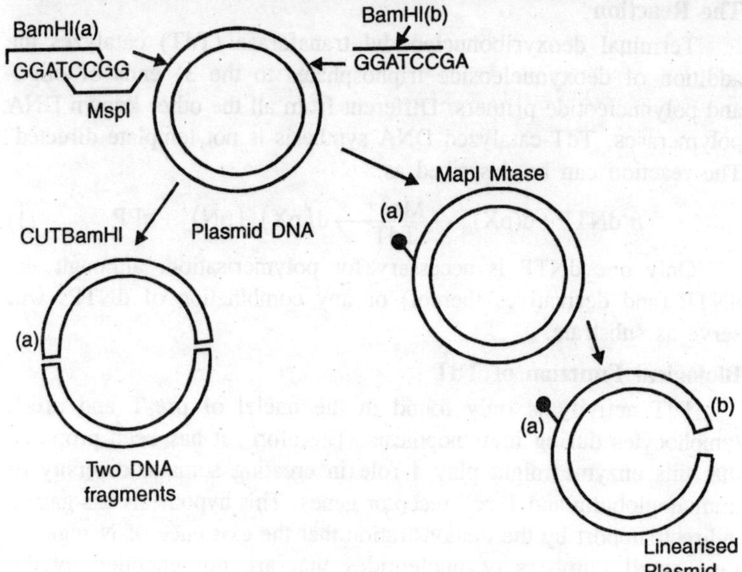

Figure 11.3 : Inactivation of restriction sites using DNA Mtase of overlapping specificity.

(CGCG), which clearly would not methylate the full complement of CpG dinucleotides in a given sequence of DNA. Unfortunately, no commercial preparations of a CpG Mtase are currently available. However, a number of purification schedules have been published for such enzymes from a range of mammalian sources. Apreparation obtained from mouse erythroleukemia cells has been well characterized and is probably the enzyme of choice for this type of work.

As before, the DNA fragment or oligonucleotide duplex to be modified is incubated with the Mtase preparation as described earlier. In order to determine the extent and sequence-specific nature of the methylation, the DNA can be subjected to strategic restriction analysis (if the DNA sequence is known). A more thorough analysis involves the comparative determination of the sequences of modified and unmodified DNA by the Maxam Gilbert method. Methylated cytosines do not react with hydrazine and, therefore, do not give bands in the C-track of the sequencing gel. Following successful methylation, the DNA can be used in comparative binding studies with transcription factors or can be used to transfect cells, and so forth.

TERMINAL DEOXYRIBONUCLEOTIDYL TRANSFERASE

The Reaction

Terminal deoxyribonucleotidyl transferase (TdT) catalyzes the addition of deoxynucleoside triphosphates to the 3' ends of oligo- and polynucleotide primers. Different from all the other known DNA polymerases, TdT-catalyzed DNA synthesis is not template directed. The reaction can be described as:

$$n\,\text{dNTP} + \text{d(pX)}_m \xrightarrow[\text{TdT}]{\text{Me}^{2+}} \text{d}(\text{pX})_m(\text{pN})_n + n\text{PP}_i \qquad (1)$$

Only one dNTP is necessary for polymerisation, although any dNTP (and derivatives thereof) or any combination of dNTPs will serve as substrate.

Biological Function of TdT

TdT activity is only found in the nuclei of pre-T and pre-B lymphocytes during immunopoiesis. Therefore, it has been proposed that this enzyme might play a role in creating somatic diversity of immunoglobulin and T-cell receptor genes. This hypothesis has gained indirect support by the demonstration that the existence of N regions, i.e., small numbers of nucleotides that are not encoded by the chromosomes, correlates with the expression of TdT. Besides this circumstantial evidence for a role as an active mutator, after 25 years of intensive research, still nothing is known about the biological function of TdT. Nevertheless, TdT is an important marker for the diagnosis and classification of pre-B-cell and pre-T-cell leukemias and differential diagnosis of myeloid leukemias. The use of TdT as a marker in leukemia diagnosis is the subject of two excellent reviews.

Applications

Because of its terminal addition properties, TdT has been widely employed for the production of synthetic homo- and heteropolymers. A wide variety of polymers can be synthesized from derivatives of dNTPs, which are N-acetylated orN-alkylated. Adetailed discussion of the mechanism of these reactions, the statistics of polymerisation, inhibitors, and practical synthetic applications has been provided by Bollum.

Homopolymeric tailing of linear duplex DNA for in vitro genetic recombination is the most common application today. In a key experiment of genetic engineering, linear bacteriophage P22 DNA was tailed with oligo(dA), and another set of linear P22 was tailed with the complementary oligo(dT). After annealing, the dA/dT tailed recombinant molecule was ligated to give covalently closed dimeric circles. Details of the tailing reaction were recently reviewed.

Another application utilizes the limited addition of rNTPs, ddNTPs, and cordycepin triphosphate for terminal additions. Since these substrates lead to chain terminations, 3' ends can be extended in a controlled manner. This has been exploited for the radioactive labeling of the 3'-hydroxyl ends of single- and double-stranded DNA. A modification of this method is the 3' labeling of DNA primers by biotin-11 -dUTP or fluorescent succinylfluorescein-labeled dideoxnucleoside triphosphates. TdT has also been used to elongate primers by only one phosphorothioate deoxynucleotide. The resulting extension products were purified by means of a mercury beaded column. This approach has been used for the introduction of mispaired primer ends for site-directed mutagenesis protocols.

THE ENZYME

Structure

TdT is a monomeric enzyme with a mol mass of about 60,000 Da, whose cDNA has been cloned and expressed in various organisms. Dependent on the biological source, the enzyme consists of 508-529 amino acids. A high degree of sequence homology (>80%) has been observed between TdTs from different species. There is also a remarkable similarity between TdT and the cellular repair polymerase, DNA polymerase β. Earlier preparations represented a proteolytically degraded form of the enzyme that consists of two subunits. The smaller 8000 Da α-subunit corresponds to amino acids 403-508 of human TdT, and the larger 24,000 Da β-subunit corresponds to amino acids 159-402 of the undegraded 60,000 Da form. The undegraded form of the enzyme has been reported to exhibit a severalfold lower turnover number than the (commercially available) degraded enzyme. It is not yet known whether proteolytic degradation changes enzymic properties other than k_{cat}.

The two-subunit form of the enzyme is commercially available from many biochemical companies. Although many firms deliver an excellent enzyme, every now and then we have observed nuclease contamination in different batches. Therefore, it is recommended that the specifications of different lots be checked, particularly for nuclease contamination, and moreover, it is wise to include experimental controls that would detect exonuclease and endonuclease activity.

Reaction Conditions

Reaction Buffer

TdT is peculiar with regard to the reaction conditions in several respects. The activity is strongly inhibited by the ammonium cation as well as chloride, iodide, and phosphate anions stimulating a quest for the optimal buffer. In general, potassium or sodium cacodylate buffers are preferred, since they were shown to be optimal for polypurine and polypyrimidine synthesis. However, cacodylate buffers suffer from important drawbacks. First, cacodylate (dimethyl arsenic acid) is toxic; second, cacodylate might be contaminated by heavy metal ions, which must be removed prior to use, e.g., by treatment with a complexing chelate resin, and third, SH-protective compounds, such as DTT, which are mandatory for an enzyme's activity, react themselves with cacodylate to yield garlic-like smelling and highly toxic sulfur-cacodylic compounds. Thus, an alternative for cacodylate buffers is highly desirable. The only systematic study for replacing cacodylate by other buffer substances revealed that TdT activity in 2(N-morpholino)ethane sulfonic acid (MES) has only one-quarter the activity of TdT in cacodylate. With Tris-HCI, only 2% of maximal activity was observed. We found that 100 mM Tris-acetate, pH 7.2, is an attractive alternative to cacodylate buffers, giving nearly the same activity as Chelex-treated potassium cacodylate.

Divalent Cations

The polymerisation reaction requires the presence of a divalent cation, with an order of efficiency of $Mg^{2+} > Zn^{2+} > Co^{2+} > Mn^{2+}$ for the elongation of oligonucleotide primers with dAMP. dGTP is also optimally added in the presence of Mg^{2+}; on the other hand, the polymerisation of pyrimidines is best in Co^{2+}-containing buffers. In genetic engineering, most tailing reactions are performed in the presence of either Co^{2+} for pyrimidine additions or Mn^{2+} for purine additions. The frequent use of Mn^{2+} for purine additions is based on the finding that Mn^{2+} permits TdT to extend duplex termini, and that contaminating nucleases are less active or even inactive in the presence of Mn^{2+}.

Parameters Influencing the Reaction

TdT binds its substrates with rather high K_m values of 100 μM for dATP and dGTP, 500 μM for dTTP and dCTP, and 1 μM for oligonucleotide primers and up to 1 mM for homopolymer primer ends. Thus, for optimal reaction rates, all the substrates should be present in highly concentrated solutions. This is most readily achieved by working in small volumes. The number of nucleotides added to a distinct primer molecule will in principle be determined by the ratio of Mol dNTPs:Mol 3'-OH-termini in the reaction mixture. Thus, if there is a 100-fold molar excess of dNTP over primer ends, nearly 100 nucleotides will be incorporated per primer with a distribution of products that ideally obeys the Poisson distribution. Such a behaviour is based on the assumption that all primer ends are equally utilized by the enzyme, which is only true for the homopolymeric extension of an isohomopolymeric primer, e.g., (dC)-tailing of $(dC)_{10}$ On the other hand, tailing of restriction fragments is far from being ideal, mainly because differences in primer binding exist that are dependent on the availability of the 3'-OH group and the nucleotide composition of the primer termini. Assume a blunt-ended DNA that should become extended by an oligo(dC) tail. The very first nucleotide incorporation step at the blunt end will be rather slow, because of the poor availability of a free, i.e., temporarily melted 3' terminus. The melting (or breathing) process of the primer end, in turn, is dependent on the relative base composition: Ends that are rich in G: C base pairs are more stable and thus less accessible to the elongation reaction than A:T rich ends. As soon as one nucleotide is inserted, the primer has a 3'-OH overhang, which will be elongated more easily. As a consequence, a few blunt-end primers will be elongated to a larger extent than initially expected. Furthermore, TdT binds about 10-fold stronger to a terminal dG or dA than to a dC or dT. This again might lead to a bias in the expected distribution of elongated termini, particularly if different primer ends are involved. All these effects lead to more or less unpredictable results in both the fraction of primers elongated and the number of nucleotides added.

For the extension of restriction fragments with subsequent insertion into vectors, very long extensions are as harmful as very short ones. Long stretches of homopolymers within a plasmid or phage vector are not very stable, and tend to delete or otherwise hinder the vector propagation. Very short stretches or a large fraction of primer ends without any addition drastically reduce the cloning efficiency. Therefore, a careful control of the lengths of tails by direct visualisation from a gel is required before the annealing reaction with a complementary tailed counterpart is performed. This is not necessary for tailing with dGMP, because this reaction ceases after the incorporation of about 20 nucleotides owing to aggregate formation of the newly generated oligo(dG) stretches. Thus, by using dG/dC-tailing, only the lengths of the dCtail must be controlled and optimized. For the extension of restriction fragments, the amount of enzyme relative to DNA ends is a critical factor. It is conceivable that at a low ratio of enzyme to 3'-OH ends the enzyme tends to elongate an already elongated primer in preference to a new one, yielding few primers containing many nucleotides and many primers without any addition. To ensure that most of the primers will be elongated, particularly when the reaction is carried out with 3'recessive or blunt ends, a 10- to 100-fold molar excess of TdT over primer ends is necessary. Terminal additions to blunt-ended DNA and 3'-recessive-ended DNA also require a partial melting of such ends. Therefore, conditions that destabilize DNA duplexes, such as low salt buffers and/or the replacement of Mg^{2+} by Co^{2+} or Mn^{2+}, are favorable. Likewise, X exonuclease can be used to trim the Sends and thereby converting a blunt or 3'-recessive end into a 3' overhang.

EXPERIMENTAL PROCEDURES

In the following, we will give some examples for the practical use of terminal transferase. We readdress the issue of homopolymer synthesis by providing data for the use of phosphorothioate derivatives of the four dNTP substrates (dNTPαS). Homopolymers containing phosphorothioate diester linkages are interesting mainly because they are expected to be more stable against nuclease degradation than phosphatediesterhomopolymers. Their thermodynamic behaviour in an annealed state is not yet known. Furthermore, $(dCS)_{28}$ has been shown to display anti-HIV-1 activity, both in vivo and in vitro.

We further describe reaction conditions for the addition of dNTPαS to 3'-hydroxyl ends of linear double-stranded DNA. The insert DNA will be tailed uniformly with $(dG)_{20}$. The less valuable linearized vector will be tailed with oligo(dC). After annealing, the dG/dC-tailed recombinant molecule will be used to transform competent *E. coli* cells. Within the bacteria, remaining single-stranded DNA will be filled up by DNA polymerase I and ligated to covalently closed circular DNA. As a sidereaction, intracellular nucleases degrade the recombinant molecule, preferably at its vulnerable single-stranded regions. This reduces transformation efficiencies. Replacing the dNTPs by their α-phosphorothioate analogs yields single-stranded DNA intermediates that are less susceptible to degradation.

Materials

1. Terminal deoxynucleotidyl transferase.
2. Cacodylic acid.
3. Solid KOH.
4. Chelex 100; 200-400 mesh, sodium form.
5. $p(dC)_{10}$.
6. 0.1 M $CoC1_2$.
7. 0.1 M $MnC1_2$.
8. 0.1 M Dithiothreitol (Boehringer, Mannheim, Germany).
9. Deoxynucleoside triphosphates (Boehringer).
10. [^{3}H]Deoxynucleoside triphosphates, 600 Ci/mmol.
11. Deoxynucleoside-5'-O-(1-thio)-triphosphates, dNTPaS (Amersham).
12. [^{35}S]Deoxynucleoside-5'-O-(1-thio)-triphosphates, 1300 Ci/mmol.
13. DE81 filter disks.
14. 0.5 M Na_2HPO_4.
15. Competent cells of E. *coli* DH5α.
16. 10 mM Tris-HC1, pH 7.2, 1 mM EDTA, 100 mM NaCl.

Methods

Preparation of the Tailing Buffer

1. Equilibrate 5 g Chelex 100 with 3M potassium acetate, pH 7. After 5 min at room temperature, remove excess liquid by passing the slurry through a glass sintered funnel by applying water aspirator vacuum. Wash the Chelex in the funnel with 10 mL distilled water by applying water aspirator vacuum.

2. Prepare a 1.2M solution of cacodylic acid. Add KOH pellets, until a pH of approx 7 is obtained. Adjust to pH 7.2 by the dropwise addition of 1M KOH. Dilute with distilled water to obtain a 1M stock solution.
3. Add the equilibrated Chelex to the potassium cacodylate solution, and stir for about 5 min. Remove the ionic exchanger by filtration.
4. Prepare a 5X C-Tailing Buffer by pipeting in the following order:
   a. 10 mL 1 M potassium cacodylate, pH 7.2
   b. 8.8 mL distilled water
   c. 0.2 mL 0.1 M DTT
   d. 1 mL 0.1 M $CoCl_2$
   Store in 500-μL aliquots at -20°C.
5. Prepare a 5X G-Tailing Buffer by pipeting in the following order:
   a. 10 mL 1M potassium cacodylate, pH 7.2
   b. 8.8 mL distilled water
   c. 0.2 mL 0.1 M *DTT*
   d. 1 mL 0.1 *M* $MnCl_2$
   Store in 500-μL aliquots at -20°C.

Preparation of Phosphorothioate-Containing Homopolymers

1. Mix:

$p(dC)_{10}$ (10 A_{260} U/mL)	1.5 μL (10 μM 3'-OH)
5X C-tailing buffer	5 μL
[α-^{35}S]dCTP (1300 Ci/mmol, 13 μCi/μL)	2 μL
5 *mM* dCTPaS	2.5μL
Distilled water	13 μL
Terminal transferase (17 U/μL)	1 μL

2. Incubate for 60 min at 37°C.
3. Spot a 2-μL aliquot onto a DE81-filter disk, dry under a fan, and count the total radioactivity in a toluene-based scintillant.
4. After measuring the total radioactivity, place the filter disk on a suction device. Rinse several times with ethanol (to remove toluene), and then dry under a fan.
5. Keep the dried filter for 5 min in a 0.5M solution of Na_2HPO_4.

Then place it on a suction device, and rinse five times with 1 mL (each) of 0.5M Na_2HPO_4, two times with 1 mL water, and two times with 1 mL of ethanol. Then dry under a fan, and count the dCMPS incorporation in a toluene-based scintillant.

6. Calculate the amount of dCS-tails added to $(dC)_{10}$ from the amount of filter-retained radioactivity:

   (cpm binding to DE81/Total cpm) ×
   (500 μM dCMPS/10 μM 3'-OH) =
   dCMPS additions/primer (2)

7. Ethanol-precipitate the remaining DNA in the presence of potassium acetate.
8. Control the tail lengths by denaturing polyacrylamide gel electrophoresis.

Tailing of Restriction Fragments with dNTPaS

Tailing Double-Stranded DNA with Oligio(DGS)

1. Prepare a reaction mixture containing:

Double-stranded insert DNA (dried pellet)	About 2.5 pmol 3' ends
5X G-tailing buffer	5 μL
[α-^{35}S]dGTP (1300 Ci/mmol, 13 μCi/μL)	2 μL
100 μM dGTPαS	6 μL
Distilled water	10 μL
Terminal transferase (17 U/μL)	2 μL

2. Incubate at 37°C for 30 min.
3. Spot a 2-μL aliquot onto a DE81-filter disk.
4. Calculate the amount of dG-tails added to the double-stranded DNA from the amount of radioactivity bound to the DE81 filter:

   (cpm binding to DE81/Total cpm) ×
   (24 μM dGMPS/20)
   = 3'-OH termini of insert DNA (μM) (3)

If the expected number of 20 for dGMPS added to each 3' terminus is not achieved or (likewise) the calculated number of 3'-OH termini of insert DNA is too low, the tailing procedure should be repeated after ethanol precipitation of the DNA. Because dG- and dGS-containing tails cannot exceed 20 residues, there is no danger of generating tails that are too long.

5. Ethanol-precipitate the remaining DNA in the presence of potassium acetate.

Tailing Linearized Vector DNA with Oligo(DCS)

1. Prepare a reaction mixture with:

Linearized vector DNA (dried pellet)	10 pmol 3' ends
5X C-tailing buffer	20 μL
0.1 mM dCTPaS	30 μL
[α-^{35}S]dCTP (1300 Ci/mmol, 13 μCi/μL)	2 μL
Distilled water	38 μL
Terminal transferase (17 U/pL)	10 μL

2. Spot a 2-μL aliquot onto a DE81 filter disk, dry under a fan, and count the total radioactivity in a toluene-based scintillant.
3. Incubate at 20°C, take 2-μL aliquots every 15 min, and determine as soon as possible the incorporated amount of dCS per 3'-OH terminus by using the DE81 filter technique. When about 20 dCS residues are incorporated per primer, stop the reaction by heating to 65°C for 10 min. Chill on ice.
4. Ethanol-precipitate DNA in the presence of potassium acetate.
5. Take an aliquot, cleave with a restriction endonuclease that cuts close to the extension, and control the tail lengths by polyacrylamide gel electrophoresis.
6. If necessary, repeat the procedure by incubating for a shorter or longer time.

Annealing of teh Tailed DNA Components and Transformation of Competent E.Coli Cells

1. Dissolve the (dGS)-tailed insert DNA and the (dCS)-tailed vector DNA in 10 μL (each) of 10 mM Tris-HCI, pH 7.2, 1 mM EDTA, and 100 mM NaCl.
2. Mix aliquots of the two solutions in an approximate molar ratio of 1:1 (3' ends) for each component.
3. Incubate at 65°C for 10 min in an incubator block.
4. Turn off the incubator block, and let slowly cool down to room temperature.
5. Take an aliquot with about 1 fmol annealed construct for the transformation of 100 μL competent *E. coli* cells.

The same protocol is applicable for tailing with unmodified deoxynucleoside triphosphates. Transformation efficiencies are about threefold higher with phosphorothioate-tailed constructs than those obtained with normal dG/dC-tailed constructs.

THE BAL 31 NUCLEASES

The extracellular nucleases commonly called the BAL 31 nuclease take their name from the designation given the marine bacterium producing them, which was originally classified as *Pseudomonas* BAL 31 and reclassified as belonging to the small genus *Alteromonas* with the species named *espejiana* after its discoverer, a Chilean microbiologist. The nuclease activities were originally found as contaminants in preparations of bacteriophage PM2 grown on this organism, but were shown to be bacterial products. Only 10-20% of the nuclease activity is found in the periplasm. The American Type Culture Collection strain of *Alteromonas espejiana* produces BAL 31 nuclease as proficiently as the strain originally obtained from its discoverer.

Some Physical Features

Two kinetically and molecularly distinct forms of the nuclease constitute the bulk of the activity in culture supernatants and are the only ones that have been partially characterized. The smaller of these single-subunit enzymes, the "slow" (S) form (mol mass 85,000 Da), is derived by proteolysis, mediated by *anAlteromonas-produced* protease, of the "fast" (F) form (mol mass 109,000 Da), which in turn derives from an even larger precursor. Conversion of F nuclease to a species indistinguishable in molecular size and catalytic properties from the S enzyme can be done by proteolysis in vitro.

The BAL 31 nucleases are remarkable for their resistance to inactivation and or denaturation in the presence of detergents, urea, or high concentrations of electrolyte, and are very stable upon extended storage in the cold. The nuclease activities are not highly resistant to inactivation at elevated temperatures, but the overall secondary and tertiary structures of the S enzyme are extremely resistant to disruption: The internally proteolytically cleaved enzyme fails to dissociate at 100°C in the presence of 1% sodium dodecyl sulfate.

Reactions Catalyzed

The nucleases have three general activities against DNA: a 3'-5' exonuclease activity that apparently removes one residue at a time from duplex structures, a 5'-3' exonuclease activity that carries out the bulk of the degradation of single-stranded DNA, and a much slower (2-3% of the bond cleavage rate for the 5'-3' exonuclease) endonuclease activity against single-stranded DNA that is also elicited by a variety of covalent and noncovalent lesions or distortions in duplex DNA. The two forms differ greatly in the rates of catalysis, at given molar concentrations of duplex ends and enzyme, for the exonuclease reaction if the DNA is double-stranded (hence the aforementioned "fast" and "slow" designations); but the two forms are much more comparable in kinetic behaviour if the substrate is singlestranded DNA. The nucleases are not sugar-specific and catalyze the terminally directed hydrolysis of duplex RNA and readily degrade RNA containing nonduplex structure. Endonucleolytic cleavage of duplex RNA in response to lesions or distortions has not been examined except that it is likely that the nucleases can cleave in response to a strand break in duplex RNA.

The combination of a 3'-5' exonuclease activity on duplex DNA, a 5'-3' mode of attack on single-stranded DNA, and relatively infrequent endonucleolytic attack in single-stranded DNA serves to reduce the length of linear duplex DNA and is referred to as the duplex exonuclease activity. Partially degraded molecules possess the expected 5'terminated single-stranded "tails," and there was no evidence for 3'-terminated tails, implying that the 5'-3' attack is limited to singlestranded DNA. An unexpected aspect of the mechanism is that the average length of the tails, for a constant number of nucleotides removed by the exonuclease action, decreases markedly to a limiting value with increasing enzyme concentration. In the absence of DNA polymerasemediated repair, ligation of partially shortened duplexes under conditions favoring the joining of fully base-paired ends is undetectable when low concentrations of nuclease are used, but becomes quite substantial (up to 50%) at nuclease concentrations that give rise to tails of minimum average length. This strongly suggests that the 5'-3' exonuclease activity, acting on 5'-terminated single-stranded tails that are generated through attack by the 3'-5' exonuclease on duplex ends, terminates at the junction between single-stranded and duplex structures to leave fully base-paired ends on a significant fraction of the molecules in a partially digested population. The length reduction of duplex RNA presumably proceeds by a similar mechanism, but has not been characterized. The BAL 31 nucleases are the most efficacious enzymes known for the controlled length reduction of duplex DNA, and are apparently the only enzymes that can mediate this reaction for duplex RNA.

The covalent lesions in duplex DNA that have been shown to elicit cleavage by BAL 31 nuclease, some of which have been examined only in the case of the S form, include strand breaks, nitrous acidinduced inter-strand crosslinks, UV irradiation-induced photoproducts, adducts from reaction with arylating and alkylating agents or with Ag^{+} and Hg^{++} ions, and apurinic sites. Quasiduplex DNA, in which there are extra nucleotides in one strand separated by 3 bp, was attacked by the BAL 31 nuclease, but not by the S 1 and *Neurospora crassa* nucleases. Noncovalent alterations in duplex structure that can cause endonucleolytic attack include those associated with a very moderate degree of negative supercoiling and very high positive supercoiling, as well as the junctions between regions of B- and Z-DNA helical structures in the same molecule.

The only known nonsubstrate nucleic acids are nonmodified, nonsupercoiled (or slightly positively supercoiled) circular duplex DNAs, which are cleaved at such slow rates that they serve as excellent controls for the detection of lesions or distorted structures in such DNAs. The strand break (nick) introduced as the initial cleavage event itself provides a substrate site, so that the other strand is cleaved, usually after the removal of several nucleotides from the originally nicked strand at the site of the nick, to produce termini that are then attacked so as to shorten the resulting duplex as previously outlined.

Uses in Manipulation of Nucleic Acids

The duplex exonuclease activity of the nucleases has been very widely exploited, because the removal of sequences from DNA termini in a controlled manner is desirable in numerous applications. Unidirectional deletions can be done for DNAs cloned into a circular vector if single unique sites for two restriction enzymes are present, one on each side of the cloned insert. Linearisation by cleavage with one of the restriction enzymes allows for unidirectional deletions in the insert, whereas subsequent cleavage with the other restriction enzyme releases the partially degraded vector DNA so that the shortened insert can be religated to an intact vector DNA for subcloning. Conversely, manipulations of the vector DNAs themselves in this manner have proven useful-an early application of this method led to a vector DNA modified so that cloned inserts abutted the DNA coding for a signal peptide of the vector, allowing secretion of the protein products of cloned genes into the periplasmic space of *E. coli*. Such unidirectional deletions in cloned inserts have been used as a means to sequence, using a single primer for the DNA polymerase-mediated extension, long segments of DNA cloned into M13 phage-derived vectors through the production of a set of progressively shorter unidirectionally deleted derivatives of the original insert. The progressive removal of sequences from duplex termini allows the determination of the restriction map of a given DNA by observing the order in which the fragments from digestion with the restriction enzyme in question disappear from gel electrophoretic patterns of progressively shortened samples. This technique is greatly enhanced if the fragment to be mapped is in a circular cloning vector as ambiguity resulting from simultaneous degradation from both ends of the fragment is eliminated. Cloning vectors containing infrequently cleaved restriction sites have been constructed in this laboratory to take advantage of BAL 31 nuclease-mediated deletions in both sequencing and restriction fragment mapping.

The blockage of the exonuclease action by nucleosomes and by interstrand crosslinks, such as can be induced by psoralen derivatives, has been used to map the locations of such structures on duplex DNAs. The detection of lesions or distortions in duplex DNA through the endonuclease activity has been alluded to earlier.

ENZYME REQUIREMENTS

pH

The near-neutral pH optima for the nuclease-catalyzed reactions is an advantage for most work. There is a difference in the pH optima for single-stranded and linear duplex substrates for both the S and F nucleases, which are in the ranges 8.5-8.8 and 7.0-8.0 for singlestranded and linear duplex DNA, respectively. Attack on supercoiled DNA, examined only for the S enzyme, was optimal at the same pH as for linear duplex DNA.

Metal Ion Cofactors

Ca^{2+} and Mg^{2+} are both required cofactors. Ca^{2+} is essential for activity, with nuclease activities on both single- and double-stranded DNA irreversibly (with respect to the readdition of excess Ca^{2+}) lost in solution if the molar concentration of EDTA exceeds that of this cation. However, activity can be recovered after electrophoresis under denaturing conditions (sodium dodecyl sulfate [SDS]-polyacrylamide gels) by incubation in a Ca^{2+}-containing buffer.

Maximum velocity against both single-stranded and linear duplex DNAs at constant [Ca^{2+}] is achieved between 10 and 15 mM Mg^{2+}.

At nominally zero (actually 0.01-0.02 mM) Mg^{2+}, there was residual duplex exonuclease activity (8 and 45% for the S and F enzymes, respectively), but virtually none for single-stranded DNA. In corresponding profiles where [Ca^{2+}] was varied, concentrations near 10 mM are needed to achieve full velocity for the length reduction of duplex DNA, but activity is maximal on single-stranded DNA at or below 1 mM.

In light of the above, a buffer containing 12.5 mM each of Ca^{2+} and Mg^{2+} is recommended, since this confers full activity with respect to both classes of substrate. The 5-mM concentrations of these ions in the B AL 31 nuclease buffers recommended by some suppliers would yield only 60-65% of the duplex exonuclease activity.

Effects of Temperature

The optimal temperature for the activity on single-stranded DNA is near 60°C, but the use of such elevated temperatures is impractical. Internal breaks would almost certainly be introduced into duplex DNA because of at least transient thermally mediated unstacking of base pairs. Moreover, the half-life of the activity on single-stranded DNA and the duplex exonuclease activity are 3-5 min at 50°C. The half-life of the activity is at least 20 h at 30°C, which is the recommended temperature for manipulations with these enzymes.

In a storage buffer containing 5 mM each of Ca^{2+} and Mg^{2+} near 4°C, the nuclease should retain most of its activity on a time scale of years as inferred from studies in which no activity was lost upon storage for several months. Most suppliers provide the enzyme in 50% (v/v) glycerol for storage at -20°C, in which it should maintain full activity indefinitely.

Effects of Ionic Strength, Protein Denaturing Agents, and Other Potential Inhibitors

Aconcentration of NaCl of 0.6M is present in the usual assay buffer. The activity on single-stranded DNA is optimal in this range, but the duplex exonuclease activity appears to increase if the ionic strength is reduced from that of the assay buffer. However, the NaCl concentration of the usual buffer will repress the activity of contaminating salt-sensitive nucleases or other factors that can cause adventitious breaks. The nuclease was used at 0.1M NaCl with no evidence of internal breaks in nonsupercoiled closed duplex DNA provided that the buffers were autoclaved; omission of this step did lead to the appearance of unexpected strand breaks. Such internal breaks must be avoided because each such lesion, after cleavage of the intact strand, gives rise to two new termini, which then serve as substrates for the duplex exonuclease activity.

Very high concentrations of 1 : 1 electrolytes are tolerated extremely well by the nucleases. With single-stranded DNA as substrate, the S enzyme displays 40 and 27% of its maximal activity in 4.5MNaC1 and 7M CsCl, respectively; in a comparison at only two concentrations, the F nuclease was 70% as active in 4.5M NaCl as in 0.6M NaCl. The duplex exonuclease activity of the F nuclease at 4.5M NaCl was 26% of that at 0.6M.

Protein denaturing agents at concentrations that normally denature proteins are unable to abolish the BAL 31 nuclease activities. Singlestranded DNA was hydrolyzed at 40% of the maximal rate in 6.5M urea, whereas crude preparations maintained activity against both single- and double-stranded substrates in the presence of 5% SDS if both Mg^{2+} and Ca^{2+} were present prior to exposure to the detergent. Preliminary experiments on the S enzyme showed that substantial activity on single-stranded DNA was observed in up to 6M guanidinium hydrochloride.

Cleavage of adducts of nonsupercoiled closed duplex DNA with Hg^{2+} or Ag^{+} by the F nuclease suggested that the enzyme is not highly sensitive to these metal ions, but it must be pointed out that, in the case of Ag^{+}, virtually all the metal ion was bound to DNA nucleotide. The only metal ion tested that appeared to be strongly inhibitory was Zn^{2+}. Inhibitors of unknown nature that can very strongly repress the duplex exonuclease activity of the BAL 31 nucleases have been observed in preparations of plasmid DNA prepared by rapid procedures, such as the alkaline lysis protocol of Birnboim and Doly, that do not involve purification by centrifugation to equilibrium in CsC1 density gradients containing intercalating compounds such as ethidium bromide. An additional purification step, using small columns of Bio-Gel A-0.5m agarose-based size exclusion resin (BioRad Laboratories, Richmond, CA), has been shown to render the DNA digestible by commercially provided BAL 31 nuclease, thus avoiding the time-consuming equilibrium ultracentrifugation procedure.

Another unexplained strong source of inhibition was found as a component of a T4 DNA polymerase repair reaction: DNA that had been thus treated, and the polymerase inactivated, without subsequent precipitation and resuspension of the DNA was virtually unaffected by the duplex exonuclease activity. The inhibitor was shown not to be the 5'-deoxynucleoside triphosphate substrates for the polymerase and was not a large macromole since the nuclease activity on the repaired DNA was restored when it was passed through a Sephadex° G25 "spun column."

The 5'-deoxynucleoside monophosphates (5'-dNMPs) resulting from nuclease action are apparently inhibitory because the velocity of the duplex exonuclease reaction decreases significantly with time, even though the substrate concentration (concentration of DNA ends) remains constant until molecules begin to be completely degraded. The nuclease is known to bind to 5'-dNMPs, because an affinity column consisting of 5'-dNMPs covalently attached to agarose is used in the purification procedure.

Systematic studies have been done on the effects of a "chaotropic" agent ($NaClO_4$) and compounds that protect cells from damage due to increases or decreases in extracellular osmolarity. However, these are not compounds that are likely to be part of a reaction mixture in general applications of these enzymes, and it is noted here only that $NaClO_4$ rapidly inactivates the nucleases at concentrations in the range of 1-2M.

Effects of Proteolysis

The proteolytic conversion of F to S nuclease that takes place in *Alteromonas* culture supernatants can be mimicked by proteolysis in vitro. However, it has also been found that extended exposure to protease of the S nuclease can preferentially remove the duplex exonuclease activity, but the bulk of the activity against single-stranded DNA survives. The ability to linearize a nicked circular DNA correlates with the remaining activity on single-stranded DNA, so that the endonuclease activity in response to lesions is not lost upon proteolysis. Since the bulk of the activity on single-stranded DNA is of the 5'-3' exonuclease variety, whereas attack on duplex DNA is through a 3'-5' mode, the data suggest that the protease-treated nuclease may lose its 3-5' exonuclease capability. The observation is interesting, because if conditions for the removal of the duplex exonuclease activity can be optimized, the remarkable power of the enzyme to cleave endonucleolytically in response to alterations in duplex DNA structure could be much better exploited to reveal imperfections, such as small mismatches in nominally duplex DNAs, since the information as to the site of cleavage would not be obliterated by the exonuclease activity.

The internal breaks that were thought to be introduced into the S nuclease by proteolysis to produce the preferential loss of duplex exonuclease activity were not revealed in denaturing polyacrylamide gels, where the proteins are heated in boiling water in the presence of 1 % SDS and 20 mM β-mercaptoethanol prior to electrophoresis. Rather, the mol mass appeared to be unaffected for samples that had clearly undergone loss of over 90% of the starting duplex exonuclease activity, but retained over half of the activity against single-stranded DNA. When urea was added to the denaturation buffer to 6M, the internal breaks were revealed by the progressively more extensive fragmentation of the nuclease with increasing extent of exposure to protease. This indicates a resistance of the breakup of the secondary and tertiary structure of the S nuclease to extreme denaturing conditions that is apparently without parallel. It is interesting that the catalytic activity is maintained near room temperature in the presence of strong denaturants, but is rather thermolabile. A relatively thermally sensitive binding/catalytic site(s) in an otherwise extremely stable protein structure is suggested.

UNIT ASSAY AND CHARACTERISATION OF THE DUPLEX EXONUCLEASE ACTIVITY

General Considerations

Most commercial preparations of BAL 31 nuclease are mixtures of the F and S forms. A problem with their use for the most common application, the controlled length reduction of duplex DNA, is the possible variance of the relative amounts of F and S nucleases from batch to batch. This is because the F enzyme is derived from a larger precursor by proteolysis, mostly in the culture supernatant, and itself is the source of the smaller S enzyme. Hence, the relative amounts of the two species can vary depending on the extent of exposure to the supernatant protease(s) before the nucleases are separated from these. Only one supplier (International Biotechnologies, Inc., New Haven, CT) offers the purified S and F nucleases, which could be expected to be reproducible in duplex exonuclease activity from batch to batch. Since the nuclease is very stable on extended storage in the cold, it seems advisable to obtain as large a sample as possible and to characterize it for duplex exonuclease activity unless the pure S or F species is obtained.

The recommended unit assay for general characterisation of a sample (e.g., in order to check its activity after long periods of storage) is using single-stranded DNA because maximum velocity conditions, desirable in any enzyme assay, are readily achieved. Examination of the velocity vs substrate concentration profile with viral ϕX174 DNA as substrate shows that 90% of the apparent maximum velocity V_{max}^{aPP} is achieved at a concentration of this single-stranded DNA near 10 μg/ mL for the S nuclease and at an even lower concentration for the F nuclease. Calculations using recent carefully determined values for the apparent kinetic parameters, which take into account the fact that the major mode of degradation of single-stranded DNA is exonucleolytic attack from the Sends, indicate that 90% of V_{max}^{aPP} for the S and F nucleases can be obtained with approx 34 and 9 μg/ mL, respectively, of a DNA the size of ϕX174. Since the denatured calf thymus DNA normally used in this assay is not expected to be larger in molecular weight that ϕX174 DNA and is routinely used at concentrations of at least several hundred μg/mL, the concentration of ends should be much more than sufficient to provide maximum velocity conditions. Data with calf thymus DNA as substrate in the velocity vs substrate concentration curve mentioned above showed that V_{max}^{app} was reached at concentrations near 30 μg/mL, which supports the notion that the standard assay provides maximum velocity conditions. However, this assay cannot be used to predict the kinetics of the duplex exonuclease reaction for mixtures of the two forms.

The use of duplex DNA to assay the nuclease through the release of nucleotides owing to the duplex exonuclease activity, as recommended by some suppliers, will not take place under close to V_{max} conditions unless short duplexes are used at high concentrations. For a DNA of 1000 bp, its concentration would have to be over 150 μg/ mL for the F nuclease to achieve 90% of maximum velocity according to the most recent estimates for the apparent K_m; for the S nuclease, the corresponding concentration would be in the 2 mg/mL range. DNA of defined molecular size is required so that the substrate concentration (molar concentration of termini) is known. If enzyme is assayed using duplex DNA for the purpose of determining the activity of a particular batch on linear duplexes, care should be taken to keep the substrate concentration in subsequent digests close to those in the pilot experiments. Also, the G + C content of the DNA in the pilot experiments should not differ significantly from that of DNA in experiments based on the pilot studies, unless the F nuclease is used. Finally, there is a DNA length dependence of the kinetic parameters.

Assay Using Single-Stranded DNA

This assay has been described in detail, but is reproduced here for the convenience of the reader and because of recent modifications in the procedure. Calf thymus DNA is dissolved at a nominal concentration of 1-2 mg/mL in BE buffer (100 mM NaCl, 20 mM Tris-HCI, 1 mM EDTA [pH 8]) by overnight stirring. NaOH (2M or higher) is added to a nominal concentration of 0.1M and, after several minutes of alkaline denaturation, the solution is neutralized with a severalfold excess of 3M sodium phosphate (pH 7). Dialysis into BE buffer is followed by adjustment to the proper concentrations of NaCl, $MgCl_2$, and $CaC1_2$ in 20 MM Tris-HCl/1 mM EDTA (pH 8) to yield, after addition of enzyme solution, the desired composition of assay buffer: 0.6M NaCl, 12.5 mM $CaC1_2$,12.5 mM$MgC1_2$ in the aforementioned Tris/EDTA buffer. If 0.1 vol of enzyme solution in CAM buffer (100 mM NaCl, 5 mM $MgCl_2$, 5 mM $CaCl_2$, 20 mM Tris-HCl, 1 mM EDTA [pH 8]) is mixed with the DNA solution in the assay, the DNA solution should contain 0.656M NaCl, 13.3 mM MgC12, and 13.3 mM $CaC1_2$, which is conveniently achieved with 5, 0.5, and 0.5M stock solutions of the respective chemicals, each containing the Tris/EDTA buffer. CAM buffer (usually containing 50% [v/v] glycerol) is used as the nuclease storage buffer by several suppliers.

After mixing 0.9 vol of the above DNA solution with 0.1 vol of enzyme solution at 30°C, aliquots of 0.4 ml, are withdrawn at such times as 5, 10, and 15 min and mixed with 40 μL of 0.5M sodium EDTA (pH 8-8.5) followed by 0.8 ml, of 10% (w/v) $HClO_4$. This is conveniently done in a 1.5-mL Eppendorf centrifuge tube for centrifugation of the precipitated nondigested DNA after the acidified mixtures have stood on ice for several minutes (but not for extended periods). Five-tenths to 1 mL of supernatant is carefully removed from each tube after centrifugation for approx 10 min, and the value of A_{260} is measured in a 1-cm path length quartz cuvet with a 4 mm wide cavity. The blank (zero time of incubation) is made up separately by mixing 40 μL of 0.5M EDTA with 0.4 mL of the denatured DNA solution before addition of the enzyme and $HClO_4$ solution. This ensures that any acid-soluble UV-absorbing material in nuclease preparations will not contribute to the A_{260} values when the blank absorbance is subtracted.

The resulting plot of A_{260} vs time (t) (in minutes), including the point (0,0), should be a good straight line (correlation coefficient from linear least squares analysis 0.995 or greater) from which the value of the slope, DA_{260}/t, is obtained. According to Vogt, the number of enzyme units in the total mixture is given by:

$$U = (\Delta A_{260}/t)(30\ 3)(V_{tot}) \quad (1)$$

where V_{tot} is the total volume of the acidified reaction mixture aliquot in mL (V_{tot} = 1.24 mL if the above volumes are used). Division by the volume of enzyme solution per aliquot (0.04 mL in the above) yields the U/mL of nuclease in the solution before dilution into the assay mixture.

For the removal of aliquots at 5-min intervals, $\Delta A_{260}/t$ should be at least 0.02 min^{-1}, which implies that the enzyme solution being assayed (before dilution into the aforementioned reaction mixture) should be in the range of 20 U/mL. Correspondingly, longer assays are needed for accurate results with more dilute nuclease solutions. Only when the absorbance of the acid-soluble released DNA nucleotides significantly exceeds 1 has nonlinearity in the plots of A_{260} vs *t* been observed.

Characterisation of the Duplex Exonuclease Activity

Kinetics of Molecular Weight Reduction

Testing of each batch of commercial BAL 31 nuclease for its ability to catalyze the duplex exonuclease reaction, where this use is to be made of the enzyme, is necessary as discussed in Section 3.1. This is conveniently done by subjecting restriction enzyme-generated fragments of a relevant unique sequence (e.g., a plasmid DNA or the cloned, isolated DNA fragment that is itself to be treated) to progressive digestion and analyzing the partially degraded fragments by agarose gel electrophoresis using (usually) the nondegraded restriction fragments as molecular size markers. It is advisable to assume that the kinetics will be those of the S nuclease, since this species usually predominates in culture supernatants, and to adjust the amount of enzyme accordingly if significantly faster degradation is observed.

For estimation of the rate of degradation, the equation:

$$v_0/[S] = V_{max}^{app} / [S] + K_m^{app} \tag{2}$$

has been used, where v_0 is the initial reaction velocity, [S] is the substrate concentration (mol/L of duplex termini), V_{max}^{app} is the apparent maximum velocity corresponding to the concentration of nuclease used, and K_m^{app} is the apparent Michaelis constant. It is important to note that the kinetic parameters are apparent values and do not necessarily have the interpretations associated with MichaelisMenten kinetic analysis of reactions in which only a single enzymesubstrate need be considered. At least two such intermediates, one corresponding to enzyme nonspecifically bound to duplex DNA away from the ends and not catalytically active, and another productive complex with enzyme bound at the termini, are needed to model this reaction.

The significance of $v_0/[S]$ is that it is the number of nucleotide residues released per DNA terminus per unit time, which is usually the desirable parameter. This can be taken as equal to one-half the number of base pairs removed per terminus for digests of more than approx 50-100 nucleotides removed per terminus at enzyme concentrations that minimize single-stranded tail length. At very low nuclease concentrations, substantial single-stranded tails may be present so that the number of residues removed is not equal to one-half the number of base pairs removed.

Data obtained in recent kinetic studies of the duplex exonuclease activity to determine the effects of the G + C content of the substrate for the S nuclease gave very large standard deviations for the values of K_m^{app} and V_m^{app} so that substitution of these values in Eq. (2) would not give reliable estimates for $v_0/[S]$. The reason proved to be that the values of K_m^{app} tend to be much larger than the largest values of [S] compatible with the photometric technique, even after cleavage of the DNA with a restriction enzyme to increase the number of termini per unit weight concentration of DNA. Hence, the substrate concentration range falls far short of the range of approx 0.33-2 K_m^{app} that is considered to yield optimally accurate results for such determin-ations. In agreement with this, the plots of v_0 vs [S] were all very good straight lines, as expected from Eq. (2) where $[S] << K_m^{app}$; the right-hand side of Eq. (2) then reduces to V_m^{app}/K_m^{app}, which will be constant at a fixed enzyme concentration and is the slope of a plot of v_0 vs [S].

Hence, it is possible, in the range of [S] that will ordinarily be used in such experiments, to calculate $v_0/[S]$ at any given enzyme concentration and data are moreover available to include the dependence on G + C content. As an example, consider a test DNA of G + C content near 50 mot percent containing 5000 by that has been cut into six fragments by a restriction endonuclease, and 2 μg are treated with BAL 31 nuclease in a vol of 100 μL. This would be convenient for removal of several aliquots containing approx 0.5 μg of DNA each for analysis (after quenching the reaction with an excess of EDTA) at different times of exposure (i.e., 5-min intervals) to nuclease in an ordinary agarose gel electrophoresis experiment. For duplex DNA of this G + C content, the value of $v_0/[S]$ is close to 0.80 residues released/terminus/min at an enzyme concentration of 1.0 U/mL. Assuming a desired rate of 10 residues removed/terminus/min, an enzyme concentration of 10/0.8 = 12.5 U/mL should be used independently of the substrate concentration.

The older data for the S nuclease give very different maximum velocities/U/L of nuclease and an estimate of K_m^{app} that is at least an order of magnitude lower than those evidenced (although with considerable error) from the more recent experiments. Because it turns out that these data were measured under conditions where the total enzyme concentration $[E]_{tot}$ actually was in excess of that of the substrate, the usual assumption of Michaelis-Menten kinetics that $[S] >> [E]_{tot}$ was not valid. The more recent data do not suffer from that limitation, allow the dependence of v_0 on G + C content to be assessed, and appear to be quite accurate as long as $[S] << K_m^{app}$ a condition that will be met unless concentrations of hundreds of micrograms per milliliter of fragments averaging several thousand base pairs in length are used. Hence, use of the more recent data is recommended.

Recent studies on the duplex exonuclease activity of the F nuclease give results much more in agreement with the older data and strongly suggest that the values for K_m^{app} are generally more than an order of magnitude lower than those for the S nuclease, which necessitates the use of Eq. (2) without the assumption that [S] can be dropped from the right-hand side. Here, values of K_m^{app} and V_{max}^{app} were detemined with good accuracy, because it was possible to access substrate concentrations that were in the range of *K.W.*

These studies revealed that both K_m^{app} and V_m^{app} decrease with increasing length of the linear duplex substrate for the F nuclease. Parallel studies could not be done for the S enzyme because of the error in determination of K_m^{app} noted above. These data are consistent with an interpretation of the mechanism in which nonspecific binding to the duplex away from the ends is followed by a "search" process to form a productive complex with nuclease bound at a terminus. The values of V_{maX}^{app}, but not those of K_m^{app}, seem to level off above a length in the range of 2000-2500 bp. It appears that reasonable kinetic parameters to use for the F nuclease, for molecules in this size range or above, are 2.14 ± 0.04 nmol/U/min for the maximum velocity/U of nuclease/L and 58 ± 2 nM for $K_m aPP$. V_{max}^{app} in Eq. (2) is obtained simply by multiplying the normalized maximum velocity above by the nuclease concentration in U/L (the velocity was shown to be linear with enzyme concentration for both the F and S enzymes over a wide range of substrate concentration). The dependence of $v_0/[S]$ on %(G + C) is much weaker than for the S enzyme.

Several comments are in order here. The DNA should be precipitated and resuspended in a small volume of Tris/EDTA buffer, and diluted into a reaction mixture that will give the desired buffer composition (same as that for the assay using single-stranded DNA), since it will be difficult to resuspend directly in the reaction buffer. The expected extent of degradation, at least in the first one or two aliquots, should not be more than approx 20% of the average fragment length so that smearing of the bands of asynchronously digested DNA will not preclude reasonable molecular size estimation from at least some of the fragments. To avoid changes in substrate concentration owing to complete digestion of fragments, a restriction enzyme that produces any very short fragments from the DNA should be avoided. Because the velocity does decrease with time, apparently because of inhibition by released 5'-dNMPs, it should not be assumed that extensive digestion of a DNA (more than 25%, for example) will proceed at the same rate as that observed early in the course of a digest.

Effect of Nuclease Concentration

The presence of substantial single-stranded tails on partially digested duplexes can be avoided if the concentration of nuclease is near 2 and 10 U/mL for the F and S enzymes, respectively, or greater. The value corresponding to S nuclease should be used for all commercial preparations not sold as separated S and F enzymes. These (or higher) concentrations are compatible with most uses of the enzymes, corresponding to 33 and 8 residues removed/terminus/min from a DNA of near 50% G + C content for the F and S nucleases, respectively, at the substrate concentration (7.3×10^{-8} M, needed only for the calculation for F nuclease) of the above example. Significantly, digestion at or above the aforementioned concentrations not only leaves short average tail lengths (approx 7 residues), but allows ligation under conditions favoring blunt-end joining of up to 50% of that observed with DNA known to possess fully base-paired ends. Hence, treatment with S1 nuclease or DNA polymerase to remove single-stranded ends and provide fully based-paired termini is unnecessary, although it is apparent that polymerase-mediated repair will enhance the fraction of ligatable termini.

Effects of Base Composition

The duplex exonuclease activity of the S form decreases markedly with increasing G + C content, and the magnitude of this effect has,been examined over the range of 37-66 mol% G + C residues. For the S nuclease, the effect is significant and leads to an approx 4.2fold change in v_0/[S] over this composition range. The dependence of v_0/[S] on G + C content over the above range can be approximated, independently of the substrate concentration as previously noted, by the equation:

$$v_0/[S] = 16.85 - 0\ 546P + 0.00456P^2 \quad (3)$$

where P is the mol% G + C, and the value of $v_0/[S]$ is that corresponding to an enzyme concentration of 1 U/mL. The desired value of $v_0/[S]$ is thus calculated by multiplication of that from Eq. (3) by the nuclease concentration (U/mL) to be used in the actual experiment. The equation should not be used to extrapolate outside of the above range of base composition ($37 \leq P \leq 66$).

The duplex exonuclease activity of the F enzyme is much less dependent on composition than for the S enzyme, with $v_0/[S]$ varying only by a factor of 1.7 over the above range at the substrate concentration of the example used earlier. Moreover, nearly all the change occurs between approx 52 and 37% G + C. Given the paucity of data, the values for the maximum velocity/U/L of nuclease and $K,_n$ °*nn* above should be used in the range of 66-52% G + C, and the maximum velocity parameter should be increased by interpolation from the value above to a maximum of 1.7 times that value in the range from 52-37% G + C.

The less pronounced effect of variances of G + C content for the F nuclease suggests its use where it is desired to minimize the effects of local sequence features, such as local regions of high G + C content, on the termination points of digestion of partially degraded samples. With samples that corresponded to S nuclease, sequence analysis of partially degraded samples showed a strong tendency for the nuclease to stop in regions of several consecutive G + C pairs with a strong preference for dG at the 5' end. When the location of "stop" sites was analyzed in this laboratory for both the S and F nucleases, this tendency was confirmed for the S enzyme and shown also to be true for the F species. This suggests that the factors accounting for the dependence of the rates of 3'-terminal nucleotide removal by the respective nuclease species on G + C content are not a direct result of the relative thermodynamic stability of the two types of base pairs. This is because the F nuclease, which removes 3'-terminal nucleotides more rapidly and shows a lesser dependence of this rate on G + C content, would otherwise be expected to show a lesser tendency to stop in "runs" of several G - C pairs. Further support for the lack of a direct role of relative thermodynamic stability of base pairs in this connection are the facts that stops tend to occur at 5' dG residues, whereas stops at both dC and dG residues should be expected if stability alone were the dominant factor and the kinetic parameters show little dependence on G + C content above about 50%, whereas the melting temperature T_m, an indicator of stability, is linearly dependent on the G + C content. Use of the F nuclease will make average extents of degradation more predictable, however, because localized changes in velocity associated with local differences in G + C content are decreased.

THE NUCLEASE AS A PROBE FOR LESIONS IN NONSUPERCOILED CLOSED DNA

It was noted earlier that the nucleases can introduce breaks in doublestranded DNA in response to a wide variety of both covalent and noncovalent distortions of duplex structure, whereas covalently closed circular duplexes that are essentially nonsupercoiled or slightly positively supercoiled (form 1° DNA) are extremely refractory to nucleasemediated attack. This provides for very stringent and general tests of altered duplex structure caused by any perturbation that does not introduce strand breaks (such breaks themselves provide substrate sites) or result in significant negative supercoiling of form I° DNA. Information as to whether duplex DNA structure is perturbed by a given reagent or treatment is valuable as a guide to further studies.

Form 1° DNA is readily produced from any supercoiled duplex species by incubation with DNA topoisomerase I. This enzyme is commercially available, but crude preparations used as described are quite adequate for this purpose. The nicking-closing reaction is routinely monitored by electrophoresis in 1 % agarose gels.

Whether the modifying treatment itself introduces strand breaks is readily ascertained by electrophoresis of the treated form I° DNA and nontreated controls in agarose gels containing a low concentration of ethidium bromide (e.g., 11). Ethidium bromide at 1 μg/mL should also be present in nuclease reaction mixtures to ensure that changes in the ionic environment and/or temperature between the conditions of the nuclease reaction and the topoisomerase treatment do not result in even slight negative supercoiling. High enzyme concentrations, on the order of 100 U/mL, should initially be used to provide a stringent test, because the rate of attack in response to various types of lesions varies widely.

The rate of disappearance of form I° DNA, as by agarose gel electrophoresis of aliquots containing equal volumes of the reaction mixture, is the parameter to be assayed, since nuclease at such concentrations will rapidly degrade, via the duplex exonuclease activity, the linear duplexes produced after the introduction of a strand break, and cleavage in the second strand. In fact, this reaction has been so rapid that it led to a novel coupled assay in one study in which the loss of fluorescence of EtdBr owing to the complete digestion of a DNA of 10,000 by was used to monitor the rate of cleavage in response to the presence of apurinic sites; this digestion was fast compared to the rate of introduction of the initial scission. In the only study where direct comparisons were made, the F nuclease was more efficient in cleaving in response to a covalent lesion than the S species.

The exonuclease activity unfortunately destroys information as to the site of the initial cleavage in reactions converting an appreciable percentage of the DNA. For a type of distortion that readily elicits cleavage, that resulting from negative supercoiling, it has proven possible to localize sites of BAL nuclease cleavage in experiments in which the overall extent of cleavage was very limited and the DNA was labeled with ^{32}P after cleavage with restriction nucleases so that the small fraction undergoing endonucleolytic attack could be examined in autoradiograms of agarose gels. The interesting possibility of using nuclease that has been treated with protease so that the bulk of the duplex exonuclease activity is eliminated in order to localize sites of attack was noted in an earlier section. This will require further experimentation to optimize the conditions for such protease treatment.

PRONASE

Pronase is the name given to a group of proteolytic enzymes that are produced in the culture supernatant of *Streptomyces griseus* K-1. Pronase is known to contain at least ten proteolytic components: five serine-type proteases, two Zn^{2+}endopeptidases, two Zn^{2+}-leucine aminopeptidases, and one Zn^{2+} carboxypeptidase. Pronase therefore has very broad specificity, and is used in cases where extensive or complete degradation of protein is required. It has been used, for example, to reveal the protein components of cell organelles by the hydrolysis of tissue slices, and as an alternative to proteinase K to remove protein during plasmid DNA, chromosomal DNA, and RNA isolation. Another use of pronase is the production of a protein hydrolysate suitable for amino acid analysis.

Traditionally, protein hydrolysates for amino acid analysis are produced by hydrolysis in 6N HC1. However, this method has the disadvantage that tryptophan is totally destroyed, serine and threonine partially (5-10%) destroyed, and most importantly, asparagine and glutamine are hydrolyzed to the corresponding acids. Digestion of the protein/peptide with pronase overcomes these problems, and is particularly useful when the concentration of asparagine and glutamine is required. The use of pronase for this application is detailed in Section 3. The use of pronase in the preparation of optically active amino acids has been described. The hydrolysis of amino amides by pronase results in high-rate significant enantioselectivity and a high degree of conversion of substrates. Pronase has also been used to study membrane topology and protein membrane translocation in the same manner as described for proteinase K, and has been used to disperse mast cells prior to purification and characterisation of these cells.

ENZYME DATA

Specificity

Because it is a mixture of exo- and endoproteases, pronase has a broad specificity, cleaving nearly all peptide bonds.

Purification

The purification of individual proteolytic components of pronase has been reported by a number of authors. In 1988, the purification and characterisation of an additional amino acid-specific endopeptidase from Pronase were described by Yoshida et al..

Molecular Mass

Molecular masses of 15,000-27,000 Da have been reported for components of pronase. These weights have usually been determined by gel filtration.

pH Optimum

Pronase has optimal activity at pH 7-8. However, individual components are reported to retain activity over a much wider pH range. The neutral components are stable in the pH range 5.0-9.0, in the presence of calcium, and have optimal activity at pH 7.0-8.0. The alkaline components are stable in the pH range 3.0-9.0, in the presence of calcium, and have optimal activity at pH 9.0-10.0. The aminopeptidase and carboxypeptidase components are stable at pH 5.0-8.0 in the presence of calcium.

Assay

The assay is normally based on the hydrolysis of casein, which is followed spectrophotometrically by reaction with Folin and Ciocalteau's reagent. One milliliter of enzyme solution is suitably diluted in 0.067M phosphate buffer, pH 7.4. A 2% casein solution, adjusted to pH 7.4 with NaOH, is also prepared. One milliliter of the 2% casein solution is added to the enzyme solution, and the mixture incubated at 40°C. After 10 min, 2 mL of protein precipitation agent are added (1.63 g trichloroacetic acid, 0.82 g sodium acetate, 0.57 mL glacial acetic acid made up to 100 mL with water). The mixture is incubated for a further 10 min at 40°C, and then centrifuged for 5 min at 3000 rpm. Five milliliters of 0.4M sodium carbonate solution and 1 mL Folin and Ciocalteau's reagent (diluted 1:5 just prior to use) are added to 1 mL of supernatant. The mixture is then incubated for a further 10 min at 40°C, and the absorbance at 660 nm then measured. One PUK (Proteolytic Units of Kaken; "Kaken" is a Japanese company) unit is defined as the amount of enzyme that produces an absorbance of 1.0 at 660 nm. Using this assay, the enzyme is usually supplied with a specific activity of about 70,000 PUK/g.

Alternatively, the enzyme is supplied with a specific activity of about 7000 PU/g where one PU (proteolytic unit) is the enzyme activity that liberates Folin-positive amino acids and peptides corresponding to 1 μcool of tyrosine within 1 min under the aforementioned assay conditions. One PUK corresponds to 50 PU. Because of its broad specificity, the enzyme is capable of hydrolyzing many peptide, amide, and ester bonds, including the majority of the specific substrates of most proteolytic enzymes. Pronase also hydrolyzes peptides involving D amino acid residues.

Stability

The calcium ion dependence for the stability of some of the components (mainly exopeptidases) was one of the earliest observations made of Pronase. Pronase is therefore normally used in the presence of 5-20 mM calcium. The addition of excess EDTA results in the irreversible loss of 70% of proteolytic activity. Two peptidase components are inactivated by EDTA, but activity is restored by the addition of Co^{2+} or Ca^{2+}. One of these components, the leucine aminopeptidase, is heat stable up to 70°C. All other components of Pronase lose 90% of their activity at this temperature. The leucine aminopeptidase is not inactivated by 9M urea, but is labile on dialysis against distilled water. Some of the other components of Pronase are also reported to be stable in 8M urea, and one of the serine proteases retains activity in 6M guanidinium chloride. Pronase retains activity in 1 % SDS (w/v) and 1 % Triton (w/v). The enzyme is stable at 4°C for at least 6 mo and is usually stored as a stock solution of 5-20 mg/mL in water at -20°C.

Inhibitors

Among the alkaline proteases, there are at least three that are inhibited by diisopropyl phosphofluoridate (DFP). In general, the neutral proteinases are inhibited by EDTA, and the alkaline proteinases are inhibited by DFP. No single enzyme inhibitor will inhibit all the proteolytic activity in a Pronase sample.

EXPERIMENTAL PROCEDURES

DNA Isolation

When used in DNA isolation, Pronase is generally prepared as a stock solution at about 5-20 mg/mL in water. Prior to storage at-20°C, the solution is heated to 56°C for 15 min followed by a 1-h incubation at 37°C. The purpose of this step is to encourage self-digestion. This eliminates contamination with DNases and RNases. For use, the enzyme is added to the DNA sample (in the presence of 0.5-1 % SDS to disrupt DNA-protein interactions) typically at 250-500 μg protein/ mL, 37°C, for 1-4 h.

Protein Hydrolysis

To prepare a protein hydrolysate, dissolve 0.2 pmol of protein in 0.2 mL of 0.05M ammonium bicarbonate buffer, pH 8.0 (or 0.2M sodium phosphate pH 7.0 if ammonia interferes with the amino acid analysis). Add Pronase to 1 % (w/w), and incubate at 37°C for 24 h. To achieve complete hydrolysis, it is usually necessary to make a further addition of aminopeptidase M at 4% (w/w), and incubate at 37°C for a further 18 h. The sample is then lyophilized and subjected to amino acid analysis. When using two enzymes in this way, there is often an increase in the background amino acids owing to hydrolysis of each enzyme. It is therefore important to carry out a digestion blank to correct for these background amino acids.

PROTEOLYTIC ENZYMES FOR PEPTIDE PRODUCTION

There are three main reasons why a protein chemist might wish to cleave a protein of interest into peptide fragments. The first reason is to generate, by *extensive* proteolysis, a large number of relatively small peptides either for peptide mapping or for purification and subsequent manual sequence determination by the dansyl-Edman method. The second reason is to generate relatively large peptides (50-150 residues) by *limited* proteolysis for automated sequence analysis, such as with the gas-phase sequencer. The third reason is to prepare, again by limited proteolysis, specific fragments for studies relating structure to function. In each case, the specificity of the enzyme used to generate the peptides is a prime consideration, since the aim is to provide high yields of discrete fragments. It can be appreciated that significantly <100% cleavage at some or all of the cleavage sites on the protein being digested will generate a far more complex mixture of a larger number of polypeptides, each in relatively low yield. It is for this reason that enzymes of high specificity, such as trypsin, which

Table 11.1 : Major Cleavage Sites of Some Proteolytic Enzymes

Enzymes	Cleavage site
Chymotrypsm	C-terminal to hydrophobic residues, e g , Phe, Tyr, Trp, Leu
Clostripain	C-terminal to Arg residues
Elastase	C-terminal to amino acids with small hydrophobic side chains
Endoproteinase Arg-C	C-terminal to Arg residues
Endoproteinase Asp-N	N-terminal to Asp and Cys
Endoprotemase Glu-C	C-terminal to Asp and Glu
Endoprotemase Lys-C	C-terminal to Lys
Pepsin	Broad specificity, preference for cleavage C-terminal to Phe, Leu, and Glu
Thermolysin	N-terminal to amino acids with bulky hydrophobic side-chains, e.g , Ileu, Leu, Val, and Phe
Trypsin	C-terminal to Lys and Arg

cleaves at the C-terminal side of arginine and lysine residues, are mainly used for peptide production. However, other proteases with considerably less specificity have also found use in peptide production, particularly when limited proteolysis is being used, or where native protein is used as the substrate when only a limited number of susceptible peptide bonds are available to the proteases (i.e., those on the surface of the protein). This chapter describes ten proteolytic enzymes that have found extensive use in peptide production in recent years. These are chymotrypsin, clostripain, elastase, endoproteinase Arg-C (submaxillary protease), endoproteinase Asp-N, endoproteinase Glu-C (V8 protease), endoproteinase Lys-C, pepsin, thermolysin, and trypsin. The specificities of these enzymes are summarized in Table elsewhere in this chapter.

ENZYME DATA

Chymotrypsin

General Information

The enzyme is initially synthesized in the pancreas as the inactive zymogen chymotrypsinogen. Chymotrypsinogen is converted to the active enzyme by cleavage by other proteolytic enzymes when pancreatic juice passes into the small intestine. Tryptic cleavage of the bond between Arg 15 and Ileu 16 results in a fully active enzyme (ir-chymotrypsin). Further chymotryptic cleavage between Leu 13 and Ser 14 liberates the dipeptide Ser 14-Arg 15, and results in the formation of S-chymotrypsin. Further chymotryptic cleavage liberates the dipeptide Thr 147-Asn 148 resulting in the commonly used and commercially available form of the enzyme, α-chymotrypsin. Chymotrypsin is known as a serine protease, since its catalytic activity results from a strongly ionized serine residue that is generated at the active site as a consequence of the microenvironment of serine residue 195. This strongly nucleophilic oxyanion catalyzes the cleavage of peptide bonds. Many inhibitors of serine proteases (such as diisopropylphosphofluoridate [DFP] and phenylmethane sulfonyl fluoride [PMSF]) function by reacting with the active site serine residue. The commercially available source of the enzyme is from bovine pancreas.

Specificity

Chymotrypsin has relatively broad specificity, cleaving on the C terminal side of hydrophobic residues, especially phenylalanine, tyrosine, tryptophan, and leucine. However, some significant cleavages have also been reported C terminal to Met, Ileu, Ser, Thr, Val, His, Gly, and Ala residues. The nature of neighboring residues can affect the rate of cleavage of a particular bond. Irrespective of the nature of X, X-Pro bonds are not cleaved.

Molecular Mass

α-Chymotrypsin (chymotrypsin A) has a mol mass of about 25,000 Da and contains 241 amino acid residues. The molecule has three polypeptide chains (A chain-13 residues; B chain-131 residues; and C chain-97 residues) linked by disulfide bridges. The amino acid sequence is known.

pH Optimum

The enzyme has a pH optimum between 7.5 and 8.5.

Assay

The assay is based on measuring the esterolytic activity of the enzyme. The decrease in absorbance caused by the hydrolysis of the ester linkage ofN-acetyl-L-tyrosine ethyl ester (ATEE) is measured spectrophotometrically at 237 nm.

ATEE (2.8 mL of 2 mM in 0.1M phosphate buffer, pH 7.0) is incubated at 25°C. At time zero, 20 μL of enzyme (diluted in 1 mM HCl) are added. The change in absorbance is monitored with time at 237 nm. The enzyme is usually supplied with a specific activity of 9000-11,000 ATEE U/ mg when this assay is used. One ATEE unit is defined as the amount of enzyme that causes a decrease in absorbance of 0.001/min.

Alternatively, an assay based on hydrolysis of Suc-$(Ala)_2$-Pro-Phe4-nitroanilide can be used. The reaction is monitored at 410 nm. Using this assay, the enzyme is usually supplied with a specific activity of 60-70 U/mg, where 1 U is defined as the amount of enzyme that hydrolyzes 1 pmol of substrate/min at pH 7.0, and 25°C.

Stability

The enzyme functions in the presence of 2M guanidine hydrochloride and 0.1 % SDS. Stock solutions (10 mg/mL) in 1 mM HCl can be stored frozen or for a number of days at 4°C. When diluted for use to 1 mg/mL in 1 mM HCl (since autolysis occurs at high pH), the presence of 2 mM Ca^{2+} helps stabilize (or possibly activate) the enzyme. The lyophilized enzyme is stable for years at 4°C.

Chymotrypsin is a basic protein, and hence, it sticks easily to glass. Therefore, dilute solutions of the enzyme should be avoided.

Inhibitors

The most commonly used serine protease inhibitors are DFP and PMSF. PMSF is preferred, since it is much less toxic than DFP. PMSF is prepared as a stock solution (1M) in *a dry* solvent, such as propan-2-ol or methanol (stable for months at 4°C). It is added to the enzyme solution with vigorous stirring to a final concentration of 1 mM. It has a half-life of about 1 h in aqueous solution. DFP (highly toxic!) is prepared as a 500 mM stock solution in propan-2-ol and added to a final concentration of 0.1 mM. It also has a half-life of 1 h in aqueous solution. Other low-mol-mass inhibitors include chymostatin (mol mass 605 Da), a microbial amino acid aldehyde, active at 10-100 μM, and tosyl phenylalanyl chloromethyl ketone (TPCK), active at 10-100 pM. TPCK reacts with histidine 57 involved in the catalytic triad. High-mol-mass inhibitors include aprotinin, a_1 antitrypsin, and a_2 macroglobulin, and they are all effective at equimolar concentrations with chymotrypsin. The enzyme is also completely inhibited in the presence of 10 mM Cu^{2+} and Hg^{2+}, and partially inhibited by 1 mM Zn^{2+}. Chymotrypsin that has been completely inhibited by Cu^{2+} can be reactivated by the addition of Ca^{2+} or Versene.

Clostripain

General Information

Clostripain (clostridiopeptidase B) is a sulfhydryl protease isolated from culture filtrates of *Clostridium histolyticum.* Sulfhydryl proteases are all characterized by a thiol group at the active site that is essential for proteolytic activity.

Specificity

The enzyme has amidase and esterase activity as well as protease activity. The enzyme primarily cleaves proteins at the C-terminal side of arginine residues, i.e., -Arg-X-bonds, although -Lys-Xbonds are also cleaved, but at a much slower rate. Thus, under conditions of controlled hydrolysis, the enzyme may be used to cause cleavage effectively at arginine residues only.

Molecular Mass

The enzyme has a mol mass of approx 50,000 Da. The amino acid composition has been published.

pH Optimum

The enzyme has a pH optimum in the range pH 7.4-7.8 in phosphate or Tris buffer.

Assay

The assay is based on the hydrolysis of benzoyl arginine ethyl ester (BAEE). The substrate solution is 0.05M sodium phosphate buffer, pH 7.8, 0.25 mM BAEE, and 2.5 mM dithiothreitol, at 25°C. The reaction is followed by monitoring the change in absorbance at 253 nm with time after addition of enzyme. A molar extinction coefficient of $1150 M^{-1}\ cm^{-1}$ is used to express the results as .tmol BAEE hydrolyzed/min at 25°C. Using this assay, the enzyme is usually supplied with a specific activity of 50-250 U/mg protein, where 1 U is defined as the amount of enzyme that hydrolyzes 1 μmol of BAEE/ min at pH 7.6 at 25°C.

Stability

Enzymic activity is rapidly lost on incubation in the presence of oxidizing agents because of oxidation of the reactive thiol group, but can be regenerated by treatment with dithiothreitol. The enzyme is active in 6M urea.

Inhibitors

The enzyme is inhibited by any thiol reactive agent (e.g., iodoacetic acid), by PMSF, and by parachloromercuribenzoate. The PMSF inhibition can be reversed by the addition of thiol reagents, such as DTT or mercaptoethanol. Since the reactive thiol group is easily oxidized, the enzyme is usually used in the presence of a reducing agent, such as dithiothreitol (1-2 mm). Some authors report that binding of calcium is an essential prerequisite of activity, and certain suppliers indicate that the enzyme requires activation prior to use, by preincubation in 1 mM calcium acetate, 2.5 mM DTT for 2-4 h at 25°C. The esterase activity is inhibited in the presence of Co^{2+}, Cu^{2+}, Cd^{2+}, Na^{+}, and K^{+}. EDTA completely inhibits activity at a concentration of 10 μM. This is probably the result of the calcium requirement of the enzyme. When performing assays, it should be noted that citrate, borate, Veronal, and Tris anions partially inhibit esterase activity.

Elastase

General Information

The commercial enzyme is normally prepared from porcine pancreas. Elastase I (pancreatopeptidase E) is a serine protease and is produced in the pancreas as the inactive zymogen proelastase. Proelastase is converted to elastase by tryptic cleavage of a single peptide bond. The isolation of porcine elastase II has also been described, but elastase I is the commercially available enzyme.

Specificity

The enzyme has rather a broad specificity cleaving C-terminal to small hydrophobic side chains, e.g., Gly, Ala, Ser, Val, Leu, and Ileu.

Molecular Mass

Elastase is a single polypeptide chain of 240 residues crosslinked by four disulfide bridges, and has a mol mass of 25,900 Da. The complete amino acid sequence is known; the gene has been cloned and mol mass confirmed.

pH Optimum

The enzyme has a pH optimum of pH 8.0-8.5.

Assay

An assay based on the hydrolysis of N-benzoyl-L-alanine methyl ester (BAME) is used. The reaction conditions consist of 10 mM substrate in 0.1M KCl and 0.0 *IM* Tris-HCI, pH 8.0, at 25°C. The enzyme solution is added and the change in absorbance with time is monitored at 253 nm.

Alternatively, an assay based on the liberation of congo red from elastin-congo red can be used. The enzyme is supplied with a specific activity of 15-20 U/mg using this assay, where 1 U is defined as the amount of enzyme that solubilizes 1 mg of elastin from elastincongo red in 60 min at 25°C in carbonate buffer, pH 8.8.

Also, hydrolysis of succinyl-(L-ala)$_3$-4-nitroanilide by elastase can be monitored at 410 nm. Using this assay, the enzyme is supplied with a specific activity of 10-15 U/mg where 1 U corresponds to the amount of enzyme that liberates 1 pmol of 4-nitroaniline/min at 25°C, pH 7.8.

Stability

The enzyme is active in 0.1 % SDS. The lyophilized enzyme is stable indefinitely at-20°C and for at least 6 mo at 4°C. The enzyme rapidly autolyzes if stored at room temperature near its pH optimum of 8.8. However, in solution at pH 5.0, proteolytic activity is slight, and solutions at this pH can be used at room temperature with minimal autolysis. The addition of 0.1 mM Ca^{2+} (or 0.1 mM Mg^{2+} to a lesser extent) stabilizes the native form of the enzyme against thermal denaturation, even if a destabilizing ion (e.g., Cu^{2+}) is present. Elastase retains activity in 2M urea. Note that the enzyme sticks to glass, and therefore, plasticware should be used where possible The addition of Triton X100 to incubation solutions at 0.05% helps minimize this adhesion.

Inhibitors

Since it is a serine protease, elastase is inhibited by DFP and PMSF. However, it is not inhibited by TPCK. It is inhibited by elastinal, a low-mol-mass (mol mass 513 Da) amino acid aldehyde at 10-100 μM, and also by a$_2$ macroglobulin (at equimolar concentrations with elastase) and by a$_1$ antitrypsin. Various sulfonyl fluorides and p-dinitrophenyl diethyl phosphate also inhibit the enzyme. The activity of elastase has been shown to be appreciably affected by salts. Sodium chloride (5070 mM) gave 50% inhibition, and similar effects were observed with potassium chloride, ammonium sulfate, and sodium cyanide. Copper sulfate (10 mM) gave 50% inhibition, but millimolar concentrations of zinc, manganese, cobalt, magnesium, or calcium had no effect. The enzyme is unaffected by EDTA.

ENDOPROTEINASE ARG-C

General Information

The enzyme is prepared from mouse submaxillary glands. It is also known as submaxillary protease.

Specificity

This enzyme cleaves specifically C terminal to arginine residues, although cleavage at some residues does not always go to completion. Other workers have found occasional bonds that are resistant to cleavage. For example, when the immunoglobulin x chain was used as substrate, cleavage occurred at most arginine residues, but one ArgVal bond and one Arg-Arg bond were resistant to cleavage. The specificity of the enzyme appears to broaden during long incubations, so it is best to use the enzyme at high ratios for short periods, e.g., 1:50 dilution for 2 h at 37°C.

Molecular Mass

The enzyme is a serine protease and has a mol mass of about 28,000 Da. The amino acid composition is also known.

pH Optimum

The enzyme has a pH optimum between pH 7.5 and 8.5.

Assay

The assay is based on the hydrolysis of N_ap-tosyl-L-arginine methyl ester (TAME). The substrate solution is 0.8M TAME in 0. 1M phosphate, pH 8.0. The reaction is started by the addition of 0.1 mL of enzyme (suitably diluted in redistilled water) to 2.9 mL of substrate solution. The change in absorbance with time at 247 nm is monitored, and enzyme activity calculated. Using this assay, the enzyme is supplied with a specific activity of 100-250 U/mg where 1 U is defined as the amount of enzyme that hydrolyzes 1 p.mol of TAME/min.

Stability

The enzyme retains 90% activity after 1 h at 25°C in 0.1% SDS. It is unstable at low pH, and the enzyme is denatured in 4M urea. Solutions of the enzyme can be stored at -20°C without loss of activity. The enzyme loses specificity during long incubations; digestions proceed more rapidly if the substrate is first denatured. The purified form of the enzyme (enzyme A) undergoes autolysis on storage at pH 8.0 overnight, resulting in the formation of enzyme D, formed by the cleavage of a fragment of mol mass 3000-5000 Da from enzyme *A*. Commercially available enzyme is a mixture of enzymes A and D.

Inhibitors

The enzyme is inhibited by low pH (<2), DFP, TLCK *(43)* *(see* Section 2.1.7. for practical details), and a_2 macroglobulin when used in equimolar ratios with the enzyme. The enzyme has also been shown to be inhibited at 1 mM concentrations by Hg^{2+} (100% inhibition), Cu^{2+} (80%), Zn^{2+} (65%), Cd^{2+} (50%), and Fe^{2+} (50%).

Endoproteinase Asp-N

General Information

This enzyme has been isolated from the culture filtrates of *Pseudomonas fragic*. The enzyme is a neutral protease and is a metallo-enzyme having an essential zinc atom at the active site.

Specificity

The enzyme cleaves protein substrate on the N-terminal side of aspartic acid residues or cysteic acid residues. Cleavage occurred for all such residues in oxidized ribonuclease, but when myoglobin was used, a similar specificity was observed, except that only four out of six aspartyl bonds present were hydrolyzed. However, occasional additional cleavage of Glu residues has been reported.

Molecular Mass

The mol mass of the enzyme is about 50,000 Da. The amino acid analysis of the enzyme has been published.

pH Optimum

The enzyme has a pH optimum of about pH 7.0, with activity dropping off rapidly above pH 9.0 and below pH 6.0.

Assay

The assay is based on the hydrolysis of azocasem. A 2% azocasein substrate in 50 mM phosphate buffer, pH 7.0, is used. Substrate (250 μL) is incubated at 25°C and 150 μL of enzyme then added. After incubation for 20 min, the reaction is terminated by addition of 1.2 mL of 10% (w/v) TCA, and the mixture allowed to stand for a further 15 min to allow complete precipitation of the remaining azocasein. The mixture is centrifuged (8000g) or filtered, and 1-2 mL of the supernatant are then combined with 1.4 mL of 1.OM NaOH. The absorbance is read at 440 nm, and the concentration of enzyme calculated. One unit of enzyme activity is defined as the amount of enzyme that produces an absorbance change of 1.0 in a 1-cm path length cuvet under the above conditions.

Stability

The enzyme is fully active in 2M urea. It is stable between pH

7.0 and 9.5, but is rapidly inactivated outside these values. The enzyme is stable at temperatures up to 40°C in the absence of Ca^{2+}. The presence of 5 mM $CaCl_2$ appears to stabilize the enzyme somewhat with residual activity still remaining at 60°C. The enzyme is stable at 4°C, stored dry. A solution in redistilled water may be used for 1 wk at maximum if stored at 4°C. Aqueous solutions of the enzyme are stable up to 1 mo at -20°C. Freeze/thawing of enzyme solutions is not recommended. The enzyme is stable in 0.01% SDS, 0.1M guanidine hydrochloride, and 10% (v/v) acetonitrile.

Inhibitors

Being a metalloenzyme, the enzyme is inhibited by chelating agents, such as EDTA, and 1, 10-phenanthrolene. Activity can be restored by the addition of zinc or cobalt ions, but not by adding calcium or magnesium ions. The enzyme is not inhibited by iodoacetic acid, PCMB, dithiothreitol, PMSF, or soyabean trypsin inhibitor.

Endoproteinase Glu-C

General Information

The enzyme is purified from the culture filtrates of *Staphylococcus aureus* strain V8. The enzyme has also been named staphylococcal protease and V8 protease.

Specificity

The enzyme cleaves C terminal to Asp and Glu residues in protein substrates when used in phosphate buffer, pH 7.8. However, when used in ammonium buffers (ammonium bicarbonate, pH 7.8, or ammonium acetate, pH 4.0), cleavage is restricted to Glu residues only. -Glu-X- or Asx-X bonds are not cleaved if X is Pro or S-carboxymethylcysteine. If *X* is a bulky hydrophobic residue, cleavage is slow. In addition to the expected cleavages at Asp and Glu, a number of other workers have noted a number of nonspecific cleavages, often of Ser-X bonds.

Molecular Mass

The enzyme is a serene protease with a mol mass of 12,000 Da. The amino acid composition has been published.

pH Optimum

The protease is active in the pH range 3.5-9, and exhibits maximum proteolytic activity at pH 4.0 and 7.8 with hemoglobin as substrate.

Assay

The assay is based on the hydrolysis of casein. The reaction conditions consist of 1 % casein in 0. 1M Tris-HCI, pH 7.8. Enzyme is added, and the mixture is incubated at 25°C for 10 min. The reaction is stopped by adding an equal volume of 10% TCA, and the mixture is allowed to stand for a further 15 min. The mixture is then filtered or centrifuged; the absorbance of the supernatant is read at 280 nm, and the enzyme activity calculated. Using this assay, the enzyme is usually supplied with a specific activity of 500 U/mg, where 1 U is the amount of enzyme that caused 0.001 A_{280} unit of change/min at 25°C, pH 7.8.

Alternatively, an assay based on the hydrolysis of N-CBZ-L-glutamyl a-phenyl ester can be used. Using this assay, the enzyme is usually supplied with a specific activity of 500-1000 U/mg, where 1 U is defined as the amount of enzyme that will hydrolyze 1 pmol of substrate/min at pH 7.8, at 37°C. (One unit is equivalent to approx 0.004 casein digestion units.)

Also, an assay based on the hydrolysis of carbobenzoxy-Phe-LeuGlu-4-nitroanilide, at pH 7.8, 410 nm, and 25°C has been described. The enzyme is supplied with a specific activity of 10-20 U/mg when this assay is used, where 1 U hydrolyzes 1 pmol of substrate/min at pH 7.8 and 25°C.

Stability

The enzyme is active in the presence of 0.2% SDS, and retains 50% activity in 4M urea. The enzyme has been shown to be equally resistant to heat denaturation over a wide range of temperatures. Seventy percent of activity remained after heating at 100°C for 3 min. Aqueous solutions are stable to freeze/thawing. The enzyme is stable at pH 4-10, but precipitates below pH 4. It retains activity in 1M guanidine hydrochloride and 10% acetonitrile. The enzyme is stable at 4°C, when stored dry. A solution in redistilled water may be used for 1-2 d at maximum, if stored at 4°C.

Inhibitors

Being a serine protease, the enzyme is inhibited by DFP. The enzyme is not inhibited by EDTA, nor does the addition of a range of divalent metal ions have any effect on enzyme activity. The enzyme is also inhibited by a_2 macroglobulin at concentrations equimolar with the enzyme.

Endoproteinase Lys-C

General Information

This enzyme is a serine protease isolated from culture filtrates of *Lysobacter enzymogenes.*

Specificity

The enzyme cleaves specifically C-terminal to lysine residues. Some minor nonspecific cleavages, notably at Asn-X bonds, have also been reported.

Molecular Mass

It has a mol mass of about 33,000 Da.

pH Optimum

The enzyme has a pH optimum in the range 8.5-8.8.

Assay

The assay is based on hydrolysis of Tosyl-Gly-Pro-Lys-4-nitroanilide (commercially sold as Chromozym PL). Chromozym PL (9 mg) is dissolved in 1 ml, of redistilled water. This solution (0.05 mL) is combined with 1.0 mL of buffer (25 *mM* Tris-HC1, pH 7.7, 1 mM EDTA) and incubated at 25°C. The reaction is started by addition of 0.05 mL of enzyme (suitably diluted with the above buffer). The increase in absorbance is monitored at 405 nm, and using an extinction coefficient of 10.4 mM^{-1} cm^{-1}, the enzyme activity of the sample is calculated. Using this assay, the enzyme is supplied with a specific activity of 25-35 U/mg, where 1 U is defined as the amount of enzyme that liberates 1 μmot 4-nitroaniline/min at 25°C from Tosyl-Gly-Pro-Lys4-nitroanilide.

Stability

The enzyme is active in 0.5% SDS, 5M urea, and in 10% acetonitrile. Aqueous solutions are stable at pH 5-12 for up to 1 d at 4°C, or can be stored at -20°C for months. The enzyme is stable at 4°C, stored dry. To avoid autolysis, the incubation temperature should not exceed 37°C.

Inhibitors

The enzyme is inhibited by DFP (1 mM), TLCK (1 mM), leupeptin (0.4 mM), and aprotinin (100 μg/mL). The enzyme is not inhibited by EDTA (2 mM), PMSF (1 mM), or a_1-antitrypsin.

Pepsin

General Information

The term "pepsin" refers to a group of gastric proteinases that are active at acid pH values (pH 1-5). These enzymes are formed by partial proteolysis of the inactive zymogens, the pepsinogens. Pepsin is very suitable for investigating arrangement of disulfide bonds in proteins owing to minimisation of disulfide interchange reactions because of low pH. The commercially available form of the enzyme is Pepsin A from porcine gastric mucosa.

Specificity

Pepsin has broad specificity with a preference for peptides containing linkages with aromatic or dicarboxylic L-amino acids. It preferentially cleaves C terminal to Phe and Leu and to a lesser extent Glu linkages. The enzyme does not cleave at Val, Ala, or Gly.

Molecular Mass

The enzyme has a mol mass of 36,000 Da. The amino acid sequence and composition have been determined.

pH

Optimum Pepsin is optimally active at pH 2-4.

Assay

The assay is based on hydrolysis of denatured hemoglobin. Five milliliters of hemoglobin solution (20 mg/mL in 60 mM HCl) and 0.1 mL enzyme in HCl (10 mM) are combined and incubated for 10 min at 37°C. Ten milliliters of TCA (5% [w/v]) are then added, and the mixture centrifuged at 8000g or filtered. Five milliliters of filtrate are combined with 10 μL of 0.5M NaOH and 3 mL of Folin and Ciocalteau's reagent; after 10 min, the absorbance is read at 578 nm. The quantity of tyrosine released can be determined from a standard curve. Using this assay, the enzyme is supplied with a specific activity of 1 *Anson* unit/gram, where 1 Anson unit is the amount of enzyme that hydrolyzes hemoglobin, such that hydrolysis products formed/ min give the same absorbance with Folin's reagent as 1 mmol of tyrosine.

Alternatively, an assay based on the hydrolysis of acetyl-Lphenylalanyl-3,5-diiodo-L-tyrosine can be used. Using this assay, the enzyme is supplied with a specific activity of approx 150 U/g, where 1 U is defined as the amount of enzyme that hydrolyzes 1 pmol of substrate/min at pH 2.0 and 37°C.

Stability

The enzyme is irreversibly inactivated at pH values >6.0. Below pH 6, it retains activity in 4M urea and in 3M guanidine hydrochloride, and is unaffected by heating to 60°C. The lyophilized enzyme is stable for months at 4°C.

Inhibitors

Pepsin is inhibited by Pepstatin A(mol mass 685.9 Da), diazoketones, phenylacyl bromide, and aliphatic alcohols.

Thermolysin

General Information

The enzyme is produced in the culture supernatants of the microorganism *Bacillus thermoproteolyticus*. It is a metalloenzyme having an absolute requirement of zinc ions for activity. The enzyme is thermostable.

Specificity

In general, thermolysin hydrolyzes peptide bonds N terminal to hydrophobic amino acids with bulky sidechains, e.g., Ileu, Leu, Val, and Phe. However, the enzyme is not absolutely specific for these residues, with other workers having identified minor cleavage sites N terminal to most other amino acids. However, the *preferred* sites of cleavage are those N terminal to the bulky hydrophobic residues, and it is this characteristic that has made thermolysin useful in protein studies. Also, thermolysin appears to be able to cleave native proteins to a greater extent than most other proteases. Bonds involving proline of the type -Pro-Ileu- are cleaved, but in sequences of the type X-Phe-Pro-, the X-Phe- bond is resistant to cleavage. Cleavage is also inhibited by an adjacent a amino or carboxyl group, so thermolysin has no exopeptidase activity.

Molecular Mass

It is a single polypeptide chain of 316 residues and has a mol mass of 34,600 Da. The complete amino acid sequence has been determined.

pH Optimum

Thermolysin has optimal activity in the pH range 7.0-8.0, but can be used up to pH 9.0 if required.

Assay

The assay is based on the hydrolysis of casein. One milliliter of borate buffer (0.1M, pH 7.2), 1 mL of substrate solution (2 g casein dissolved in water and 10 mL of 1 M NaOH, pH adjusted to 7.2, and solution made up to 100 mL), and 0.02 mL of enzyme solution are incubated for 10 min at 37°C; 2.0 mL of TCA (0.1M) and 0.2 mL $CaCl_2$ (2 mM) are then added; the mixture is incubated for a further 10 min at 37°C, and then centrifuged or filtered. Filtrate (1.5 mL) is combined with 5 mL of Na_2CO_3 (0.4M) and 1 mL of Folin and Ciocalteau's reagent (commercial reagent diluted 1:3). This mixture is then incubated for 20 min at 37°C. The absorbance is read at 578 nm, and using standard solutions, the concentration of the enzyme is calculated. The enzyme is usually supplied with a specific activity of 4 U/mg, where 1 U is defined as the amount of enzyme that liberates 1 pmol Folinpositive amino acids and peptides (calculated as tyrosine)/ min at pH 7.2 and 37°C.

Stability

The enzyme is thermostable, and at 80°C has a half-life of 1 h. There is no loss of activity at 60°C over 1 h. A zinc atom at the active site is essential for activity, but the enzyme also has an absolute requirement for calcium ions for stability, and this should therefore be included in digestion buffers, although most preparations contain sufficient Ca^{2+} to be active without the addition of extra Ca^{2+}. The enzyme is stable in 8M urea, and also in 20% methanol or ethanol. However, these conditions enhance thermal denaturation and digestion should then be carried out at 20°C. Aqueous solutions of the enzyme can be stored frozen for weeks without loss of activity. The enzyme is also active in 1 % SDS, and 0.1M NaCl. The lyophilized enzyme is stable at 4°C for many months.

Inhibitors

Because of the requirement for a zinc atom at the active site, chelating agents, such as EDTA and 1,10 phenanthroline, are strong inhibitors. Complete inhibition occurs in 5 mM EDTA at 40°C in 3 min.

Oxalate, citrate, and phosphate also inhibit the enzyme, so phosphate buffers should be avoided. Mercuric chloride and silver nitrate at 5 mM also cause complete inactivation. The enzyme is also inhibited by a_2 macroglobulin and diethyl pyrocarbonate. The enzyme is not inhibited by DFP, TPCK, soyabean trypsin inhibitors, cysteine, or sodium cyanide.

Trypsin

General Information

Trypsin is a serine protease that is formed by cleavage of the inactive precursor trypsinogen. Activation is achieved by limited proteolytic cleavage at a single peptide bond, Lys_6-Ile_7, near the N terminus of the zymogen. The activation process is catalyzed by a variety of enzymes, including enterokinase, mold proteases, and trypsin itself. The commercial form of the enzyme is from bovine pancreas.

Specificity

Trypsin is a highly specific endopeptidase whose protease activity is restricted to the positively charged side chains lysine and arginine. Cleavage occurs C terminal to these residues. The enzyme also hydrolyzes ester and amide linkages of synthetic derivatives of these amino acids. No hydrolysis at X-Pro bonds occurs, and reduced hydrolysis occurs if X is preceded or followed by an acidic residue. Some nonspecific cleavage may occur because of the presence of yr-trypsin in some commercial preparations. Therefore, results from a *tryptic* digest should be interpreted carefully, since the number of peptides may not add up to the total number of arginine and lysine residues.

Molecular Mass

The bovine enzyme has a mol mass of approx 24,000 Da. The enzyme has been sequenced, and the amino acid composition has been reported.

pH Optimum

The enzyme has a pH optimum of 7-9.

Assay

The assay is based on hydrolysis ofN-benzoyl-L-arginine ethyl ester. The substrate solution is 46.7 mM Tris-HCl, pH 8.0, 0.9 mM BAEE, and 19 mM $CaCl_2$. Following the addition of enzyme, the change in absorbance at 255 nm at 25°C is measured, and the enzyme concentration calculated using an extinction coefficient of 0.081 m M^{-1} cm^{-1}. The enzyme is supplied with a specific activity of 6000-10,000 U/mg when this assay is used, where 1 U is defined as the amount of enzyme that produces an increase in absorbance at 255 nm of 0.001/min at pH 7.6 and at 25°C.

Stability

Trypsin is stable at pH 3 at low temperatures and can be stored for weeks without loss of activity. The lyophilized enzyme is stable for months at 4°C. Trypsin is reversibly denatured above pH 11. An increase in temperature above pH 8 results in irreversible denaturation. At pH 2-3, it is stable up to 40°C.

Trypsin is fully active in 6.5M urea, but is reversibly denatured by 8M urea. Urea-denatured trypsin regains activity by dialysis against 10^{-3}M HCl.

The enzyme retains activity in 30% ethanol, 0.1% SDS, 1M guanidine hydrochloride, and 10% acetonitrile. Trypsin has a tendency to autolyze at pH values >5. The addition of calcium ions (20 mM) to trypsin solutions retards autolysis.

Activators and Inhibitors

Trypsin is activated by calcium ions and lanthanides. The enzyme is inhibited by DFP, TLCK, soybean trypsin inhibitors, aprotinin (mol mass 6500 Da), leupeptin (an amino acid aldehyde, mol mass 426.6 Da), and a-macroglobulin. Inhibition by benzamidine is reversible.

EXPERIMENTAL PROCEDURES-PRACTICAL DETAILS

Choice of Buffer

It is generally preferable to use a volatile buffer, particularly if the peptides are to be separated by ion-exchange chromatography or paper chromatography/electrophoresis, where the buffer salts may interfere, particularly if the digested sample is concentrated or dried down at the end of the digestion. Salt concentration is, however, not a problem if reverse-phase HPLC is to be used to separate the peptides. The majority of the enzymes described above have pH optima in the range 7.58.5, and the most commonly used volatile buffers in this range are ammonium bicarbonate (100 mM gives a pH of 8.0) or N-ethylmorpholine acetate (pH 8.5). For pepsin, 5% acetic acid or 10 mM HCl is used. Ammonium acetate (0.1M) can be used in the pH range 4.55.5. If samples are to be taken for SDS gel electrophoresis, potassium ions should be avoided, since these precipitate with dodecyl sulfate and can aggregate peptides in the electrophoresis wells.

Condition of the Substrate

The physical state of the substrate is of great importance in determining the degree of digestion. Cleavage of the native protein will often result in large proteolytic fragments, since only these cleavage sites on the surface of the protein will be accessible to the enzyme. However, if it is required that the enzymic digestion go to completion, then the substrate protein should be denatured and disulfide bridges reduced, thus making all potential cleavage sites accessible to the enzyme. Denaturation can be achieved by boiling the protein solution, precipitation in 5% trichloroacetic acid, or treatment in 8M urea or 6M guanidium hydrochloride. Denatured proteins are often insoluble, and although proteolytic enzymes will cleave insoluble substrates, it is desirable that the precipitate be as finely divided as possible to allow maximum access of the enzyme. This can be achieved by dissolving the precipitate in 8M urea and then removing the urea by dialysis against buffer, or dissolving the precipitate at low pH and then slowly titrating back to the required pH. Many of the enzymes described above remain active in the presence of SDS or urea, and therefore, these agents may be used to denature the protein substrate (which remains soluble) and then enzyme may be added directly. For glycoproteins, the sugar backbone can sterically hinder enzyme access to the polypeptide backbone and severely reduce digestion. Deglycosylation prior to enzymic cleavage can be achieved using appropriate enzymes.

Reaction Conditions

When attempting to achieve complete digestion, an enzyme to substrate ratio of 1-2% (w/w) is commonly used, usually at 37°C for 24 h. However, longer times (24-48 h) with repeated addition of enzyme may have to be used with particularly resistant substrates. In these cases, one should be aware of the increased possibility of generating proteolytic fragments from the enzyme being used. Bacterial growth and protein cleavage by released enzymes can also be a problem in extended hydrolysis. In all cases, therefore, it is wise to carry out a preliminary small-scale trial digestion to determine the minimum digestion time necessary to generate a stable peptide pattern. For limited proteolysis experiments, much lower enzyme-to-substrate ratios (0.1 % or less), lower temperatures (10-20°C), and shorter times (15-30 min) may all be used in conjunction to reduce the amount of proteolysis that occurs. Again, it is necessary to carry out preliminary small-scale trial digests to determine the optimum conditions.

Monitoring of Digestion

Limited Digestion

Conditions for controlled proteolysis are far more difficult to predict than the rather obvious extreme conditions used for total proteolysis and are highly dependent on the experimental conditions chosen by the user. Since relatively large peptides are to be generated, the production of peptides in trial digests is best monitored by SDS gel electrophoresis. Following addition of the enzyme (1:100 to 1:1000 [w/w]) to either the native or denatured protein at 30-37°C, remove samples containing about 5 tg of protein at regular intervals (say every 15 min for 4 h) and add to SDS sample buffer. Since many proteolytic enzymes (e.g., clostripain, thermolysm, endoproteinase Glu-C, elastase, papain, and chymotrypsin) are still active in 0.1 % SDS, it is essential to boil the sample immediately to inactivate the protease. When all samples have been taken, they are run on the gel and the time of digestion that gives the required pattern of peptides used for the large-scale digest. If digestion was too rapid, a repeat trial should be carried out at a lower enzyme-to-substrate ratio and/or a lower temperature. However, if cleavage of the native protein is being attempted, considerably *higher* enzyme-to-substrate ratios may be needed, possibly as high as 1:1 (w/w).

Total Digestion

Following addition of the enzyme (1:10 to 1:100 [w/w]) at 37°C, remove samples containing about 5.tg of protein at 0, 1, 2, 4, and 8 h. A further addition of protease is made (to replace enzyme lost by autolysis), and incubation continued overnight. A further sample is then taken, a final addition of enzyme made, and incubation continued for a further 8 h, when a final sample is taken. Samples are either immediately frozen for later analysis by reverse-phase HPLC or dried directly onto a cellulose TLC plate (which effectively stops the reaction), and when fully loaded, the plate is run in an appropriate solvent, such as butanol/pyridine/acetic acid/ water (15:12:3:10) and peptides identified by spraying with ninhydrin or fluorescamine. Whichever method is used, the time of digestion that gives an unchanging pattern of peptides is used for the large-scale digestion.

Termination of the Digestion

The simplest method of termination is freezing followed by lyophilisation. Such a procedure will retain biological activity of limited proteolysis products where necessary. Where total digestion has taken place, samples may be easily (and more quickly) rotary evaporated to dryness, since one is not attempting to retain any tertiary structures in the proteolysis products. Alternatively, the pH digest may be altered (normally by the addition of acid) to a pH value at which the enzyme is no longer active, or the sample boiled rapidly to denature the enzyme. The reaction can also be stopped by the addition of a specific inhibitor for the enzyme. In general, serine proteases are inhibited by the addition of PMSF to 1 mM. (PMSF is prepared as a stable 1M stock solution in propan-2-*ol* and added to the reaction with vigorous mixing. PMSF *cannot* be prepared as an *aqueous* stock solution, since it is rapidly hydrolyzed in water with a half-life of about 1 h. Metalloenzymes can be inhibited by the addition of disodium EDTA to a concentration of 10 mM (from a 1M stock solution), and sulfhydryl proteases inhibited by the addition of iodoacetic acid to a concentration just in excess of that of the sulfhydryl reagent added to activate the enzyme.

Typical Digestion Conditions

The conditions given below are suitable for producing total digestion, e.g., suitable for peptide mapping.

Chymotrypsin

Dissolve the protein substrate at a concentration up to 10 mg/mL in 0.1M ammonium bicarbonate (pH 8.0). Add enzyme solution to give a ratio of 2% (w/w), and incubate for 4 h at 37°C.

Clostripain

Note that the enzyme may need to be activated prior to use. Dissolve the substrate protein at 1-10 mg/mL in 50 mM ammonium bicarbonate, 0.2 mM calcium acetate, and 2.5 mM DTT. Add clostripain at 1-2% (w/w), and incubate for 2-3 h at 37°C. More extensive incubations can result in some cleavage at lysine residues.

Elastase

Dissolve the substrate at 1-10 mg/mL in 100 mM ammonium bicarbonate at 37°C. Add elastase at 1-2% (w/w), and incubate for 4 h.

Endoproteinase Arg-C

Dissolve the substrate in 1 % ammonium bicarbonate, pH 8.0. Add enzyme at 2% (w/w), and incubate at 37°C for 8 h.

Endoproteinase Asp-N

Dissolve the substrate in 50 mM phosphate buffer, pH 7.0, at 37°C. Add enzyme at 5% (w/w) in the same buffer, and incubate for 4-18 *h*.

Endoproteinase Glu-C

For cleavage at glutamate residues, dissolve the proteins substrate in 0.1M ammonium bicarbonate, pH 8.0, at a concentration of about 10 mg/mL, and add enzyme at 3% (w/w). Incubate at 37°C for 2-18 h. To extend the cleavage to aspartate residues, perform the same incubation, but using 0.1M sodium phosphate buffer, pH 7.8.

Endoproteinase Lys-C

Dissolve the protein substrate in 0. 1M ammonium bicarbonate, pH 8, at 10 mg/mL. Add enzyme at 2% (w/w), and incubate at 37°C for 2 h. The same amount of enzyme is again added, and after 22 h, the digestion is stopped.

Pepsin

Dissolve the substrate in 0.01M HCI, and adjust the pH to 2.0.

Add pepsin to 1 % (w/w), and incubate for 1 h at 25°C.

Thermolysin

Dissolve the substrate protein in 0.1M ammonium bicarbonate containing 5 mM $CaCl_2$. Add enzyme to 2% (w/w), and incubate at 37°C for 1 h.

Trypsin

Dissolve the substrate in 0.1M ammonium bicarbonate. Add enzyme to 1-2% (w/w), and incubate at 37°C for 1-4 h.

INDEX

D

E

F

G

H

N

O

P

Q

R

S

T

U

V

X

Y

Z